Lecture Notes in Computer Science 10310

Commenced Publication in 1973
Founding and Former Series Editors:
Gerhard Goos, Juris Hartmanis, and Jan van Leeuwen

More information about this series at http://www.springer.com/series/7412

Pasquale Foggia · Cheng-Lin Liu
Mario Vento (Eds.)

Graph-Based Representations in Pattern Recognition

11th IAPR-TC-15 International Workshop, GbRPR 2017
Anacapri, Italy, May 16–18, 2017
Proceedings

 Springer

Editors
Pasquale Foggia (ID)
Università degli Studi di Salerno
Fisciano
Italy

Mario Vento (ID)
Università degli Studi di Salerno
Fisciano
Italy

Cheng-Lin Liu
Chinese Academy of Sciences
Beijing
China

ISSN 0302-9743   ISSN 1611-3349 (electronic)
Lecture Notes in Computer Science
ISBN 978-3-319-58960-2   ISBN 978-3-319-58961-9 (eBook)
DOI 10.1007/978-3-319-58961-9

Library of Congress Control Number: 2017940628

LNCS Sublibrary: SL6 – Image Processing, Computer Vision, Pattern Recognition, and Graphics

Printed on acid-free paper

This Springer imprint is published by Springer Nature
The registered company is Springer International Publishing AG
The registered company address is: Gewerbestrasse 11, 6330 Cham, Switzerland

Preface

This volume contains the papers presented at GbR2017: the 11th IAPR-TC15 Workshop on Graph-Based Representations in Pattern Recognition, held during May 16–18, 2017, in Anacapri, Italy.

GbR2017 was the 11th edition of a series of workshops organized every two years by Technical Committee 15 of the International Association for Pattern Recognition (IAPR). This workshop series traditionally provides a forum for presenting and discussing research results and applications in the intersection of pattern recognition, image analysis on one side and graph theory on the other side. In addition, given the avenue of new structural/graphical models and structural criteria for solving computer vision problems, GbR2017 organization encourages researchers in this more general context to actively participate in the workshop. Furthermore, the application of graphs to pattern recognition problems in other fields like computational topology, graphic recognition systems, and bioinformatics is also highly welcome at the workshop.

The present volume contains 25 papers. Each accepted paper was reviewed by two Program Committee members. The program of GbR2017 also included two invited talks by Prof. Luc Brun (Ecole Nationale Superieure d'Ingenieurs de Caen, France) and Prof. Vladimir Batagelj (University of Ljubljana, Slovenia).

We want to thank the International Association for Pattern Recognition for making GbR2017 an IAPR-sponsored event. We also thank the University of Salerno and the Department of Information and Electrical Engineering and Applyed Mathematics (DIEM) for sponsoring the workshop, and the Municipality of Anacapri for endorsing GbR2017.

May 2017

Pasquale Foggia
Cheng-Lin Liu
Mario Vento

Organization

Conference Chairs

Pasquale Foggia   Università degli Studi di Salerno, Italy
Cheng-Lin Liu   Institute of Automation of Chinese Academy of Sciences,
                           China
Mario Vento   Università degli Studi di Salerno, Italy

Program Committee

Nicole M. Artner   PRIP, TU Wien, Austria
Isabelle Bloch   ENST, CNRS UMR 5141 LTCI, France
Sebastien Bougleux   Normandie University, UNICAEN, France
Luc Brun   GREYC, ENSICaen, France
Vincenzo Carletti   Università degli Studi di Salerno, Italy
Ananda S. Chowdhury   Jadavpur University, India
Donatello Conte   Ecole Polytechnique de l'Université François Rabelais
                           de Tours, France
Guillaume Damiand   CNRS, LIRIS, Université de Lyon, France
Francisco Escolano   University of Alicante, Spain
Pasquale Foggia   Università degli Studi di Salerno, Italy
Benoit Gaüzere   Normandie Université, INSA de Rouen, LITIS, France
Rosalba Giugno   University of Catania, Italy
Rocio Gonzalez-Diaz   University of Seville, Spain
Edwin Hancock   University of York, UK
Yll Haxhimusa   Vienna University of Technology, Austria
Pierre Héroux   Université de Rouen, LITIS EA 4108, France
Xiaoyi Jiang   University of Münster, Germany
Walter G. Kropatsch   TU Wien, Austria
Cheng-Lin Liu   Institute of Automation of Chinese Academy of Sciences,
                           China
Josep Llados   Computer Vision Center, Universitat Autònoma
                           de Barcelona, Spain
Bin Luo   Anhui University, China
Jean-Marc Ogier   University of La Rochelle, Laboratoire L3i, France
Marcello Pelillo   Università Ca' Foscari, Italy
Kaspar Riesen   University of Applied Sciences and Arts Northwestern
                           Switzerland, Switzerland
Alessia Saggese   Università degli Studi di Salerno, Italy
Francesc Serratosa   Universitat Rovira i Virgili, Spain
Christine Solnon   LIRIS CNRS UMR 5205, INSA Lyon, France
Salvatore Tabbone   Université de Lorraine, France

Andrea Torsello	Università Ca' Foscari, Italy
Seiichi Uchida	Kyushu University, Japan
Ernest Valveny	Computer Vision Center, Universitat Autònoma de Barcelona, Spain
Mario Vento	Università degli Studi di Salerno, Italy
Richard Wilson	University of York, UK

Organizing Committee

Alessia Saggese	Università degli Studi di Salerno, Italy
Vincenzo Carletti	Università degli Studi di Salerno, Italy
Antonio Greco	Università degli Studi di Salerno, Italy
Luca Greco	Università degli Studi di Salerno, Italy
Pierluigi Ritrovato	Università degli Studi di Salerno, Italy

Invited Talks

Approaches to Analysis of Large Networks

Vladimir Batagelj[1,2]

[1] Department of Theoretical Computer Science, Institute of Mathematics, Physics
and Mechanics, Jadranska 19, 1000 Ljubljana, Slovenia
vladimir.batagelj@uni-lj.si
[2] University of Primorska, Andrej Marušič Institute,
Muzejski trg 2, Koper, Slovenia

Large networks are networks with some thousands up to billions of nodes that can be entirely stored in computer's memory. Most of large networks are sparse – their number of links is of the same order as their number of nodes (Dunbar's number). This allows us to develop very efficient (subquadratic) algorithms for analysis of large networks. To support the analysis of large networks started in 1996 to develop a program Pajek (De Nooy et al. 2011). Besides basic graph theory algorithms, such as weak and strong connectivity, condensation, topological ordering, etc., Pajek contains also several specific network analysis algorithms: 3-rings and 4-rings weights, SPC weights, (generalized) cores, fragment (motif) searching, network multiplication, cuts, islands, clustering and blockmodeling (Doreian et al. 2004) and others. For details see Batagelj et al. (2014).

References

Batagelj, V., Doreian, P., Ferligoj, A., Kejžar, N.: Understanding Large Temporal Networks And Spatial Networks: Exploration, Pattern Searching, Visualization And Network Evolution. Wiley Series in Computational and Quantitative Social Science. Wiley (2014)

De Nooy, W., Mrvar, A., Batagelj, V.: Exploratory Social Network Analysis with Pajek. Revised and Expanded Second Edition. Structural Analysis in the Social Sciences. Cambridge University Press (2011)

Doreian, P., Batagelj, V., Ferligoj, A.: Generalized Blockmodeling. Structural Analysis in the Social Sciences. Cambridge University Press (2004)

Graph Edit Distance: Basics and History

Luc Brun[1,2]

[1] Normandy University, ENSICAEN, CNRS, Caen, France
luc.brun@ensicaen.fr
[2] University of Caen Normandy, GREYC (UMR 6072), Caen, France

Abstract. Defining a metric between objects is a basic step of any pattern recognition algorithm. Using graphs, this notion of distance is not straightforward. Among the different distances between graphs that one may imagine, the Graph Edit Distance has progressively become a standard tool within the structural pattern recognition framework. Indeed, this distance allows to take into account fine differences between graphs, may be easily tuned and may satisfy all the axioms of a distance. Basically, the most common definition of the graph edit distance is based on the notion of edit path. An edit path between two graphs G_1 and G_2 is a sequence of node/edge removal/substitution or insertion operations transforming G_1 into G_2. Each edit path may be associated to a cost hence defining the Graph Edit Distance between G_1 and G_2 as the minimal cost of all edit paths between these two graphs. Within this survey, we will first review some definitions and properties of the Graph Edit Distance in order to set up a framework which will allow us to review the main families of methods used to compute the graph edit distance. Among them, we may cite methods based on a tree search or methods based on integer programming.

Contents

Image and Shape Analysis

Saliency Detection via A Graph Based Diffusion Model 3
  Zhouqin He, Bo Jiang, Yun Xiao, Chris Ding, and Bin Luo

Shape Simplification Through Graph Sparsification 13
  *Francisco Escolano, Manuel Curado, Silvia Biasotti,
  and Edwin R. Hancock*

Reeb Graphs of Piecewise Linear Functions 23
  Barbara Di Fabio and Claudia Landi

Learning and Graph Kernels

Learning from Diffusion-Weighted Magnetic Resonance Images
Using Graph Kernels 39
  *Sylvain Takerkart, Gottfried Berton, Nicole Malfait,
  and François-Xavier Dupé*

Learning Graph Matching with a Graph-Based Perceptron
in a Classification Context 49
  *Romain Raveaux, Chloé Martineau, Donatello Conte,
  and Gilles Venturini*

A Nested Alignment Graph Kernel Through the Dynamic Time
Warping Framework 59
  Lu Bai, Luca Rossi, Lixin Cui, and Edwin R. Hancock

Graph Applications

GERoMe – A Novel Graph Extraction Robustness Measure 73
  Dominik Drees, Aaron Scherzinger, and Xiaoyi Jiang

Speeding-Up Graph-Based Keyword Spotting in Historical
Handwritten Documents 83
  Michael Stauffer, Andreas Fischer, and Kaspar Riesen

Detecting Alzheimer's Disease Using Directed Graphs 94
  Jianjia Wang, Richard C. Wilson, and Edwin R. Hancock

Graph Matching

Error-Tolerant Coarse-to-Fine Matching Model for Hierarchical Graphs   107
  Pau Riba, Josep Lladós, and Alicia Fornés

A Hungarian Algorithm for Error-Correcting Graph Matching   118
  Sébastien Bougleux, Benoit Gaüzère, and Luc Brun

Introducing VF3: A New Algorithm for Subgraph Isomorphism   128
  Vincenzo Carletti, Pasquale Foggia, Alessia Saggese, and Mario Vento

Large Graphs and Social Networks

Node Matching Computation Between Two Large Graphs in Linear
Computational Cost.   143
  Pep Santacruz, Shaima Algabli, and Francesc Serratosa

Measuring Vertex Centrality Using the Holevo Quantity   154
  Luca Rossi and Andrea Torsello

On the Interplay Between Strong Regularity and Graph Densification   165
  *Marco Fiorucci, Alessandro Torcinovich, Manuel Curado,
  Francisco Escolano, and Marcello Pelillo*

Mining and Clustering

Mining Frequent Patterns in 2D+t Grid Graphs for Cellular
Automata Analysis   177
  Romain Deville, Elisa Fromont, Baptiste Jeudy, and Christine Solnon

Density Normalization in Density Peak Based Clustering   187
  Jian Hou and Hongxia Cui

Fast Nearest Neighbors Search in Graph Space Based
on a Branch-and-Bound Strategy.   197
  Zeina Abu-Aisheh, Romain Raveaux, and Jean-Yves Ramel

Graph Edit Distance

Exact Computation of Graph Edit Distance for Uniform and Non-uniform
Metric Edit Costs   211
  David B. Blumenthal and Johann Gamper

Improved Graph Edit Distance Approximation with Simulated Annealing . . .   222
  Kaspar Riesen, Andreas Fischer, and Horst Bunke

An Edit Distance Between Graph Correspondences   232
Carlos Francisco Moreno-García, Francesc Serratosa,
and Xiaoyi Jiang

A Survey on Applications of Bipartite Graph Edit Distance   242
Michael Stauffer, Thomas Tschachtli, Andreas Fischer,
and Kaspar Riesen

Graphs and Information Theory

Minimising Entropy Changes in Dynamic Network Evolution.   255
Jianjia Wang, Richard C. Wilson, and Edwin R. Hancock

Synchronization Over the Birkhoff Polytope for Multi-graph Matching   266
Michele Schiavinato and Andrea Torsello

Adaptive Feature Selection Based on the Most Informative
Graph-Based Features   276
Lixin Cui, Yuhang Jiao, Lu Bai, Luca Rossi, and Edwin R. Hancock

Author Index   289

Image and Shape Analysis

Saliency Detection via A Graph
Based Diffusion Model

Zhouqin He[1], Bo Jiang[1(✉)], Yun Xiao[1], Chris Ding[2,1], and Bin Luo[1]

[1] School of Computer Science and Technology, Anhui University, Hefei 230601, China
hezhouqin@foxmail.com, {jiangbo,xiaoyun,luobin}@ahu.edu.cn
[2] CSE Department, University of Texas at Arlington, Arlington, TX 76019, USA
chqding@uta.edu

Abstract. This paper proposes a graph based diffusion method for image saliency detection problem by adopting random walk with restart (RWR) model. Our method begins with computing background and foreground priors respectively for the input image. Based on these priors, we then adopt RWR method to obtain more reasonable and accurate background and foreground measurements by further considering the local structure of image. At last, we combine both background and foreground measurements together to obtain a more accurate saliency estimation. Experimental evaluations on four benchmark datasets demonstrate the benefits and effectiveness of the proposed method.

Keywords: Saliency detection · Background prior · Foreground prior · Random walk with restart

1 Introduction

Image saliency detection aims to automatically identify the salient or interesting regions of an image that attract human attention [5,12,13]. It is an important and fundamental problem in pattern recognition and computer vision area. The salient regions of the input image usually indicate the main objects or discriminative features contained in this image. Thus, saliency detection techniques can be widely used in many computer vision and image processing tasks such as image segmentation [34], image retrieval [4] and visual object tracking [23].

In recent years, many methods have been proposed for image saliency detection problem. Generally, these methods can be roughly categorized into three types, i.e., bottom-up methods, top-down methods and combination of top-down and bottom-up methods [9,33,39,40]. Bottom-up methods are usually unsupervised while top-down methods are commonly supervised. In this paper, we focus on bottom-up methods. In the past decade, many bottom-up methods have been proposed [14,21,27,37]. Among them, one kind of commonly used methods is to use graph models and methods [8,10,14]. For example, Gopalakrishnan et al. [8] adopted a random walk model for image salient object location. Yang et al. [39] used graph manifold ranking method to obtain image saliency. Zhu et al.

© Springer International Publishing AG 2017
P. Foggia et al. (Eds.): GbRPR 2017, LNCS 10310, pp. 3–12, 2017.
DOI: 10.1007/978-3-319-58961-9_1

[40] provided a general optimization model to obtain a more accurate saliency detection results by combining background and foreground priors at the same time. Li et al. [20] proposed to use a regularized random walks ranking model to achieve object saliency estimation. Wang et al. [34] also presented a saliency detection approach by combing both local graph structure and background priors together.

In this paper, we propose a new graph based diffusion method for image saliency detection problem by adopting random walk with restart (RWR) model [17,19]. First, we compute both background and foreground priors respectively for the input image. Based on these priors, we then adopt RWR method to obtain more accurate measurements of background and foreground. At last, we combine these two cues together to obtain more accurate saliency estimation. Comprehensive evaluations on benchmark datasets indicate that our method performs better than state-of-the-art methods.

2   Brief Review of Random Walk with Restart

Random walk with restart model has been widely used in many computer vision problems, such as image saliency and segmentation [18,19], object tracking [17] and data mining [26]. Here, we give a brief review of random walk with restart model.

Given a graph $G(V, E)$ with V and E denoting nodes and edges, respectively. A transition matrix $\mathbf{A}$ is first defined, in which $\mathbf{A}_{ij}$ denotes the probability that a walker moves from node v_i to node v_j. In random walk with restart (RWR), starting at a node, the walker have two options at each step, i.e., moving to a randomly chosen neighbor with probability α or jumping to a specified node with probability $(1 - \alpha)$ [17,26]. Formally, in RWR, it iteratively computes the probability distribution $\mathbf{r}^{(t)}$ as,

$$\mathbf{r}^{(t+1)} = \alpha \mathbf{A} \mathbf{r}^{(t)} + (1 - \alpha)\mathbf{p} \tag{1}$$

where $\alpha \in [0, 1]$ is a restarting probability, and $\mathbf{p} = (\mathbf{p}_1, \mathbf{p}_2, \cdots \mathbf{p}_N)$ is a restarting distribution. It is known that regardless of any initialization $\mathbf{p}^{(0)}$, as the iteration t increases, the diffusion process will converge to the stationary distribution $\mathbf{p}^*$. Moreover, since the stationary distribution $\mathbf{p}^*$ satisfies the following,

$$\mathbf{r}^* = \alpha \mathbf{A} \mathbf{r}^* + (1 - \alpha)\mathbf{p}. \tag{2}$$

Thus, the stationary distribution $\mathbf{p}^*$ can also be solved directly as

$$\mathbf{r}^* = (1 - \alpha)(\mathbf{I} - \alpha \mathbf{A})^{-1}\mathbf{p} \tag{3}$$

where $\mathbf{I}$ is an identity matrix.

From diffusion aspect, the above RWR process provides a kind of diffusion. That is, the restarting distribution $\mathbf{p}$ are propagated or diffused throughout the graph $G(V, E)$ and matrix $(1 - \alpha)(\mathbf{I} - \alpha \mathbf{A})^{-1}$ can be regarded as a kind of diffusion matrix [22].

3   Saliency Detection

In this section, we present our saliency detection method based on RWR model. Our method consists of three main steps which are introduced below.

3.1   Graph Construction

Given an input image $\mathcal{I}$, we first segment it into a set of non-overlapping super-pixels via the simple linear iterative clustering (SLIC) approach [2]. Then, we construct a k-regular graph $G(V, E)$ as follows [39]. Each node $v_i \in V$ represents a superpixel and an edge $e_{ij} \in E$ exists between node v_i and v_j if node v_i and v_j are either neighbour or having common boundaries with their neighboring nodes. For each superpixel, we use the average CIE LAB color of all pixels in this superpixel as the feature descriptor [39], and then compute the weight $\mathbf{W}_{ij}$ between node v_i and v_j as,

$$\mathbf{W}_{ij} = exp(-\eta \left\| \mathbf{c}_i - \mathbf{c}_j \right\|^2) \tag{4}$$

where $\mathbf{c}_i$ and $\mathbf{c}_j$ are the mean Lab color feature of superpixel s_i and s_j. The η is set to $\eta = 0.1$ in this paper.

In order to conduct the diffusion process via RWR model, we need to transform graph weight matrix $\mathbf{W}$ to transition matrix $\mathbf{A}$ by normalizing each column of weight matrix $\mathbf{W}$ to 1, i.e.,

$$\mathbf{A}_{ij} = \frac{\mathbf{W}_{ij}}{\sum_{i=1}^{N} \mathbf{W}_{ij}}. \tag{5}$$

Based on transition matrix $\mathbf{A}$, we can conduct diffusion process for both background and foreground as follows.

3.2   Diffusion with Background Prior

Given any background prior $\mathbf{p}_b$, we can use the above RWR model to conduct the diffusion of background prior as follows,

$$\mathbf{b}^{(t+1)} = \alpha \mathbf{A} \mathbf{b}^{(t)} + (1 - \alpha)\mathbf{p}_b \tag{6}$$

where $t = 0, \cdots n$ and $\mathbf{p}_b$ is the restart term. As iteration t increases, $\mathbf{b}^{(t+1)}$ will converge to a stationary distribution $\mathbf{b}^*$ regardless of any initialization $\mathbf{b}^{(0)}$. Comparing with original background prior $\mathbf{p}^b$, the diffusion stationary distribution $\mathbf{b}^*$ maintains more local smooth constraint and thus provides a kind of more reasonable background measurement.

For image saliency detection problem, there exist many background prior computation methods [15,37,39,40]. In this paper, we compute the background prior $\mathbf{p}_b$ using the boundary connectivity. Formally, for any superpixel s, the boundary connectivity (BC) [40] of superpixel s is defined as [40],

$$BC(s) = \frac{\sum_{s_i \in \mathcal{B}} S(s, s_i)}{\sqrt{\sum_{i=1}^{N} S(s, s_i)}} \tag{7}$$

where $\mathcal{B}$ denotes the boundary area of image and $S(s, s_i)$ is the similarity between superpixel s and s_i. N is the number of superpixels. Based on the above boundary connectivity, the background prior $\mathbf{p}_b(s)$ for superpixel s is computed as [40],

$$\mathbf{p}_b(s) = 1 - \exp\left(-\frac{BC(s)^2}{2\sigma_b^2}\right) \tag{8}$$

where σ_b is set to 1 in our experiments.

3.3 Diffusion with Foreground Prior

Similar to background diffusion process, we can also conduct diffusion for foreground prior. Given any foreground prior $\mathbf{f}_b$, such as contrast [5,40], objectness [15] and center prior [34], we can use the above RWR model to obtain the foreground measurement $\mathbf{f}$ as follows,

$$\mathbf{f}^{(t+1)} = \alpha \mathbf{A} \mathbf{f}^{(t)} + (1 - \alpha)\mathbf{p}_f \tag{9}$$

where $t = 0, \cdots n$ and $\mathbf{p}_f$ is the restart term. Comparing with original foreground prior $\mathbf{p}_f$, the diffusion stationary result $\mathbf{f}^*$ maintains more smooth consistent of image object and thus provides a more reasonable foreground measurement.

In this paper, we use background weighted contrast [5,40] to define the foreground prior $\mathbf{p}_f$ which incorporates background information in its definition, i.e.,

$$\mathbf{p}_f = \sum_{i=1}^{N} \mathbf{b}^*(s_i) d_a(s, s_i) \exp\left(-\frac{d_s^2(s, s_i)}{2\sigma_f^2}\right) \tag{10}$$

where $d_a(s, s_i)$ and $d_s(s, s_i)$ denote the Lab color feature distance and spatial distance respectively between superpixel s and s_i. The weight $\mathbf{b}^*$ is the background measurement obtained from the above background diffusion. Parameter σ_f is set to 0.25 in our experiments.

3.4 Combination

After obtaining the background measurement $\mathbf{b}^*$ and foreground measurement $\mathbf{f}^*$ based on RWR model, we then combine these two cues together to compute the final saliency result $\mathbf{r}^*$.

Note that large background measurement (probability) $\mathbf{b}_i^*$ of superpixel s_i encourages superpixel s_i to take a small saliency value $\mathbf{r}_i$, while large foreground prior $\mathbf{b}_i^*$ encourages superpixel s_i to take a large saliency value. This observation can be achieved by minimizing the following energy function [34,40],

$$\min_{\mathbf{r}} \frac{1}{2} \sum_{i=1}^{N} \sum_{j=1}^{N} \mathbf{W}_{ij}(\mathbf{r}_i - \mathbf{r}_j)^2 + \gamma \sum_{i=1}^{N} \mathbf{b}_i^* \mathbf{r}_i^2 + \beta \sum_{i=1}^{N} \mathbf{f}_i^*(\mathbf{r}_i - 1)^2. \tag{11}$$

where γ, β are two balancing parameters. Matrix $\mathbf{W}$ is the weight matrix of graph $G(V, E)$, as mentioned above. This optimization problem has a closed-form

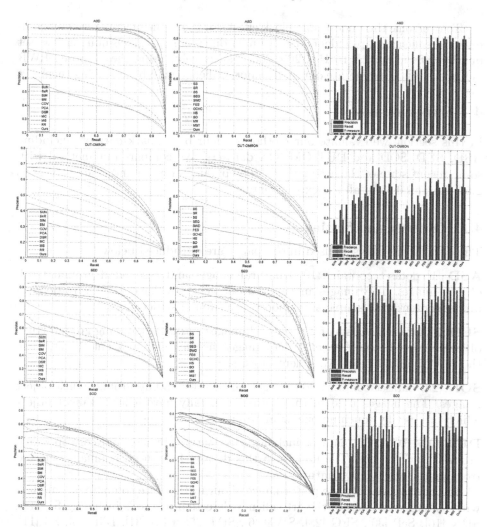

Fig. 1. Quantitative PR-curve and F-measure evaluation of recent approaches on four benchmark datasets. The rows from top to bottom correspond to ASD, DUT-OMRON, SED and SOD dataset respectively. The first two columns show the comparison of PR-curves and the last column shows the comparison of F-measure values using an adaptive threshold.

solution and the optimal solution $\mathbf{r}^*$ can be obtained by setting the first derivative of the energy function w.r.t. variable $\mathbf{r}$ to $\mathbf{0}$, i.e.,

$$\mathbf{r}^* = \beta \left(\mathbf{D} - \mathbf{W} + \gamma \mathbf{B} + \beta \mathbf{F} \right)^{-1} \mathbf{f}^* \tag{12}$$

where $\mathbf{D}$ is a diagonal matrix with $\mathbf{D}_{ii} = \sum_{j=1}^{N} \mathbf{W}_{ij}$. Matrix $\mathbf{B}$ and $\mathbf{F}$ are diagonal matrices with $\mathbf{B}_{ii} = \mathbf{b}_i^*$ and $\mathbf{F}_{ii} = \mathbf{f}_i^*$.

To overcome the undesirable block effect, for the input image $\mathcal{I}$, we first compute the saliency maps independently under different superpixels numbers (we set number of superpixels to 150, 200, 250, respectively) using the proposed RWR method, and then generate our final smooth saliency map by averaging them.

4   Experiments

In this section, we evaluate our method on four benchmark datasets including ASD, SED, SOD and DUT-OMRON datasets. We compare our method with some recent state-of-the-art methods. The parameter γ , β and α are setting to 2.5 , 0.5 and 0.35 in all the experiments.

4.1   Datasets and Settings

The benchmark datasets used in our paper are introduced below.

ASD [1] dataset contains 1000 natural images and associated ground truth object images. The images in this dataset generally have large variations in image content with simple and smooth background structures.

SED [3] dataset consists of 200 natural images and corresponding ground truth object images. In this dataset, half of images involve a single salient object, while the rest of images have two salient objects.

SOD [24] dataset contains 300 images and corresponding ground truth object images. The images used in this dataset generally have a large salient object which are usually touching one or more image boundaries.

DUT-OMRON [39] dataset includes 5168 natural images and corresponding ground truth images. The images in this dataset have been manually selected from more than 140,000 natural images.

In order to evaluate the performance of our method, we compare our method with some recent state-of-the-art saliency detection methods including SR [12], SUN [16], SeR [30], SEG [28], SWD [6], SIM [25], FES [31], BM [35], SS [11], COV [7], PCA [29], BS [36], DSR [21], GCHC [38], HS [37], MC [14], MR [39], MS [32], SO [40], RR [20] and MST [33]. The codes or executables of all these compared methods are available and we use them to obtain the results. For performance evaluation, we use standard precision-recall curves (PR-curve) and F-measure evaluation metrics. To obtained the precision-recall (PR) curve, we need to binarize the normalized saliency map with thresholds ranging from 0 to 255, respectively and then compute the precision and recall values respectively. In addition to PR curve, we also use the F-measure metric for evaluation. The F-measure is the weighted average of precision and recall, which is defined as follows,

$$F_\lambda = \frac{(1 + \lambda)\text{Precision} \times \text{Recall}}{\lambda\text{Precision} + \text{Recall}} \tag{13}$$

We set $\lambda = 0.3$ in all the experiments to give more weight to precision than recall, as suggested in [20, 39].

4.2 Results

Figure 2 shows some visual comparison examples of the different saliency methods. Intuitively, one can note that, the proposed method generally highlights the salient object in desirable and can preserve the finer object boundaries more clearly than other compared methods, which demonstrates the proposed

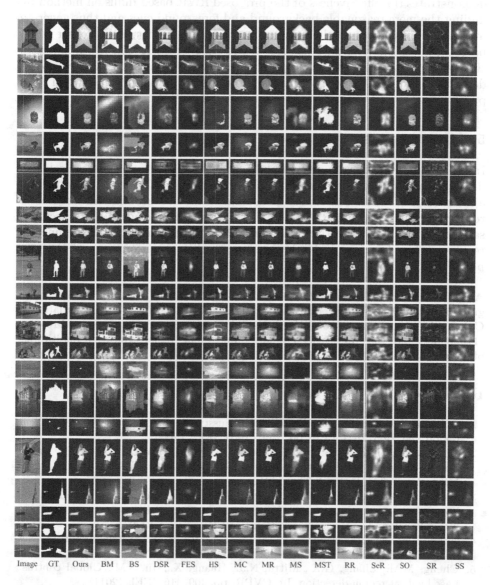

Image GT Ours BM BS DSR FES HS MC MR MS MST RR SeR SO SR SS

Fig. 2. Sample saliency maps of the compared methods. Intuitively, our saliency maps are generally more close to the ground truth results which demonstrates the more accurate saliency detection results obtained by our method.

method can detect the saliency region more accurate than some other methods. Figure 1 shows the precision-recall curves and F-measure of all methods on different datasets, respectively. Overall, the results demonstrate that the proposed method outperforms the state-of-the-art methods on all the four public benchmark datasets. More detailly, we can note that: (1) Comparing with the traditional SO [40] method, our method can obtain better results, which clearly demonstrates the effectiveness of the proposed RWR based diffusion method by finding the more reasonable background and foreground cues and thus generate more accurate saliency results. (2) Our method performs better than other graph based diffusion methods such as MR [39], MC [14] and RR [20], which indicates the more effectiveness of the proposed method on conducting saliency object detection task. (3) Our method generally obtains better performance when compared to some recent saliency methods such as MST [33] and RR [20].

5   Conclusions

In this paper, we propose a new graph based diffusion method for image saliency detection problem by using random walk with restart (RWR) model. Our method aims to derive a more reasonable measurement for background and foreground respectively by adopting RWR model, which can thus lead to more accurate saliency estimation by further using an optimization framework. Experimental results on four benchmark databases show the better performance of the proposed method.

Acknowledgement. This study was funded by the National Key Basic Research Program of China (973 Program) (2015CB351705); National Nature Science Foundation of China (61602001, 61572030, 61671018); Natural Science Foundation of Anhui Province (1708085QF139); Natural Science Foundation of Anhui Higher Education Institutions of China (KJ2016A020).

References

1. Achanta, R., Hemami, S., Estrada, F., Susstrunk, S.: Frequency-tuned salient region detection. In: CVPR, pp. 1597–1604. IEEE (2009)
2. Achanta, R., Shaji, A., Smith, K., Lucchi, A., Fua, P., Süsstrunk, S.: SLIC superpixels compared to state-of-the-art superpixel methods. IEEE Trans. Pattern Anal. Mach. Intell. **34**(11), 2274–2282 (2012)
3. Alpert, S., Galun, M., Basri, R., Brandt, A.: Image segmentation by probabilistic bottom-up aggregation and cue integration. In: 2007 IEEE Conference on Computer Vision and Pattern Recognition (CVPR), pp. 1–8 (2007)
4. Chen, T., Cheng, M.M., Tan, P., Shamir, A., Hu, S.M.: Sketch2photo: internet image montage. ACM Trans. Graph. (TOG) **28**(5), 124 (2009)
5. Cheng, M.M., Zhang, G.X., Mitra, N.J., Huang, X., Hu, S.M.: Global contrast based salient region detection. In: CVPR, pp. 409–416. IEEE (2011)
6. Duan, L., Wu, C., Miao, J., Qing, L., Fu, Y.: Visual saliency detection by spatially weighted dissimilarity. In: 2011 IEEE Conference on Computer Vision and Pattern Recognition (CVPR), pp. 473–480. IEEE (2011)

7. Erdem, E., Erdem, A.: Visual saliency estimation by nonlinearly integrating features using region covariances. J. Vis. **13**(4), 1–20 (2013)
8. Gopalakrishnan, V., Hu, Y., Rajan, D.: Random walks on graphs for salient object detection in images. IEEE Trans. Image Process. **19**(12), 3232–3242 (2010)
9. Han, B., Zhu, H., Ding, Y.: Bottom-up saliency based on weighted sparse coding residual. In: International Conference on Multimedia 2011, Scottsdale, AZ, USA, November 28 - December, pp. 1117–1120 (2011)
10. Harel, J., Koch, C., Perona, P.: Graph-based visual saliency. In: Advances in Neural Information Processing Systems, pp. 545–552 (2006)
11. Hou, X., Harel, J., Koch, C.: Image signature: highlighting sparse salient regions. IEEE Trans. Pattern Anal. Mach. Intell. **34**(1), 194–201 (2012)
12. Hou, X., Zhang, L.: Saliency detection: a spectral residual approach. In: 2007 IEEE Conference on Computer Vision and Pattern Recognition (CVPR), pp. 1–8. IEEE (2007)
13. Itti, L., Koch, C., Niebur, E., et al.: A model of saliency-based visual attention for rapid scene analysis. IEEE Trans. Pattern Anal. Mach. Intell. **20**(11), 1254–1259 (1998)
14. Jiang, B., Zhang, L., Lu, H., Yang, C., Yang, M.H.: Saliency detection via absorbing Markov chain. In: ICCV, pp. 1665–1672 (2013)
15. Jiang, P., Ling, H., Yu, J., Peng, J.: Salient region detection by UFO: uniqueness, focusness and objectness. In: Proceedings of the IEEE International Conference on Computer Vision, pp. 1976–1983 (2013)
16. Kanan, C., Tong, M.H., Zhang, L., Cottrell, G.W.: Sun: top-down saliency using natural statistics. Vis. Cognit. **17**(6–7), 979–1003 (2009)
17. Kim, H.U., Lee, D.Y., Sim, J.Y., Kim, C.S.: SOWP: spatially ordered and weighted patch descriptor for visual tracking. In: Proceedings of the IEEE International Conference on Computer Vision, pp. 3011–3019 (2015)
18. Kim, J.S., Sim, J.Y., Kim, C.S.: Multiscale saliency detection using random walk with restart. IEEE Trans. Circ. Syst. Video Technol. **24**(2), 198–210 (2014)
19. Kim, T.H., Lee, K.M., Lee, S.U.: Generative image segmentation using random walks with restart. In: Forsyth, D., Torr, P., Zisserman, A. (eds.) ECCV 2008. LNCS, vol. 5304, pp. 264–275. Springer, Heidelberg (2008). doi:10.1007/978-3-540-88690-7_20
20. Li, C., Yuan, Y., Cai, W., Xia, Y., Feng, D.D.: Robust saliency detection via regularized random walks ranking. In: CVPR, pp. 2710–2717 (2015)
21. Li, X., Lu, H., Zhang, L., Ruan, X., Yang, M.H.: Saliency detection via dense and sparse reconstruction. In: ICCV, pp. 2976–2983 (2013)
22. Lu, S., Mahadevan, V., Vasconcelos, N.: Learning optimal seeds for diffusion-based salient object detection. In: Proceedings of the IEEE Conference on Computer Vision and Pattern Recognition, pp. 2790–2797 (2014)
23. Mahadevan, V., Vasconcelos, N.: Saliency-based discriminant tracking. In: IEEE Conference on Computer Vision and Pattern Recognition, pp. 1007–1013 (2009)
24. Movahedi, V., Elder, J.H.: Design and perceptual validation of performance measures for salient object segmentation. In: Computer Society Conference on Computer Vision and Pattern Recognition-Workshops, pp. 49–56. IEEE (2010)
25. Murray, N., Vanrell, M., Otazu, X., Parraga, C.A.: Saliency estimation using a non-parametric low-level vision model. In: 2011 IEEE Conference on Computer Vision and Pattern Recognition (CVPR), pp. 433–440. IEEE (2011)

26. Pan, J.Y., Yang, H.J., Faloutsos, C., Duygulu, P.: Automatic multimedia cross-modal correlation discovery. In: Proceedings of the Tenth ACM SIGKDD International Conference on Knowledge Discovery and Data Mining, pp. 653–658. ACM (2004)

27. Perazzi, F., Krähenbühl, P., Pritch, Y., Hornung, A.: Saliency filters: contrast based filtering for salient region detection. In: CVPR, pp. 733–740. IEEE (2012)

28. Rahtu, E., Kannala, J., Salo, M., Heikkilä, J.: Segmenting salient objects from images and videos. In: Daniilidis, K., Maragos, P., Paragios, N. (eds.) ECCV 2010. LNCS, vol. 6315, pp. 366–379. Springer, Heidelberg (2010). doi:10.1007/978-3-642-15555-0_27

29. Ran, M., Tal, A., Zelnikmanor, L.: What makes a patch distinct? In: IEEE Conference on Computer Vision and Pattern Recognition, pp. 1139–1146 (2013)

30. Seo, H.J., Milanfar, P.: Static and space-time visual saliency detection by self-resemblance. J. Vis. **9**(12), 1–27 (2009)

31. Rezazadegan Tavakoli, H., Rahtu, E., Heikkilä, J.: Fast and efficient saliency detection using sparse sampling and kernel density estimation. In: Heyden, A., Kahl, F. (eds.) SCIA 2011. LNCS, vol. 6688, pp. 666–675. Springer, Heidelberg (2011). doi:10.1007/978-3-642-21227-7_62

32. Tong, N., Lu, H., Zhang, L., Ruan, X.: Saliency detection with multi-scale super-pixels. IEEE Signal Process. Lett. **21**(9), 1035–1039 (2014)

33. Tu, W.C., He, S., Yang, Q., Chien, S.Y.: Real-time salient object detection with a minimum spanning tree. In: 2016 IEEE Conference on Computer Vision and Pattern Recognition (CVPR), pp. 2334–2342 (2016)

34. Wang, Q., Zheng, W., Piramuthu, R.: Grab: visual saliency via novel graph model and background priors. In: 2016 IEEE Conference on Computer Vision and Pattern Recognition (CVPR), pp. 535–543 (2016)

35. Xie, Y., Lu, H.: Visual saliency detection based on Bayesian model. In: International Conference on Image Processing, vol. 263, no. 4, pp. 645–648 (2011)

36. Xie, Y., Lu, H., Yang, M.H.: Bayesian saliency via low and mid level cues. IEEE Trans. Image Process. **22**(5), 1689–1698 (2013)

37. Yan, Q., Xu, L., Shi, J., Jia, J.: Hierarchical saliency detection. In: CVPR, pp. 1155–1162 (2013)

38. Yang, C., Zhang, L., Lu, H.: Graph-regularized saliency detection with convex-hull-based center prior. IEEE Signal Process. Lett. **20**(7), 637–640 (2013)

39. Yang, C., Zhang, L., Lu, H., Ruan, X., Yang, M.H.: Saliency detection via graph-based manifold ranking. In: CVPR, pp. 3166–3173 (2013)

40. Zhu, W., Liang, S., Wei, Y., Sun, J.: Saliency optimization from robust background detection. In: CVPR, pp. 2814–2821 (2014)

Shape Simplification Through Graph Sparsification

Francisco Escolano[1]([✉]), Manuel Curado[1], Silvia Biasotti[2],
and Edwin R. Hancock[3]

[1] Department of Computer Science and AI,
University of Alicante, 03690 Alicante, Spain
{sco,mcurado}@dccia.ua.es
[2] CNR-IMATI, Via de Marini, 6 (Torre di Francia), 16149 Genova, Italy
silvia.biasotti@ge.imati.cnr.it
[3] Department of Computer Science, University of York, York YO10 5DD, UK
erh@cs.york.ac.uk

Abstract. In this paper, we draw on Spielman and Srivastava's method for graph sparsification in order to simplify shape representations. The underlying principle of graph sparsification is to retain only the edges which are key to the preservation of desired properties. In this regard, sparsification by edge resistance allows us to preserve (to some extent) links between protrusions and the remainder of the shape (e.g. parts of a shape) while removing in-part edges. Applying this idea to alpha shapes (abstract representations which have a huge number of edges) opens up a way of introducing a hierarchy of the edge strength, thus being relevant for shape analysis and interpretation.

Keywords: Graph sparsification · Shape simplification · Alpha shapes

1 Introduction

1.1 Shape Representations: Triangulations vs Alpha Shapes

The traditional problem addressed by shape reconstruction is to recover a digital representation of a physical shape that has been scanned, where the scanned data contain a wide variety of defects or the representation of data acquired by different diagnostic equipments such as angiography, Computed Tomography (CT) and Magnetic Resonance (MR). To encode the data in a digital model different geometric representations have been explored in detail. The work reported in [11] organizes them into a spectrum with respect to the achieved trade-off between verbosity and complexity. Voxel grids are at one extreme of the spectrum, since they are the simplest, but the most verbose and less accurate representation. Although, in principle, the use of arbitrarily fine grids could achieve any level of approximation, the practical limit comes from constraints on the resolution. At the other end of the spectrum, the functional representations - using smooth functions to specify the continuous of points that make up the shape - provide

© Springer International Publishing AG 2017
P. Foggia et al. (Eds.): GbRPR 2017, LNCS 10310, pp. 13–22, 2017.
DOI: 10.1007/978-3-319-58961-9_2

an accurate and complex representation. Piecewise linear representations are at the center of the spectrum.

The most popular representations in the piecewise class are the simplicial complexes [12], including triangular meshes that have become the *de-facto* standard in graphics accelerators [5] and tetrahedral meshes that are used to represent volumes and are used for the simulation of deformable models, such as organs or tissues. A generalization of the concept of triangulation are the so-called alpha (α-) shapes, that are families of piecewise linear simple curves in the Euclidean space associated with a dense and unorganized set of data points. An alpha shape is demarcated by a frontier, which is a linear approximation of the original shape. First introduced in the 2D plane by Edelsbrunner et al. [8], they were extended to 3D spaces [10] and higher dimensions [6]. In the case of 2D, an alpha shape consists of vertices, edges and triangles, while for 3D there are also tetrahedra. In our graph representation, we consider the 1-skeleton of both triangulations and alpha-shapes, i.e., the set of vertices and edges of the complex.

Fig. 1. From left to right: a point set, a triangulation and a sequence of three alpha shapes with increasing values of α.

Alpha shapes depend on the parameter α used as radius of spheres centered on the points that determine the connection among the neighbourhoods. A very small value will generate many isolated points and the alpha shape degenerates to the point cloud when $\alpha \to 0$. On the other hand, a large value of α will consider many points inside the spheres and therefore the size of the 1-skeleton considerably increases. The limit of the alpha shape when $\alpha \to \infty$ is the convex hull of the point cloud and the 1-skeleton of the alpha shape becomes the complete graph.

The main application of alpha shapes is the reconstruction of objects which have been sampled by points. How to determine the best value of α is not obvious and in practice α is found using a trial-and-error strategy. This leads the computation of quite large families of alpha shapes and the 1-skeleton increases as long as the value of α increases. Moreover, there are point-sets for which there is no unique α value, for instance because small α values capture local characteristics while larger ones determine large connectivity. For instance, this is the case when a point cloud is not uniformly sampled or the point cloud is supposed to represent either small or large features (for instance, it contains both thin and long handles like the examples shown in Fig. 1). Low density sampling requires a rather large radius to build a connected representation.

But a large value of α will unfortunately close some handles. In practice, a large value of α results in (among possibly other things) a closure of handles, connection of multiple components and joints (e.g., sharp turns) being destroyed. For this reasons and because the general size of the 1-skeleton, the approach proposed in this work is able to simplify connections without destroying the global topology of the alpha shape.

1.2   Contributions

Spielman and Srivastava [15] have developed an efficient method for graph sparsification based on edge resistance (which is proportional to the commute time of its end nodes). The method is based on the observation that the probability of an edge appearing in a random spanning tree of a graph is equal to its effective resistance. Drawing on Spielman and Teng's approximately linear solver [16], they show how to efficiently compute resistance, and hence sample edges for the purposes of sparsification.

Herein, we present a unified view of resistance sparsification through sampling. In addition, we exploit such a sampling for retaining edges (both in triangulations and alpha spaces) that are key to the preservation of the topological properties of the input shape. Our experiments show high compression rates as the allowed error ϵ increases. However each shape is sensitive to a different value of ϵ. It is the persistence of a given edge as ϵ increases what will provide us with is the relative importance of a vertex. This characterisation is pivotal for subsequent tasks such as efficient shape matching and shape representation.

2   Graph Sparsification

2.1   Definition and Ingredients

Graph sparsification [16] is the principled study of how to significantly decrease the number of edges of an input graph G so that the output, H, preserves some of the structural properties of G.

Benczúr and Karger [4] showed that every cut in $G = (V, E)$ can be approximated in $H = (V, E')$, with $E' \subseteq E$, so that every cut in H has a value within $(1 \pm \epsilon)$ times its value in G. For instance, a K_n (complete) graph with n vertices and $O(n^2)$ edges can be approximated by a random d–regular graph, i.e. a graph with $O(dn)$ edges. This means that for every subset $S \subset V$ the ratio between the value of a cut in K_n and that of the same cut in the random d–regular graph H is n/d. This link between sparsification and random graphs is useful (to some extent). For instance, if an edge in G is included in H with probability p, we must set $p \gg 1/c$ where c is the value of the minimal cut. As a result, if we have m edges in G we can only have $O(m/c)$ edges in H.

This limitation leads to *non-uniform sampling*, i.e. to associate a different probability p_e to each edge $e \in E$. The edge e it is included in E' with probability p_e and it is given a weight $1/p_e$ if it is included. This inverse weighting ensures that the expected weight of e in H is unity.

The *choice of a suitable value of* p_e is the first step in graph sparsification. For cut sparsification, the choice of p_e relies on the *strong* connectivity c_e of e. The strong connectivity c_e is the maximum value of a cut in a connected component including e. This quantity is upper bounded by the standard connectivity of e (the minimal value of a cut separating its endpoints), but it is hard to find. However lower bounds $c'_e \leq c_e$ can be founds through sparse certification (see details in [4]). In this way we have that $p_e = \rho/c'_e \geq \rho/c_e$, where ρ is the compression factor, is a good choice for p_e. The *compression factor* ρ has complexity $O(c(d+2)(\log n)/\epsilon^2)$ and it is in turn inversely proportional to the squared error ϵ^2. The setting $p_e = \min\{1, \rho/c'_e\}$ then ensures the correctness of the approximation with probability $1 - n^{-d}$.

The above rationale leads to the second ingredient of sparsification, namely the *minimal number of samples* required to correctly sparsify the graph with high probability. For cut sparsification, we have that taking $O(n\rho)$, i.e. $O(n \log n/\epsilon^2)$, samples will suffice. This can be proved by means of the Chernoff bound, which is a standard information-theoretic tool for limiting the number of samples.

2.2 Spectral Formulation and Effective Resistances

An alternative approach to the the sparsification problem consists of enforcing the preservation of structural properties by bounding the quadratic form associated with the graph Laplacian of the sparsified graph H with respect to that of the input graph G (see the survey in [3]). Therefore, given $G = (V, E, w)$ we must obtain $H = (V, E', w')$ by taking $O(n \log n/\epsilon^2)$ independent samples, so that we satisfy (with probability at least $1/2$) the following constraint

$$\forall\, x \in \mathbb{R}^n : (1 - \epsilon) \leq x^T L_G x \leq x^T L_H x \leq (1 + \epsilon)x^T L_G, \qquad (1)$$

where $\epsilon > 0$, $n = |V|$, and L_G, L_H are the respective Laplacian matrices of G and H. Recall that $L_G = D - W$ where D is the diagonal degree matrix and W is the weighted adjacency matrix, and that $x^T L_G x = \sum_{(u,v) \in E}(x(u) - x(v))^2 w_{uv}$ and similarly for L_H.

Since Laplacian matrices are Semidefinite Positive (SDP), which is denoted by $L_G \succeq 0$, we can reformulate Eq. 1 in terms of circumventing the hyper ellipsoid associated with L_H with that associated with L_G, i.e. one must satisfy

$$(1 - \epsilon)L_G \preceq L_H \preceq (1 + \epsilon)L_G, \text{ or equivalently } L_G \preceq L_H \preceq \kappa L_G, \qquad (2)$$

with $\kappa = \frac{1+\epsilon}{1-\epsilon}$. This implies that all of the eigenvalues λ'_i of L_H satisfy $\lambda'_i \leq \kappa \lambda_i$, where λ_i is the corresponding eigenvalue of L_G. In addition, since Eq. 2 is invariant under rescaling, we have that

$$L_G^{-1/2} L_G L_G^{-1/2} \preceq L_G^{-1/2} L_H L_G^{-1/2} \preceq \kappa L_G^{-1/2} L_G L_G^{-1/2} \qquad (3)$$

i.e.

$$I \preceq L_G^{-1/2} L_H L_G^{-1/2} \preceq \kappa I, \qquad (4)$$

where I is the identity matrix and $L_G^{-1/2} L_H L_G^{-1/2}$ is the so called *relative Laplacian*. This leads to locating L_H so that the relative Laplacian is properly contained between I and κI. In this regard, the structure of L_H is determined by a weighted sum of outer products: $L_H = \sum_{e \in E} w'_e b_e b_e^T$, where w'_e are the unknown weights, $b_e = \delta_u - \delta_v = b_{uv}$, δ_u is the unit vector with a 1 at u and zeros elsewhere (similarly for v), and $e = (u, v)$ is the edge. In this regard, since $E' \subseteq E$, an edge of E not included in E' will have $w'_e = 0$. We define the random variables s_e (our unknowns) so that $w'_e = s_e w_e$ where $\mathbb{E}(s_e) = 1$ for all $e \in E$. Then, Eq. 4 can be rewritten as follows

$$I \preceq L_G^{-1/2} \left(\sum_{e \in E} s_e w_e L_G^{-1/2} b_e b_e^T \right) L_G^{-1/2} \preceq \kappa I. \tag{5}$$

It is well known that the Laplacian matrix L cannot be inverted since it contains the zero eigenvalue. Expressions including the inverse must be computed using the pseudo-inverse L^+ instead. The pseudo inverse plays a key role in defining the effective resistance across $e = (u, v)$ (the scaled commute time) R_e, which is given by

$$R_e = (\delta_u - \delta_v)^T L^+ (\delta_u - \delta_v) = b_e^T L^+ b_e. \tag{6}$$

Then, combining Eqs. 5 and 6 we obtain

$$I \preceq \sum_{e \in E} s_e w_e v_e v_e^T \preceq \kappa I, \tag{7}$$

where $v_e = L_G^{-1/2} b_e$, i.e., the squared norm of v_e is

$$\|v_e\|^2 = (L_G^{-1/2} b_e)^T (L_G^{-1/2} b_e) = (b_e^T L_G^{-1/2})(L_G^{-1/2} b_e) = b_e^T L_G^+ b_e = R_e. \tag{8}$$

This squared norm allows us to treat $\sum_{e \in E} s_e w_e v_e v_e^T$ in Eq. 5 as a quadratic form quite close to the identity matrix I. This is extremely important since: (i) the relative Laplacian relies on the effective resistances of G, and (ii) we can pose the sparsification problem in terms of finding the sampling probabilities p_e so that the constraint in Eq. 5 is satisfied. To this end, Batson et al. [2] exploited the following fact:

$$\sum_{e \in E} \tilde{v}_e \tilde{v}_e^T = I, \tag{9}$$

where $\tilde{v}_e = w_e^{1/2} v_e$. This can be proved by using the $m \times n$ incidence matrix of G, i.e. B, with elements $B(e, v) = 1$ if v is e's head, $B(e, v) = -1$ if v is e's tail, and $B(e, v) = 0$ otherwise. Then the Laplacian matrix of G is given by $L_G = B^T W_e B$, where W_e is the diagonal $m \times m$ matrix where $W_e(e, e) = w_e$. Since the vectors $v_e = L_G^{-1/2} b_e$ rely on the columns of B^T, we have that vectors $\tilde{v}_e = v_e w_e^{1/2}$ are the columns of a $n \times m$ matrix $\tilde{V} = L_G^{-1/2} B^T W_e^{1/2}$. Then

$$\sum_{e \in E} \tilde{v}_e \tilde{v}_e^T = \tilde{V} \tilde{V}^T = L_G^{-1/2} B^T W_e^{1/2} W_e^{1/2} B L_G^{-1/2} = L_G^{-1/2} L_G L_G^{-1/2} = I.$$

In addition we have that

$$||\tilde{v}_e||^2 = (w_e^{1/2}L_G^{-1/2}b_e)^T(w_e^{1/2}L_G^{-1/2}b_e) = w_e(b_e^T L_G^+ b_e) = w_e R_e, \qquad (10)$$

i.e. we obtain weighted effective resistances. The identity $||\tilde{v}_e||^2 = w_e R_e$ suggests to sample E with probabilities p_e proportional to $w_e R_e$.

Let $y_1, y_2, \ldots, y_q$ vectors drawn independently with replacement from the distribution

$$y = \frac{1}{\sqrt{p_e}}\tilde{v}_e \text{ with probability } p_e. \qquad (11)$$

Then, the expectation of yy^T (which contains the effective resistances) is

$$\mathbb{E}\left[yy^T\right] = \sum_{e \in E} p_e \frac{1}{p_e}\tilde{v}_e \tilde{v}_e^T = I. \qquad (12)$$

In addition, the shape of each of the q samples $y_i = \tilde{v}_e/\sqrt{p_e}$ leads to

$$\frac{1}{q}\sum_{i=1}^{q} y_i y_i^T = \frac{1}{q}\sum_{i=1}^{q} \#e \frac{\tilde{v}_e}{\sqrt{p_e}} \cdot \frac{\tilde{v}_e^T}{\sqrt{p_e}} = \frac{1}{q}\sum_{i=1}^{q} \#e \frac{\tilde{v}_e \tilde{v}_e^T}{p_e} = \sum_{e \in E} s_e \tilde{v}_e \tilde{v}_e^T, \qquad (13)$$

where $\#e$ is the number of times that e is sampled, and $s_e = \#e/qp_e$. Then, we obtain

$$\frac{1}{q}\sum_{i=1}^{q} y_i y_i^T = \sum_{e \in E} s_e \tilde{v}_e \tilde{v}_e^T = \sum_{e \in E} s_e w_e v_e v_e^T, \qquad (14)$$

i.e. a proper sampling process leads to the relative Laplacian. This is ensured insofar $\frac{1}{q}\sum_{i=1}^{q} y_i y_i^T$ and $\mathbb{E}yy^T$ conform the Chernoff bound for matrices [14]:

$$\mathbb{E}\left[\left|\left|\frac{1}{q}\sum_{i=1}^{q} y_i y_i^T - \mathbb{E}\left[yy^T\right]\right|\right|\right] \leq min\left(CM\sqrt{\frac{\log q}{q}}, 1\right), \qquad (15)$$

where $||\left[\mathbb{E}yy^T\right]|| < 1$ and $sup_y||y|| \leq M$. The first norm condition is verified since $\mathbb{E}\left[yy^T\right] = I$. For verifying the second norm condition we must set the link between $w_e R_e$ (weighted effective resistances) and p_e (sampling probabilities). In order to do so, Spielman and Srivastava [15] exploit the fact that $\sum_e w_e R_e = n - 1$. Therefore, we may set

$$p_e = \frac{w_e R_e}{n - 1} \text{ so that } ||y|| = \frac{1}{\sqrt{p_e}}\sqrt{w_e R_e} = \sqrt{\frac{n-1}{w_e R_e}}\sqrt{w_e R_e} = \sqrt{n - 1}. \qquad (16)$$

Therefore, taking $q = 9C^2 n \log n/\epsilon^2$ yields

$$\mathbb{E}\left[\left|\left|\frac{1}{q}\sum_{i=1}^{q} y_i y_i^T - \mathbb{E}\left[yy^T\right]\right|\right|\right] \leq C\sqrt{\epsilon^2 \frac{\log(9C^2 n \log n/\epsilon^2)(n-1)}{9C^2 n \log n}} \leq \epsilon/2, \qquad (17)$$

for n large enough and $\epsilon \geq 1/\sqrt{n}$.

Summarising, the resistance-based sparsifier [15] consists in five steps:

1. Given the input graph $G = (V, E, w)$, estimate the effective resistances R_e for each $e \in E$.
2. Set an error tolerance ϵ. Set $E' = \emptyset$, $w' = \emptyset$, define $H = (V, E', w')$ and set $\#e = 0$ for all edges in E.
3. Make $q = 9C^2 n \log n / \epsilon^2$ independent samples (with replacement) with probability $p_e \propto w_e R_e$. Each sample is associated with an edge e.
4. If e is selected from a cumulative sum test, then increment $\#e$ and add e to E' with weight $1/p_e$.
5. For all $e \in E'$ set $w'_e = \frac{\#e}{q p_e}$.

Finally, the computation R_e can be accomplished using exact spectral methods [13]. However, this step takes $O(n^3)$ steps and the eigenvalues are ill conditioned if the graph G has several connected components. This is why Spielman and Srivastava [15] propose to approximate the computation of effective resistances by exploiting the Achlioptas version [1] of the Johnson-Lindenstrauss (JL) Lemma. This lemma states that if we project the original vectors (for instance those belonging to the effective resistance embedding) onto a subspace spanned by $O(\log n)$ random vectors, the distances between the projected vectors and the original ones are preserved, and then to some extent are given by ϵ.

3   Experiments

We have performed several experiments on the reduction of the 1-skeleton of both triangulations and alpha shapes. As previously mentioned, triangulations are the standard *de-facto* representations of the surface of 3D objects.

Triangulations are sets of triangles and vertices and are fully described by their 1-skeleton. All vertices of a triangulation have the same importance. For instance, it is not possible to distinguish peaks, pits or passes from other structures. Moreover, connections are all represented without any relations with their importance (for instance from shape outliers or dense regions). For this reason, it is necessary to derive more abstract, high level shapes. In this sense, sparsification can act has a tool able to determine a hierarchy between the vertex connections. It may therefore determine a relative importance of the vertices.

Alpha shapes provide a family of shape *representations* that is very useful when performing shape reconstruction. The reason for this is that they connect vertices with all neighbourhoods that are enclosed in a ball of radius alpha. In general, alpha shapes generalize triangulations and their importance is mainly theoretical. In our experiments on 3D point clouds, triangulations represent the external boundary connections, while alpha shapes encode spatial (volumetric) relationships.

Figure 2 shows five triangulations used in our experiments. These 3D models correspond to an abstract shape, a cactus, a deer, a cup and a cow model, respectively. Most of these models contain features that can be considered to be at a small scale (for instance the small handles in the abstract shape, the details of the cow and deer models, etc.) or to a larger scale, such as the handles and the elongated parts (in the cactus, the deer and the cup models). The results

Fig. 2. Examples of triangulations used in our experiments.

Table 1. Statistics on the number of edges of the 1-skeleton of some triangulations when the parameter ϵ increases.

Triangulation	$\epsilon = 0$	$\epsilon = 0.25$	$\epsilon = 0.75$	$\epsilon = 1.25$	$\epsilon = 1.75$	$\epsilon = 2.25$
Model in Fig. 2(a)	3906	3905	3837	3090	2098	1438
Model in Fig. 2(b)	4623	4622	4550	3643	2543	1782
Model in Fig. 2(c)	15012	15011	14757	11871	80623	5506
Model in Fig. 2(d)	18837	18836	18504	17146	10392	7290
Model in Fig. 2(e)	21759	21757	21415	17201	11677	7989

in Table 1 report the number of edges when sparsification is performed and how they vary when the value of the ϵ parameter increases[1].

Similarly Fig. 3 shows the 1-skeletons of five alpha shapes that were constructed over various point clouds, also varying the α value. These correspond to two different versions of the abstract shape already shown in Fig. refFig:triangulations(a), two alpha shapes of the deer model in Fig. 2(c) and an alpha shape from the cow point set that correspond to Fig. 2(e). The choice of these alpha shapes is motivated by the presence of small and larger handles and features that alpha shapes have difficulty capturing with a single choice of the parameter α, as previously discussed. The results in Table 2 report the number of edges of the 1-skeleton of the alpha shape when the value of the ϵ parameter increases. From these experiment, we think that with sparsification would be possible to overcome the limitations of alpha shapes in the sense that we hope that it will be possible to commence from a quite large value of the parameters α and then to remove the redundant edges by using sparsification, thus implementing a connected, progressive, geometrical-topological peeling of the shape.

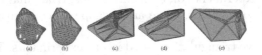

Fig. 3. Examples of alpha shapes used in our experiments.

---

[1] In this paper, the parameter ϵ controls the number of samples needed by the process, whereas the weight for choosing the edges is given by effective resistances.

Fig. 4. Degradation of topological properties as ϵ increases. 2D projections of the blob alpha shape. From left to right: $\epsilon = 0$, $\epsilon = 0.75$, $\epsilon = 1.0$ and $\epsilon = 1.25$.

Table 2. Statistics on the number of edges of the 1-skeleton of some alpha shapes when the parameter ϵ increases.

Alpha-shape	α	$\epsilon = 0$	$\epsilon = 0.25$	$\epsilon = 0.75$	$\epsilon = 1.25$	$\epsilon = 1.75$	$\epsilon = 2.25$
Model in Fig. 3(a)	3	8492	8491	6960	4167	2453	1621
Model in Fig. 3(b)	10	9526	9525	7643	4316	2563	1663
Model in Fig. 3(c)	1	39224	39223	33083	19476	11596	7531
Model in Fig. 3(d)	10	39707	39705	33416	19440	11664	7502
Model in Fig. 3(e)	10	55598	55596	48044	28769	17304	11235

Finally, Fig. 4 shows the potential degradation of the topological properties of the simplified shape as ϵ increases. For the abstract shape (models (a), (b) in Fig. 3) we observe that the shape of the graph is preserved up to $\epsilon = 1.2$. However, for $\epsilon = 1.25$ the representation collapses to the most important connected component. This is partially due to the fact that the link between the original resistances R_e ($\epsilon = 0$) and their sampled counterparts R'_e is governed by $R'_e = (1 \pm \epsilon)^2 R_e$ according to the JL lemma, if we do not compute them by spectral means. In addition, as ϵ increases we reduce the number of samples $q = 9C^2 n \log n / \epsilon^2$ ($C = 1$ in this paper). This leads to an increment of entropy, which in turn flattens the importance of certain key edges. Therefore, the critical value of ϵ is larger for shapes with an increasing number of nodes. For instance, for the deer alpha shapes we have that the critical value of ϵ is in the range $[1.4, 1.45]$ whereas for the cow alpha shape we have that it is in the range $[1.4, 1.5]$. For triangulations the values are similar but larger. For the blob the critical value is close to 1.4, and for the remaining ones is the range $[1.4, 1.5]$.

4 Conclusions

In this paper, we have shown that graph sparsification leads to a principled way of simplifying shapes. Experiments on both triangulations and alpha shapes show promising preliminary results. In particular, it introduces a hierarchy (and therefore a priority queue) of the edge strength. It is relevant for shape analysis and interpretation. We plan to further develop these ideas, in particular, in relation to the filtrations induced by the theory of topological persistence [7,9].

Acknowledgments. F. Escolano and M. Curado are funded by Project TIN2015-69077-P of the Spanish Government.

References

1. Achlioptas, D.: Database-friendly random projections. In: Buneman, P. (ed.) Proceedings of the Twentieth ACM SIGACT-SIGMOD-SIGART Symposium on Principles of Database Systems, PODS, 21–23 May 2001, Santa Barbara, California, USA. ACM (2001)
2. Batson, J.D., Spielman, D.A., Srivastava, N.: Twice-ramanujan sparsifiers. SIAM J. Comput. **41**(6), 1704–1721 (2012)
3. Batson, J.D., Spielman, D.A., Srivastava, N., Teng, S.: Spectral sparsification of graphs: theory and algorithms. Commun. ACM **56**(8), 87–94 (2013)
4. Benczúr, A.A., Karger, D.R.: Approximating s-t minimum cuts in $\tilde{O}(n^2)$ time. In: Proceedings of the Twenty-Eighth Annual ACM Symposium on the Theory of Computing, STOC, Philadelphia, Pennsylvania, USA, 22–24 May 1996, pp. 47–55 (1996)
5. Berger, M., Tagliasacchi, A., Seversky, L.M., Alliez, P., Guennebaud, G., Levine, J.A., Sharf, A., Silva, C.T.: A survey of surface reconstruction from point clouds. Comput. Graph. Forum **36**, 301–329 (2016)
6. Edelsbrunner, H.: Alpha shapes - survey. In: Tessellations in the Sciences: Virtues, Techniques and Applications of Geometric Tilings. Springer Verlag (2011)
7. Edelsbrunner, H., Harer, J.: Computational Topology: An Introduction. American Mathematical Society, Providence (2010)
8. Edelsbrunner, H., Kirkpatrick, D., Seidel, R.: On the shape of a set of points in the plane. IEEE Trans. Inf. Theory **29**(4), 551–559 (1983)
9. Edelsbrunner, H., Letscher, D., Zomorodian, A.: Topological persistence and simplification. Discrete Comput. Geom. **28**, 511–533 (2002)
10. Edelsbrunner, H., Mücke, E.P.: Three-dimensional alpha shapes. ACM Trans. Graph. **13**(1), 43–72 (1994)
11. Naylor, B., Bajaj, C., Edelsbrunner, H., Kaufman, A., Rossignac, J.: Computational representations of geometry. In: SIGGRAPH 1996 Course Notes (1996)
12. Paoluzzi, A., Bernardini, F., Cattani, C., Ferrucci, V.: Dimension-independent modeling with simplicial complexes. ACM Trans. Graph. **12**(1), 56–102 (1993)
13. Qiu, H., Hancock, E.R.: Clustering and embedding using commute times. IEEE Trans. Pattern Anal. Mach. Intell. **29**(11), 1873–1890 (2007)
14. Rudelson, M., Vershynin, R.: Sampling from large an approach through geometric functional analysis. J. ACM **54**(4), 21 (2007)
15. Spielman, D.A., Srivastava, N.: Graph sparsification by effective resistances. SIAM J. Comput. **40**(6), 1913–1926 (2011)
16. Spielman, D.A., Teng, S.: Spectral sparsification of graphs. SIAM J. Comput. **40**(4), 981–1025 (2011)

Reeb Graphs of Piecewise Linear Functions

Barbara Di Fabio and Claudia Landi[✉]

Università di Modena e Reggio Emilia DISMI, Pad. Morselli,
Via Amendola 2, 42122 Reggio Emilia, Italy
claudia.landi@unimore.it

Abstract. The Reeb graph is a popular tool in the field of computational topology for shape analysis. The Reeb graph is usually thought of as a transform from shapes, viewed as spaces endowed with functions, to graphs. It finds its roots in the classical Morse theory, where the Reeb graph transform is granted to produce a graph, but it finds its applications mostly in Computer Graphics. Therefore it is usually applied on objects that are not smooth but polyhedral. While the definition of the Reeb graph perfectly makes sense also in the polyhedral case, it is not straightforward to see that the output of the transform in this case is a graph. This paper is devoted to provide a formal guarantee of this fact.

1 Introduction

The Reeb graph is defined for shapes modeled as spaces endowed with scalar functions. It is obtained by shrinking each connected component of a level set of the function to a single point. Often, vertices of the Reeb graph are labeled by the value of the function at the corresponding level set.

Reeb graphs have been initially studied in pure mathematics where spaces are assumed to be differentiable and functions to be Morse [14]. These assumptions guarantee that the Reeb graph is actually a graph, that is a structure consisting of vertices connected by edges.

Since [16,17], the Reeb graph construction gained popularity in the Computer Graphics community as an effective tool for shape analysis and description tasks. Application fields related to the use of Reeb graphs are: surface analysis and understanding [1,17]; identification of topological quadrangulations [11]; data simplification [5]; animation [12]; human body segmentation [18]; surface parameterization [13,19]; sub-part correspondence [6].

A number of characteristics of the Reeb graph have contributed to make it useful for specific application domains. For example, with regard to utilization of the Reeb graph as a search query for 3D objects, it is interesting that there is a natural link between the function and the shape, and the possibility of adopting different functions for describing different aspects of shapes have led to a wide use of Reeb graphs for similarity evaluation, shape matching and retrieval. If the function is constructed from geometric information, such as a height function or a distance function, the Reeb graph captures both topological and geometric features of a shape, thus combining global and local information.

© Springer International Publishing AG 2017
P. Foggia et al. (Eds.): GbRPR 2017, LNCS 10310, pp. 23–35, 2017.
DOI: 10.1007/978-3-319-58961-9_3

Also, by defining the function appropriately, it is possible to construct a Reeb graph that is invariant to translation and rotation, or even more complicated isometries of the shape. A more complete account on these aspects can be found in the survey paper [4].

In this paper, we focus on another characteristic of the Reeb graph that is very important independently of the application: the property of having a one-dimensional structure without any higher dimension components. This is an interesting property that in applications makes the Reeb graph sometimes preferable to other similar transforms such as the medial axis, where instead degenerate surfaces can occur.

The property of being a graph is of the utmost importance not only in applications, but also in the development of the theory of Reeb graphs. For example, the graph structure plays a central role in the results concerning the stability of Reeb graphs, i.e. the property that small perturbations on the input data still produce small perturbations in the Reeb graph. In the literature concerning the stability of the Reeb graph it is usually assumed that the data produce a Reeb graph that is one-dimensional [2, 3, 7].

That a Reeb graph has a combinatorial structure is important also for those algorithms conceived to handle Reeb graphs for specific tasks, like maintaining the Reeb graph through time [9], or computing the persistent Reeb graph homology for a sequence of Reeb graphs defined on a filtered space [8]. In [10], where the Reeb graph is generalized by using several functions on the same space at the same time, the authors prove that every point of the obtained structure has a neighborhood that is a cone over a Reeb space of dimension one less. Hence, the polyhedral nature of this structure depends on that of the Reeb graph.

On the other hand, to the best of our knowledge, the literature lacks of a acknowledgeable result in this sense in the piecewise linear case. Indeed, the one-dimensional nature of the Reeb graph is usually accepted as a matter of fact in the transition from pure mathematics, where the Reeb graph is built under differentiability and genericity assumptions so that the Reeb graph is actually guaranteed to be a graph [14], to applications, where spaces and functions are usually only piecewise linear.

This paper aims at providing a guarantee that the Reeb graph is actually a graph also in the piecewise linear setting. In our opinion, this will allow researchers that develop the theory and applications of the Reeb graph to found their results on a more solid basis, filling a gap in the literature.

The paper is organized as follows. In Sect. 2, we review the general definition of the Reeb graph, and the basic properties common to all the frameworks where it may be defined, let it be the smooth setting or the piecewise linear setting. In Sect. 3, we show some properties of polyhedra that turn out to be useful to prove that the Reeb graph in this context is actually a graph. Section 4 is devoted to the main result of this paper. Few comments in Sect. 5 conclude this work.

2   The Topological Reeb Graph and Its Properties

In this section, we review the definition and properties of the Reeb graph.

The more general setting in which the definition of Reeb graph makes sense is when X is a compact topological space endowed with a continuous function $f : X \to \mathbb{R}$. This setting is so general that it encompasses the case when spaces and functions are smooth as well when they are piecewise linear. The basic idea is that of shrinking each connected component of a level set of the function f to a single point. Technically, this is done via a quotient.

Definition 1. *The* topological Reeb graph *of f is the quotient space $X/\sim_f$ where, for every $x, x' \in X$, $x \sim_f x'$ if and only if x and x' belong to the same connected component of $f^{-1}(f(x))$.*

In other words, the points of the topological Reeb graph correspond to connected components of the level sets of f. Examples of Reeb graphs in the smooth and piecewise linear cases are displayed in Fig. 1.

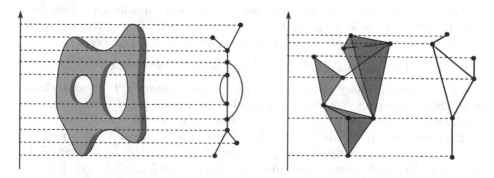

Fig. 1. Left: the Reeb graph of a Morse function on a smooth manifold. Right: the Reeb graph of a PL function on a polyhedron.

In the case when the space is a smooth manifold and the function is generic enough, technically a simple Morse function, it is well-known that this construction is guaranteed to yield a graph.

Theorem 2 ([14]). *Let M be a compact n-dimensional manifold and f a simple Morse function defined on M. The quotient space $M/\sim_f$ has the structure of a finite graph whose vertices bijectively correspond to the critical points of f.*

In the next sections, we will prove that we get a graph also when the space is piecewise linear.

3   Preliminary Facts on Polyhedra

This section is initially devoted to reviewing the definitions of a polyhedron and a piecewise linear function. Next, we focus on properties of level and inter-level sets of piecewise linear functions.

Let $A, B \subset \mathbb{R}^n$. The *join* of A and B is the set

$$A * B = \{\lambda \cdot a + \mu \cdot b : a \in A, b \in B, \lambda, \mu \in \mathbb{R}, \lambda, \mu \geq 0, \lambda + \mu = 1\}.$$

If $A = \{a\}$, we say that the join of A and B, denoted simply by $a * B$ is a *cone* with *vertex* a and *base* B.

Following [15], a subset $P \subset \mathbb{R}^n$ is called a *polyhedron* if each point $p \in P$ has a cone neighborhood $N(p) = p * L(p)$ in P, where $L(p)$ is compact. In this case, $p, N(p)$ and $L(p)$ are called the *vertex*, the *star* and the *link* of the cone, respectively. Moreover, a map $f : P \to Q$ between polyhedra is said to be *piecewise linear* (abbreviated PL) if each point $p \in P$ has a star $N(p) = p * L(p)$ in P such that $f(\lambda \cdot p + \mu \cdot \ell) = \lambda \cdot f(p) + \mu \cdot f(\ell)$, with $\ell \in L(p)$, $\lambda, \mu \geq 0, \lambda + \mu = 1$, i.e. if f is linear on each cone in P.

From now on, we let P be a compact polyhedron (and hence a Hausdorff space), and $f : P \to \mathbb{R}$ a PL-function. Because the definition of Reeb graph is based on level sets, we now consider the level and inter-level sets of f.

Lemma 3. *For every $c \in f(P)$, $f^{-1}(c)$ is a compact sub-polyhedron of P.*

Proof. By the continuity of f, $f^{-1}(c)$ is a compact subset of P because closed in P, that is compact. We want to show that each point $p \in f^{-1}(c)$ has a cone neighborhood $N(p) = p * L(p)$ in $f^{-1}(c)$, where $L(p)$ is compact.

Because P is a polyhedron, $p \in P$ has a cone neighborhood $N'(p) = p * L'(p)$ in P, where $L'(p)$ is compact. We prove that $N'(p) \cap f^{-1}(c)$ is the star of the cone in $f^{-1}(c)$ with vertex p and link the compact set $L'(p) \cap f^{-1}(c)$, i.e. that $N'(p) \cap f^{-1}(c) = p * (L'(p) \cap f^{-1}(c))$.

Let us show that $N'(p) \cap f^{-1}(c) \subseteq p * (L'(p) \cap f^{-1}(c))$. If $q \in N'(p) \cap f^{-1}(c)$, then $f(q) = c$ and, by definition of $N'(p)$, $q \in p * L'(p)$. In other words, $f(q) = c$, and $q = \lambda \cdot p + \mu \cdot \ell$, with $\lambda, \mu \geq 0$, $\lambda + \mu = 1$, $\ell \in L'(p)$. In the case $\mu = 0$, it follows that $q = p$, and hence the claim holds true. In the case $\mu \neq 0$, because f is PL, we deduce the following equalities:

$$c = f(q) = f(\lambda \cdot p + \mu \cdot \ell) = \lambda \cdot f(p) + \mu \cdot f(\ell) = \lambda \cdot c + \mu \cdot f(\ell). \qquad (1)$$

From equalities (1), we get $(1 - \lambda) \cdot c = \mu \cdot f(\ell)$, with $\mu \neq 0$, i.e. $f(\ell) = c$. Thus we can conclude that $\ell \in L'(p) \cap f^{-1}(c)$, and, therefore, $q \in p * (L'(p) \cap f^{-1}(c))$.

Let us prove that $p * (L'(p) \cap f^{-1}(c)) \subseteq N'(p) \cap f^{-1}(c)$. If $q \in p * (L'(p) \cap f^{-1}(c))$, then $q = \lambda \cdot p + \mu \cdot \ell$, with $\lambda, \mu \geq 0$, $\lambda + \mu = 1$, $\ell \in L'(p) \cap f^{-1}(c)$. Therefore $q \in p * L'(p) = N'(p)$ and it holds that

$$f(q) = f(\lambda \cdot p + \mu \cdot \ell) = \lambda \cdot f(p) + \mu \cdot f(\ell),$$

with $f(p) = f(\ell) = c$ by assumption, implying $f(q) = c$. In conclusion, $q \in N'(p) \cap f^{-1}(c)$. $\qquad \square$

Lemma 4. *For every $a, b \in f(P)$, with $a < b$, $f^{-1}([a,b])$ is a compact sub-polyhedron of P.*

Proof. By the continuity of f, $f^{-1}([a,b])$ is a compact subset of P because closed in P, which is compact. We want to show that each point $p \in f^{-1}([a,b])$ has a cone neighborhood $N(p) = p * L(p)$ in $f^{-1}([a,b])$, with $L(p)$ a compact set. Let us distinguish the following two cases: (i) $p \in f^{-1}((a,b))$ and (ii) $p \in f^{-1}(a)$ or $p \in f^{-1}(b)$.

(i) If $p \in f^{-1}((a,b))$, the claim immediately follows because $f^{-1}((a,b))$ is open in P and hence it is a sub-polyhedron of P (see [15, p. 4]).

(ii) If $p \in f^{-1}(a)$ (the case $p \in f^{-1}(b)$ is analogous), then we observe that, in particular, $p \in P$, with P a polyhedron. This implies that p has a cone neighborhood $N'(p) = p * L'(p)$ in P, with $L'(p)$ compact. We want to prove that $N'(p) \cap f^{-1}([a,b])$ is the star of the cone with vertex p and link the compact set $L'(p) \cap f^{-1}([a,b])$, i.e. that $N'(p) \cap f^{-1}([a,b]) = p * (L'(p) \cap f^{-1}([a,b]))$. Without loss of generality, we can assume that $f(\ell) < b$ for every $\ell \in L'(p)$.

Let us show that $N'(p) \cap f^{-1}([a,b]) \subseteq p * (L'(p) \cap f^{-1}([a,b]))$. If $q \in N'(p) \cap f^{-1}([a,b])$, then $a \leq f(q) \leq b$ and, by definition of $N'(p)$, $q \in p * L'(p)$. In other words, $a \leq f(q) \leq b$, and $q = \lambda \cdot p + \mu \cdot \ell$, with $\lambda, \mu \geq 0$, $\lambda + \mu = 1$, $\ell \in L'(p)$. In the case $\mu = 0$, it follows that $q = p$, and hence the claim. In the case $\mu \neq 0$, let us consider the following equalities:

$$f(q) = f(\lambda \cdot p + \mu \cdot \ell) = \lambda \cdot f(p) + \mu \cdot f(\ell) = \lambda \cdot a + \mu \cdot f(\ell),$$

where $a \leq f(q) \leq b$ and $f(\ell) < b$ by assumption. Thus $\lambda \cdot a + \mu \cdot f(\ell) \geq a$, with $\mu \neq 0$, that yields $f(\ell) \geq a$. Hence, we get $\ell \in L'(p) \cap f^{-1}([a,b])$, and, therefore, $q \in p * (L'(p) \cap f^{-1}([a,b]))$.

Let us prove that $p*(L'(p) \cap f^{-1}([a,b])) \subseteq N'(p) \cap f^{-1}([a,b])$. If $q \in p*(L'(p) \cap f^{-1}([a,b]))$, then $q = \lambda \cdot p + \mu \cdot \ell$, with $\lambda, \mu \geq 0$, $\lambda + \mu = 1$, $\ell \in L'(p) \cap f^{-1}([a,b])$. Therefore, $q \in p * L'(p) = N'(p)$, and it holds that

$$f(q) = f(\lambda \cdot p + \mu \cdot \ell) = \lambda \cdot f(p) + \mu \cdot f(\ell),$$

with $f(p) = a$ and $a \leq f(\ell) < b$ by assumption. Thus, $q \in N'(p)$ and, on the one hand, $\lambda \cdot a + \mu \cdot f(\ell) \geq \lambda \cdot a + \mu \cdot a$, that yields $f(q) \geq a$; on the other hand, $\lambda \cdot a + \mu \cdot f(\ell) < \lambda \cdot a + \mu \cdot b$, that yields $f(q) < b$. In conclusion, $q \in N'(p) \cap f^{-1}([a,b])$.

□

Lemma 5. *For every $c \in f(P)$, there exist a real value $\varepsilon = \varepsilon(c) > 0$ and a PL-retraction $r : f^{-1}([c-\varepsilon, c+\varepsilon]) \to f^{-1}(c)$ such that $f^{-1}(c)$ is a strong deformation retract of $f^{-1}([c - \varepsilon, c + \varepsilon])$ through r.*

Proof. Because P is a polyhedron, for every $p \subset f^{-1}(c)$ there exists a cone neighborhood of p in P, $N(p) = p * L(p)$, with $L(p)$ compact. Moreover $f^{-1}(c) \subseteq \bigcup_{p \in f^{-1}(c)} N(p)$.

Let us start by proving that there exists a real value $\bar{\delta} = \bar{\delta}(c) > 0$ such that, for every $0 \le \delta \le \bar{\delta}$, $f^{-1}([c - \delta, c + \delta]) \subseteq \bigcup_{p \in f^{-1}(c)}$. By contradiction, suppose that, for every $\delta > 0$, there exists a point $p_\delta \in f^{-1}([c - \delta, c + \delta])$ with $p_\delta \notin \bigcup_{p \in f^{-1}(c)} N(p)$. Hence, for $\delta = \frac{1}{k}$, and $k \in \mathbb{N}$, there is a point $p_k \in f^{-1}([c - \frac{1}{k}, c + \frac{1}{k}])$ such that $p_k \notin \bigcup_{p \in f^{-1}(c)} N(p)$. By the compactness of P, we can extract a convergent subsequence from (p_k), say (p_{k_j}). Let $\bar{p} = \lim_{j \to +\infty} p_{k_j}$. We have $f(\bar{p}) = \lim_{j \to +\infty} f(p_{k_j}) = c$ because f is continuous, and $p_{k_j} \in f^{-1}([c - \frac{1}{k_j}, c + \frac{1}{k_j}])$. Therefore, $\bar{p} \in f^{-1}(c) \subseteq \bigcup_{p \in f^{-1}(c)} N(p)$. This means that there exists a $\bar{j}$ such that, for every $k_j > k_{\bar{j}}$, $p_{k_j} \in \bigcup_{p \in f^{-1}(c)} N(p)$, yielding a contradiction.

Now we proceed with the construction of the map r. We observe that, for every $0 \le \delta \le \bar{\delta}$, $f^{-1}(c)$ is a sub-polyhedron of $f^{-1}([c - \delta, c + \delta])$, and $f^{-1}([c - \delta, c + \delta])$ is a sub-polyhedron of P by virtue of Lemmas 3 and 4. Let us fix a value $0 < \delta \le \bar{\delta}$. By [15, Theorem 2.11], P, $f^{-1}([c - \delta, c + \delta])$ and $f^{-1}(c)$ are the underlying polyhedra of simplicial complexes K, L and M, respectively. We write

$$P = |K|, \quad f^{-1}([c - \delta, c + \delta]) = |L|, \quad f^{-1}(c) = |M|.$$

Since $|M| \subset |L| \subset |K|$, we can use [15, Addendum 2.12] to ensure the existence of simplicial subdivisions $K' \triangleleft K$, $L' \triangleleft L$, $M' \triangleleft M$ such that $M' \subset L' \subset K'$.

Let us denote by $St_{L'}(M')$, and call it the star of the simplicial complex M' in L', the set of simplexes of L' that have a face in M':

$$St_{L'}(M') = \{\beta \in L' : \alpha \le \beta, \ \alpha \in M'\}.$$

We also denote by $\overline{St}_{L'}(M')$, and call it the closed star of M' in L', the set

$$\overline{St}_{L'}(M') = \{\gamma \in L' : \gamma \le \beta, \ \beta \in St_{L'}(M')\}.$$

Finally, we denote by $Lk_{L'}(M')$, and call it the the link of M' in L', the set of simplexes of the closed star of M' in L' that have no faces in M':

$$Lk_{L'}(M') = \overline{St}_{L'}(M') - St_{L'}(M').$$

We observe that $\overline{St}_{L'}(M')$ and $Lk_{L'}(M')$ are simplicial sub-complexes of L', and $|M'|, |Lk_{L'}(M')| \subseteq |\overline{St}_{L'}(M')| \subseteq |L'|$.

For any real number $0 < \varepsilon < \delta$ such that

$$|Lk_{L'}(M')| \subseteq f^{-1}([c - \delta, c + \delta]) \setminus f^{-1}([c - \varepsilon, c + \varepsilon]) \tag{2}$$

we construct the map:

$$r : f^{-1}([c - \varepsilon, c + \varepsilon]) \to f^{-1}(c)$$

as follows:

- For every $p \in f^{-1}(c)$, we define $r(p) = p$;
- For every $p \in f^{-1}([c - \varepsilon, c + \varepsilon])$, with $f(p) \neq c$, we observe that p belongs to the interior of exactly one simplex α of $\overline{St}_{L'}(M')$. Indeed, $p \in \overline{St}_{L'}(M')$ and it cannot be a vertex because of (2). Denote by σ the maximal face of α in M' and by γ the maximal face of α in $Lk_{L'}(M')$. It holds that $p \in \sigma * \gamma$. Thus there exists a unique point a in σ and a unique point b in γ such that $p = (1 - t) \cdot a + t \cdot b$, $t \in [0, 1]$. We define $r(p) = b$.

We observe that r is a well-defined map because each point $p \in f^{-1}([c - \varepsilon, c + \varepsilon])$, with $f(p) \neq c$, belongs to exactly one such segment, and two different segments in $\overline{St}_{L'}(M') \cap f^{-1}([c - \varepsilon, c + \varepsilon])$ are either disjoint or meet at a common point $p' \in f^{-1}(c)$. Moreover, the map r is continuous, and linear on each simplex, hence a PL retraction. Finally, $f^{-1}(c)$ is a strong deformation retract of $f^{-1}([c - \varepsilon, c + \varepsilon])$ via the homotopy

$$F : f^{-1}([c - \varepsilon, c + \varepsilon]) \times [0, 1] \to f^{-1}([c - \varepsilon, c + \varepsilon]), \ F(p, s) = (1 - s) \cdot p + s \cdot r(p).$$

□

4   The Reeb Graph is a Graph Also in the PL Case

In this section we assume X to be a compact polyhedron and f to be piecewise linear.

We start this section with a preliminary result stating that the points of the Reeb graph of a piecewise linear model are separable in the same way as points in $\mathbb{R}^n$ are (technically, the space is Hausdorff).

We denote by π_f the natural projection of X onto $X/\!\sim_f$ induced by $\sim_f$. The topological Reeb graph $X/\!\sim_f$ is naturally endowed with a continuous function $\tilde{f} : X/\!\sim_f \to \mathbb{R}$ defined by setting $\tilde{f}(\pi_f(x)) = f(x)$, so that the following diagram commutes:

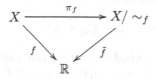

Proposition 6. *The topological Reeb graph $X/\!\sim_f$ is a Hausdorff and compact space.*

Proof. We use the fact that $X/\!\sim_f$ is Hausdorff if and only if

$$\ker \pi_f = \{(x, x') \in X \times X : \pi_f(x) = \pi_f(x')\}$$

is closed. We want to show that, for every $x, x' \in X$ such that $\pi_f(x) \neq \pi_f(x')$, there exists an open neighborhood U of (x, x') in $X \times X$ that does not intersect

$\ker \pi_f$. This yields that $(X \times X) - \ker \pi_f$ is open, and hence the claim. By Definition 1, we have to consider the following two cases: (i) $f(x) \neq f(x')$, and (ii) $f(x) = f(x')$ with x and x' belonging to different components of $f^{-1}(f(x))$.

(i) If $f(x) \neq f(x')$, let $0 < \varepsilon < |f(x) - f(x')|/2$. Then it is sufficient to take $U = f^{-1}((f(x) - \varepsilon, f(x) + \varepsilon)) \times f^{-1}((f(x') - \varepsilon, f(x') + \varepsilon))$. Indeed, it is an open neighborhood of (x, x') disjoint from $\ker \pi_f$.

(ii) If $f(x) = f(x') = c$, let $\varepsilon > 0$ be such that the connected components C and C' of $f^{-1}((c - \varepsilon, c + \varepsilon))$ that contain x and x', respectively, are disjoint. Such ε exists because, by Lemma 5, $f^{-1}((c - \varepsilon, c + \varepsilon))$ retracts onto $f^{-1}(c)$, and we are assuming that x and x' belong to different connected components of $f^{-1}(c)$. The sets C and C' are open in X. Indeed, by the continuity of f, $f^{-1}((c - \varepsilon, c + \varepsilon))$ is open in X, and therefore a sub-polyhedron of X. Hence, it is a locally path-connected space because each of its points has a cone neighborhood. This implies that the connected components of $f^{-1}((c - \varepsilon, c + \varepsilon))$ are open in $f^{-1}((c - \varepsilon, c + \varepsilon))$, and, hence, in X. From the properties that C and C' are disjoint, open and connected, it follows that $C \times C'$ is an open subset of $X \times X$ that contains (x, x') and does not intersect $\ker \pi_f$.

Finally, $X/\sim_f$ is compact because X is compact and π_f is continuous. $\square$

Proposition 7. $X/\sim_f$ *is an abstract polyhedron of dimension* 0 *or* 1.

Proof. As seen in the proof of Lemma 5, for every $c \in f(X)$ there is an $\varepsilon = \varepsilon(c) > 0$ and a retraction r such that $f^{-1}([c - \varepsilon, c + \varepsilon])$ retracts onto $f^{-1}(c)$ and is contained in the set $\bigcup_{x \in f^{-1}(c)} N(x)$, with $N(x) = x * L(x)$ a cone neighborhood of x in X. By the compactness of X, there exists a finite sequence of values $c_1 < c_2 < \cdots < c_k$ in $f(X)$ such that, setting $\varepsilon_j = \varepsilon(c_j)$, $X \subseteq \bigcup_{j=1}^k f^{-1}((c_j - \varepsilon_j, c_j + \varepsilon_j))$. Without loss of generality, we can assume that the cover is minimal, i.e. for no $\bar{\jmath} = 1, \ldots, k$, $f^{-1}((c_{\bar{\jmath}} - \varepsilon_{\bar{\jmath}}, c_{\bar{\jmath}} + \varepsilon_{\bar{\jmath}})) \subseteq \bigcup_{j \neq \bar{\jmath}} f^{-1}((c_j - \varepsilon_j, c_j + \varepsilon_j))$, and that $c_j < c_{j+1} - \varepsilon_{j+1} < c_j + \varepsilon_j < c_{j+1}$, for $1 \leq j \leq k - 1$. For $j = 1, \ldots, k$, denote by r_j the retraction of $f^{-1}([c_j - \varepsilon_j, c_j + \varepsilon_j])$ onto $f^{-1}(c_j)$ defined as in the proof of Lemma 5.

Let C_α be a connected component of $f^{-1}([c_j - \varepsilon_j, c_j + \varepsilon_j])$ with $1 \leq j \leq k$. Our goal is to define a PL function

$$h_\alpha : C_\alpha \to \mathbb{R}^2$$

whose image is a compact polyhedron of dimension one.

To this end, let $\partial^- C_\alpha = f^{-1}(c_j - \varepsilon_j) \cap C_\alpha$ and $\partial^+ C_\alpha = f^{-1}(c_j + \varepsilon_j) \cap C_\alpha$. We assume that $\partial^- C_\alpha$, if non-empty, is the union of m components, $C^-_{\alpha,1}, \ldots, C^-_{\alpha,m}$, and, similarly, $\partial^+ C_\alpha$, if non-empty, is the union of n components, $C^+_{\alpha,1}, \ldots, C^+_{\alpha,n}$, with $m, n \geq 1$. Now we observe that $\partial^- C_\alpha$ is empty if and only if the set $C_\alpha \cap f^{-1}((c_j - \varepsilon_j, c_j))$ is empty. Moreover, if $x \in C_\alpha \cap f^{-1}((c_j - \varepsilon_j, c_j))$, then there is a unique $i \in \{1, \ldots, m\}$ such that the line passing through x and $r_j(x)$ intersects $C^-_{\alpha,i}$. In that case, we denote by $s^-_i(x)$ the intersection point. Similarly for $\partial^+ C_\alpha$. Thus, in order to define the function $h_\alpha : C_\alpha \to \mathbb{R}^2$ we proceed as follows:

- For every $x \in f^{-1}(c_j) \cap C_\alpha$, we set

$$h_\alpha(x) = O = (0,0)$$

- If $\partial^- C_\alpha$ is non-empty
  - for every $x \in C^-_{\alpha,i}$, $i = 1,\ldots,m$, we set

$$h_\alpha(x) = A^-_{\alpha,i} = (i, -\varepsilon_j)$$

  - for every $x \in C_\alpha$ with $c_j - \varepsilon_j < f(x) < c_j$, letting $i \in \{1,\ldots,m\}$ be the index such that the line passing through x and $r_j(x)$ intersects $C^-_{\alpha,i}$ at $s^-_i(x)$, so that $x = (1-t) \cdot s^-_i(x) + t \cdot r_j(x)$, with $t \in [0,1]$, we set

$$h_\alpha(x) = (1-t) \cdot h_\alpha(s^-_i(x)) + t \cdot h_\alpha(r_j(x)) = (1-t) \cdot A^-_{\alpha,i}$$

- If $\partial^+ C_\alpha$ is non-empty,
  - for every $x \in C^+_{\alpha,i}$, $i = 1,\ldots,n$, we set

$$h_\alpha(x) = A^+_{\alpha,i} = (i, \varepsilon_j)$$

  - for every $x \in C_\alpha$ with $c_j < f(x) < c_j + \varepsilon_j$, letting $i \in \{1,\ldots,n\}$ be the index such that the line passing through x and $r_j(x)$ intersects $C^+_{\alpha,i}$ at $s^+_i(x)$, so that $x = (1-t) \cdot s^+_i(x) + t \cdot r_j(x)$, with $t \in [0,1]$, we set

$$h_\alpha(x) = (1-t) \cdot h_\alpha(s^+_i(x)) + t \cdot h_\alpha(r_j(x)) = (1-t) \cdot A^+_{\alpha,i}$$

It is easily seen that h_α is well-defined, continuous and PL. Moreover, by construction, $h_\alpha(C_\alpha)$ is a compact polyhedron of dimension 1 in $\mathbb{R}^2$:

$$h_\alpha(C_\alpha) = N_\alpha(O) = O * L_\alpha(O)$$

with $L_\alpha(O) = \{A^-_{j,1},\ldots,A^-_{j,m}, A^+_{j,1},\ldots,A^+_{j,n}\}$. We also observe that if $x_1, x_2 \in C_\alpha$ are such that $h_\alpha(x_1) = h_\alpha(x_2)$, then $f(x_1) = f(x_2)$.

Now we use the maps h_α in the same way as in [15, Example 2.27(3)] to prove that $X/\sim_f$ is an abstract polyhedron. To do so, for every connected component C_α of $f^{-1}([c_j - \varepsilon_j, c_j + \varepsilon_j])$, with $1 \leq j \leq k$, we construct a continuous injection

$$\eta_\alpha : N_\alpha(O) \to X/\sim_f$$

so that the maps η_α are PL related, i.e. $\eta^{-1}_\alpha \circ \eta_\beta$ is PL whenever it is defined.

Let us consider the following diagram:

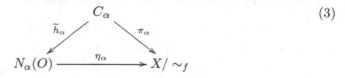

$$(3)$$

where $\tilde{h}_\alpha = h_\alpha^{|N_\alpha(O)}$, $\pi_\alpha = \pi_{f|C_\alpha}$, and $\eta_\alpha(y) = \pi_\alpha(\tilde{h}^{-1}_\alpha(y))$ for every $y \in N_\alpha(O)$. It holds that:

- η_α is a well-defined map making Diagram (3) commute.
  Let $y \in N_\alpha(O)$. Let $x_1, x_2 \in h_\alpha^{-1}(y)$. As already remarked, $f(x_1) = f(x_2)$. Moreover, by construction, x_1 and x_2 belong to the same connected component of $f^{-1}(f(x_1))$. Hence, $\pi_f(x_1) = \pi_f(x_2)$, implying the claim.
- η_α is continuous.
  Let D be a closed set in the image $\operatorname{im}\eta_\alpha$, of η_α. By the commutativity of Diagram (3), $D \subseteq \operatorname{im}\pi_\alpha$. By the continuity of π_α, $\pi_\alpha^{-1}(D)$ is closed in C_α. Moreover, since $\widetilde{h}_\alpha$ is a continuous map from a compact space to a Hausdorff space, by the closed map lemma we see that $\widetilde{h}_\alpha$ is closed, and, hence, $\widetilde{h}_\alpha(\pi_\alpha^{-1}(D)) = \eta_\alpha^{-1}(D)$ is a closed set in $N_\alpha(O)$.
- η_α is injective.
  Let $y_1, y_2 \in N_\alpha(O)$ such that $\eta_\alpha(y_1) = \eta_\alpha(y_2)$. Because $\widetilde{h}_\alpha$ is surjective, there exists $x_1, x_2 \in C_\alpha$ such that $\widetilde{h}_\alpha(x_1) = y_1$ and $\widetilde{h}_\alpha(x_2) = y_2$. By the commutativity of Diagram (3), $\eta_\alpha(\widetilde{h}_\alpha(x_1)) = \eta_\alpha(\widetilde{h}_\alpha(x_2))$ if and only if $\pi_\alpha(x_1) = \pi_\alpha(x_2)$. In turn, the latter equality occurs if and only if $f(x_1) = f(x_2)$ and x_1, x_2 belong to the same connected component of $f^{-1}(f(x_1))$. In particular, if $f(x_1) = f(x_2) = c_j$, then, by the definition of h_α, we have $\widetilde{h}_\alpha(x_1) = \widetilde{h}_\alpha(x_2) = O$, i.e. $y_1 = y_2 = O$. Otherwise, if $f(x_1) = f(x_2) < c_j$, then

$$x_1 = (1 - t_1) \cdot s_i^-(x_1) + t_1 \cdot r_j(x_1)$$

and

$$x_2 = (1 - t_2) \cdot s_{i'}^-(x_2) + t_2 \cdot r_j(x_2)$$

with $t_1 = t_2 \in [0,1]$ and $i, i' \in \{1, \ldots, m\}$. Necessarily, it holds that $i = i'$ because x_1, x_2 belong to the same component of $f^{-1}(f(x_1))$. Thus, $\widetilde{h}_\alpha(x_1) = \widetilde{h}_\alpha(x_2)$, i.e. $y_1 = y_2$. The case $f(x_1) = f(x_2) > c$ is analogous.
- η_α is an embedding.
  Since every continuous injection from a compact to a Hausdorff space is an embedding, it is sufficient to apply Proposition 6 to obtain the claim.
- If C_α is a connected component of $f^{-1}([c_j - \varepsilon_j, c_j + \varepsilon_j])$ and C_β is a connected component of $f^{-1}([c_{j'} - \varepsilon_{j'}, c_{j'} + \varepsilon_{j'}])$ such that $C_\alpha \cap C_\beta \neq \emptyset$, it holds that

$$\eta_\alpha^{-1} \circ \eta_{\beta|\widetilde{h}_\beta(C_\alpha \cap C_\beta)} : \widetilde{h}_\beta(C_\alpha \cap C_\beta) \to \widetilde{h}_\alpha(C_\alpha \cap C_\beta)$$

is PL.
By the minimality of the cover $\{f^{-1}((c_j - \varepsilon_j, c_j + \varepsilon_j))\}$, we can assume that $j' = j + 1$. From the commutativity of the diagram

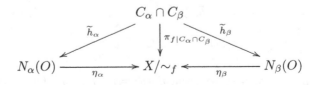

we see that, for every $x \in C_\alpha \cap C_\beta$, we have

$$\eta_\alpha^{-1} \circ \eta_\beta(\widetilde{h}_\beta(x)) = \widetilde{h}_\alpha(x).$$

Moreover, since we are assuming that $c_j < c_{j+1} - \varepsilon_{j+1} < c_j + \varepsilon_j < c_{j+1}$, if $x \in C_\alpha \cap C_\beta$, there are $s_i^+(x) \in \partial^+(C_\alpha)$ and $s_{i'}^-(x) \in \partial^-(C_\beta)$ such that

$$x = \lambda \cdot s_{i'}^-(x) + \mu \cdot s_i^+(x)$$

for $\lambda, \mu > 0, \lambda + \mu = 1$. We observe that $s_i^+(x)$ and $s_{i'}^-(x)$ belong to $C_\alpha \cap C_\beta$. By definition of $\widetilde{h}_\alpha$ and $\widetilde{h}_\beta$,

$$\widetilde{h}_\alpha(s_i^+(x)) = (i, \varepsilon_j) = A_i^+, \quad \widetilde{h}_\beta(s_{i'}^-(x)) = (i', \varepsilon_{j+1}) = B_{i'}^-$$

We set $\hat{A}_i^+ = \widetilde{h}_\beta(s_i^+(x))$ and $\hat{B}_{i'}^- = \widetilde{h}_\alpha(s_{i'}^-(x))$. Now we use the fact that $\widetilde{h}_\alpha$ and $\widetilde{h}_\beta$ are PL to deduce that

$$\widetilde{h}_\alpha(x) = \lambda \cdot \hat{B}_{i'}^- + \mu \cdot A_i^+$$

and

$$\widetilde{h}_\beta(x) = \lambda \cdot B_{i'}^- + \mu \cdot \hat{A}_i^+$$

Thus,

$$\eta_\alpha^{-1} \circ \eta_\beta(\lambda \cdot B_{i'}^- + \mu \cdot \hat{A}_i^+) = \eta_\alpha^{-1} \circ \eta_\beta(\widetilde{h}_\beta(x)) = \widetilde{h}_\alpha(x)$$
$$= \lambda \cdot \eta_\alpha^{-1} \circ \eta_\beta(B_{i'}^-) + \mu \cdot \eta_\alpha^{-1} \circ \eta_\beta(\hat{A}_i^+)$$

proving the piecewise linearity of the map $\eta_\alpha^{-1} \circ \eta_\beta$. $\square$

Corollary 8. *The topological Reeb graph $X/\!\sim_f$ embeds into a polyhedron R_f of dimension 0 or 1, via a homeomorphism $\xi : X/\!\sim_f \to R_f$ such that the map $\hat{\pi}_f = \xi \circ \pi_f$,*

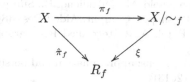

is PL. Moreover, the function $\hat{f} : R_f \to \mathbb{R}$ that makes the following diagram commute

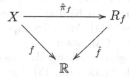

is also PL.

Proof. By Proposition 7, $X/\!\sim_f$ is the identification space of the family of 1-dimensional polyhedra $\{N_\alpha(O)\}$ under the family of PL related maps $\{\eta_\alpha\}$. Hence it can be embedded as a polyhedron R_f in $\mathbb{R}^n$ for some n in such a way that $\hat{\pi}_f$ is PL. The piecewise linearity of $\hat{f}$ follows from the surjectivity of $\hat{\pi}_f$ and the fact that f and $\hat{\pi}_f$ are PL. $\square$

5   Conclusions

In this paper we have provided a proof that the Reeb graph of a polyhedron is itself a polyhedron of dimension 0 or 1, i.e. it is a graph. It is interesting to notice that while in the smooth case the Reeb graph is guaranteed to be a graph under a genericity condition for the function defined on a manifold, in the PL case we neither need the space to be a manifold nor the function to be generic. Thus, in the PL case the result is much more general than in the differentiable case.

References

1. Attene, M., Biasotti, S., Spagnuolo, M.: Shape understanding by contour driven retiling. Vis. Comput. **19**(23), 127–138 (2003)
2. Bauer, U., Di Fabio, B., Landi, C.: An edit distance for Reeb graphs. In: Ferreira, A., Giachetti, A., Giorgi, D. (eds.) Eurographics Workshop on 3D Object Retrieval. The Eurographics Association (2016)
3. Bauer, U., Ge, X., Wang, Y.: Measuring distance between Reeb graphs. In: Proceedings of the Thirtieth Annual Symposium on Computational Geometry, SOCG 2014, New York, NY, USA, pp. 464–473. ACM (2014)
4. Biasotti, S., De Floriani, L., Falcidieno, B., Frosini, P., Giorgi, D., Landi, C., Papaleo, L., Spagnuolo, M.: Describing shapes by geometrical-topological properties of real functions. ACM Comput. Surv. **40**(4), 1–87 (2008)
5. Biasotti, S., Falcidieno, B., Spagnuolo, M.: Extended Reeb graphs for surface understanding and description. In: Borgefors, G., Nyström, I., Baja, G.S. (eds.) DGCI 2000. LNCS, vol. 1953, pp. 185–197. Springer, Heidelberg (2000). doi:10.1007/3-540-44438-6_16
6. Biasotti, S., Marini, S., Spagnuolo, M., Falcidieno, B.: Sub-part correspondence by structural descriptors of 3D shapes. Comput. Aided Des. **38**(9), 1002–1019 (2006)
7. de Silva, V., Munch, E., Patel, A.: Categorified Reeb graphs. Discrete Comput. Geom. **55**(4), 1–53 (2016)
8. Dey, T.K., Wang, Y.: Reeb graphs: approximation and persistence. Discrete Comput. Geom. **49**(1), 46–73 (2013)
9. Edelsbrunner, H., Harer, J., Mascarenhas, A., Pascucci, V.: Time-varying Reeb graphs for continuous space-time data. In: Proceedings of the Twentieth Annual Symposium on Computational Geometry, SCG 2004, pp. 366–372. ACM (2004)
10. Edelsbrunner, H., Harer, J., Patel, A.K.: Reeb spaces of piecewise linear mappings. In: Proceedings of the Twenty-fourth Annual Symposium on Computational Geometry, SCG 2008, pp. 242–250. ACM (2008)
11. Hétroy, F., Attali, D.: Topological quadrangulations of closed triangulated surfaces using the Reeb graph. Graph. Models **65**, 131–148 (2003). Special Issue: Discrete Topology and Geometry for Image and Object Representation
12. Kanongchaiyos, P., Shinagawa, Y.: Articulated Reeb graphs for interactive skeleton animation, pp. 451–467. World Scientific (2011)
13. Patané, G., Spagnuolo, M., Falcidieno, B.: Para-graph: graph-based parameterization of triangle meshes with arbitrary genus. Comput. Graph. Forum **23**(4), 783–797 (2004)

14. Reeb, G.: Sur les points singuliers d'une forme de Pfaff complétement intégrable ou d'une fonction numérique. Comptes Rendus de L'Académie ses Sciences **222**, 847–849 (1946). (French)
15. Rourke, C.P., Sanderson, B.J.: Introduction to piecewise-linear topology. Ergebnisse der Mathematik und ihrer Grenzgebiete, Band 69. Springer, New York (1972)
16. Shinagawa, Y., Kunii, T.L.: Constructing a Reeb graph automatically from cross sections. IEEE Comput. Graphics Appl. **11**(6), 44–51 (1991)
17. Shinagawa, Y., Kunii, T.L., Kergosien, Y.L.: Surface coding based on Morse Theory. IIEEE Comput. Graph. Appl. **11**(5), 66–78 (1991)
18. Werghi, N., Xiao, Y., Siebert, J.P.: A functional-based segmentation of human body scans in arbitrary postures. IEEE Trans. Syst. Man Cybern. Part B **36**(1), 153–165 (2006)
19. Zhang, E., Mischaikow, K., Turk, G.: Feature-based surface parameterization and texture mapping. ACM Trans. Graph. **24**(1), 1–27 (2005)

Learning and Graph Kernels

Learning from Diffusion-Weighted Magnetic Resonance Images Using Graph Kernels

Sylvain Takerkart[1]([✉]), Gottfried Berton[1], Nicole Malfait[1],
and François-Xavier Dupé[2]

[1] Aix Marseille Univ, CNRS, INT, Inst Neurosci Timone, Marseille, France
`sylvain.takerkart@univ-amu.fr`
[2] Aix Marseille Univ, CNRS, Centrale Marseille, LIF, Marseille, France

Abstract. Diffusion-weighted magnetic resonance imaging (DWI) is a scanning procedure that allows infering the anatomical connectivity of the brain non invasively. DWI can be used to segment the brain into a set of relevant sub-regions, yielding what is called a parcellation in the neuroimaging literature. In this paper, we introduce a generic framework that allows building predictive models using parcellations obtained on a single individual. It consists in constructing attributed region adjacency graphs to represent the parcellations and using suitable graph kernels to exploit the versatility of kernel methods. We demonstrate the relevance of this framework on real data, by showing that we can predict the age range of an individual from the connectivity structure of its corpus callosum, the main hub of connections between the left and right hemispheres of the brain. Furthermore, we study the behavior of different graph kernels for this task. This work opens new opportunities to identify DWI-based biomarkers of neurodegenerative and psychiatric diseases.

Keywords: Region-adjacency graphs · Graph kernel · Magnetic resonance imaging · Brain connectivity · Brain parcellation

1 Introduction

In recent years, the use of neuroimaging-based predictive models has seen a fast development. In most cases, these machine learning models are designed in order to build diagnosis or prognosis tools to help clinicians deal with neurological or psychiatric disorders [23]. But they also prove valuable in the field of basic research aiming at a better understanding of the organization of the brain [26]. Amongst the different neuroimaging modalities used in this context, diffusion-weighted magnetic resonance imaging (DWI) is under-exploited, with only a limited number of published studies (see for instance [7,9,11,16,17]). Yet it is attracting a growing interest.

DWI is a magnetic resonance imaging procedure that uses the diffusion of water molecules to reveal the micro-architecture of a physiological tissue. The white matter of the brain (WM) is of particular interest for neuroscientists and

© Springer International Publishing AG 2017
P. Foggia et al. (Eds.): GbRPR 2017, LNCS 10310, pp. 39–48, 2017.
DOI: 10.1007/978-3-319-58961-9_4

can be studied using DWI. Indeed, the WM mostly contains myelinated axons, the neuronal fibers, which are organized into large groups of axons called *bundles*. The structure of these bundles alters the free motion of the water molecules, and the resulting anisotropy can be captured using DWI. In the nervous system, the role of the axons is to transmit information between neurons, connecting different brain regions. Thanks to its ability to characterize the axon bundles of the WM, DWI has developed into a major tool to study *brain connectivity*. Furthermore, because a large number of brain disorders involve abnormal connectivity [6], DWI has allowed to gain unprecedent insights about these pathologies. It is therefore crucial to further design new DWI-based predictive models in order to identify non-invasive biomarkers of connectivity-based neurological syndromes.

One of the major questions to be addressed when designing predictive models based on DWI data is the representation of the data itself. A commonly used high-level summary representation of the diffusion data consists in segmenting the domain of interest – either the full brain or a given region of interest – into homogeneous sub-regions, often denoted as *parcels* in the neuroimaging community. The literature focuses on estimating parcellation models that are valid for a population, for two main reasons: first, because one of the main aims of neuroscience is to find invariants across individuals; secondly, because it is not possible to produce inference at the population level using parcellations estimated on data from a single individual using the statistical tools that are most commonly used in neuroimaging. However, because such parcellations simply are the results of a segmentation, the field of pattern recognition offers a wide range of methods which could help overcome this challenge. Our objective in this paper is therefore to demonstrate that pattern recognition tools can make it possible to build predictive models based on DWI-based parcellations estimated on a single individual, in order to make inference at the population level.

The framework we propose starts by the construction of attributed graphs from such individual parcellations and then relies on kernel methods [21], using appropriate graph kernels, to build a predictive model. We demonstrate the effectiveness of this framework on a real DWI dataset, by showing, for the first time, that we can estimate the age category of a subject from the connectivity structure of his/her corpus callosum, the main hub of connections between the two hemispheres of the brain. Furthermore, we study the influence of the choice of the kernel on the performance of the model, by comparing the expressivity of several classical graph kernels. In Sect. 2, we describe the processing workflow necessary to obtain a parcellation from the raw DWI data, the processes used to construct attributed graphs and the different graph kernels. Then, in Sect. 3, we provide a full description of the data itself, the neuroscientific question addressed in our experiments and the obtained results. Finally, we discuss these results and future research directions in Sect. 4.

2   Methods

2.1   Constructing DWI-Based Parcellations

The processing of DWI data involves a complex analysis workflow to obtain
a parcellation from the raw data, which is a set of three-dimensional volumes
that each provides, at every brain voxel, a measurement of the amount of water
diffusion in a given direction – the set of directions being a priori chosen to
isotropically span the entire sphere. After correction of the image distortions and
compensation of the between-volume displacements, we first estimate the fibre
orientation distribution at each voxel. We use the so-called *ball-and-stick* model,
which assumes that the orientation diffusion function includes an isotropic component – the ball – and one or several directions that follows the neuronal fibers
passing through this voxel – the stick(s). This model is fit at each voxel using
a sampling-based Bayesian technique [2]. Secondly, a probabilistic *tractography*
algorithm allows to estimate the probability of connection $p(m, n)$ between two
brain voxels/regions m and n [2]. For the domain $\mathcal{D}$ to be parcellated and a
set of target brain regions $\mathcal{T}$, we can thus compute the *connectivity matrix*
$C = [p(m, n)]_{m \in \mathcal{T}, n \in \mathcal{D}} \in \mathbb{R}^{M \times N}$, where M and N are respectively the number of
regions in $\mathcal{T}$ and the number of voxels in $\mathcal{D}$. Note that the vector $[p(m, n)]_{m \in \mathcal{T}}$
is called the tractogram of n. The segmentation of $\mathcal{D}$ into parcels is obtained
using a clustering of the columns of C. Here, we use Ward's hierarchical clustering, with an added spatial constraint, to identify contiguous clusters of voxels

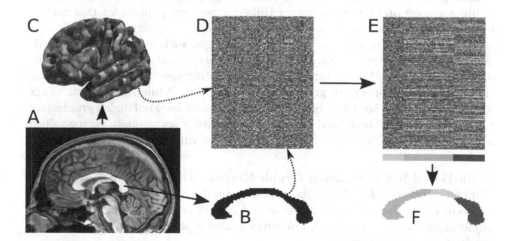

Fig. 1. Processing workflow of the DWI data. A: a structural MRI depicting the shape
of the brain. B: extraction of the domain $\mathcal{D}$ to be parcellated (here, the corpus callosum). C: definition of the set of target regions $\mathcal{T}$ for the tractography (here, a set
of anatomically defined regions of the cortex). D: the connectivity matrix C contains
the probability of connection of any given voxel of $\mathcal{D}$ to any of the target regions. E:
the re-ordered connectivity matrix after clustering of the tractograms/columns. F: the
parcellation V, where voxels are colored with the corresponding cluster label.

that have similar tractograms. Each of these clusters forms a parcel v_i, i.e. a sub-region with a homogeneous connectivity structure, as estimated using DWI data. This workflow is illustrated on Fig. 1.

2.2   Graph Construction

Given a parcellation V of the domain of interest $\mathcal{D}$, i.e. a set of R sub-regions/parcels $\{v_i\}_{i=1}^{i=R}$ that fully cover $\mathcal{D}$ and do not intersect, we would like to obtain an adequate representation of the connectivity structure within $\mathcal{D}$. We use region adjacency graphs to encode the topographical character of the information contained in the parcellation. We therefore define a node of the graph at each parcel $v_i \in V$. The set of edges $E \subset V \times V$ is given by the spatial adjacency of the parcels, i.e. $e = v_i v_j \in E \iff$ parcels v_i and v_j are spatially adjacent. We then add two vectorial attributes to each graph node: i) the average tractogram of all voxels of the parcel, and ii) the coordinates of the barycenter of the parcel, in a coordinate system that is comparable across participants. We note $\varphi_1 : V \to \mathbb{R}^M$ and $\varphi_2 : V \to \mathbb{R}^3$ the functions respectively giving the tractogram and coordinate attributes of a node, and $\varphi : V \to \mathbb{R}^M \times \mathbb{R}^3$ so that $\varphi(v) = (\varphi_1(v), \varphi_2(v))$. The connectivity structure of a given brain region is therefore represented by the attributed graph defined as $G = (V, E, \varphi)$.

2.3   Learning from Graphs

Learning from graphs is a difficult task and graph kernels, which provide an indirect projection of a graph onto a Hilbert space, are popular for that matter because they render accessible a vast array of machine learning methods. Several strategies can be employed to design graph kernels, such as instantiating an R-convolution kernel [8] (see e.g. the random walk kernel [10,27], kernels over sets of paths [24], trees [13] or graphlets [22]), exploiting spectral graph decompositions [28] or the concept of graph edit distance [18]. While most recent works focus on improving kernel scalability for unlabeled or weakly-labeled graphs (see e.g. [1,5,12,22]), we here need kernels that can easily accomodate vector-valued attributes. We describe below the kernels that we will use for our experiments.

Walk-Based R-Convolution Graph Kernels. Given a graph G, we denote V its set of vertices, $E \subset V \times V$ its set of edges and $\varphi : V \to \mathbb{R}^L$ the vector-valued function giving the attributes of the nodes. A walk w is defined as a sequence of adjacent nodes in the graph, and we suppose that a positive definite walk kernel, denoted K_{walk}, is available.

The first kernel we will use is the classical random walk graph kernel [10]. For two graphs G_1 and G_2, it is defined as

$$K_{random}(G_1, G_2) = \sum_{w_1 \in G_1} \sum_{w_2 \in G_2} K_{walk}(w_1, w_2) p(w_1|G_1) p(w_2|G_2), \qquad (1)$$

where $p(w|G)$ is the probability of walk w in G. It compares the density of walks between graphs, thus taking into account both local and global information.

Our second kernel is the bag of paths kernel [4,24] – paths are walks with unique nodes. Let P_1 and P_2 be bags of paths respectively associated with G_1 and G_2. We denote by $|.|$ the number of paths inside a bag. The mean bag of paths kernel is constructed by averaging the walk kernel over all couples of paths from each bag:

$$K_{mean}(P_1, P_2) = \frac{1}{|P_1|}\frac{1}{|P_2|}\sum_{h \in P_1}\sum_{h' \in P_2} K_{walk}(h, h'). \tag{2}$$

In practice, we use bags of paths of constant size, i.e. $P_{\mathcal{P}}(G) = \{w \in G \mid |w| \in \mathcal{P}\}$.

At the core of these graph kernels, the walk kernel K_{walk} measures the similarity between two walks. Clearly most of the expressivity of the whole kernel relies on how we compare walks with K_{walk}. Here, we use the classical kernel proposed by [10], where we only compare walks of the same size by making a direct alignment of both nodes and edges. The considered walk kernel writes as follow for two walks $h = v_1 v_2 \ldots v_{|h|}$ and $h' = v_1' v_2' \ldots v_{|h'|}'$,

$$K_{walk}(h, h') = \begin{cases} \prod_{i=1}^{|h|} K_{node}(\varphi(v_i), \varphi(v_i')) & \text{if } |h| = |h'| \\ 0 & \text{otherwise} \end{cases}, \tag{3}$$

where K_{node} is a kernel on the attributes of the nodes, usually defined with a combination of Gaussian kernels. Given our definition $\varphi = (\varphi_1, \varphi_2)$ given in Sect. 2.2, we use the following node kernel,

$$K_{node}(\varphi(v_1), \varphi(v_2)) = \exp\left(\frac{-\|\varphi_1(v_1) - \varphi_1(v_2)\|^2}{2\sigma_1^2}\right) \exp\left(-\frac{\|\varphi_2(v_1) - \varphi_2(v_2)\|^2}{2\sigma_2^2}\right), \tag{4}$$

where $\sigma_1 > 0$ and $\sigma_2 > 0$ are two hyper-parameters.

Graph Edit Distance Kernels. Graph edit distances [14,20] are an attractive way of comparing graphs as they provide a set of editions (e.g. node/edge substitution, suppression, addition...) with its cost. In [18], the authors proposed to build a graph kernel from such distances by applying them inside a Gaussian kernel. Let d_E be a graph edit distance, the corresponding kernel is defined by:

$$K_{edit}(G_1, G_2) = \exp\left(\frac{-d_E(G_1, G_2)^2}{2\sigma_E^2}\right), \tag{5}$$

where $\sigma_E > 0$ is the hyper-parameter of the kernel.

In order to avoid the computational burden associated with computing d_E, we use the approximation proposed in [20], which relies on the Munkres assignment algorithm and requires comparing nodes attributes. For this, given that $\varphi = (\varphi_1, \varphi_2)$, we use a combination of euclidean norms:

$$d_{node}(v_1, v_2) = \|\varphi_1(v_1) - \varphi_1(v_2)\| + \gamma\|\varphi_2(v_1) - \varphi_2(v_2)\|, \tag{6}$$

where γ is an equilibrium parameter.

3   Experiments and Results

The objective of our experiments is to demonstrate that our framework makes it possible to perform predictions from an individual's DWI-based parcellation. In the following, we present the addressed prediction task and its neuroscientific motivation, the dataset used in our experiments, the experimental procedure and the quantitative results that we obtained.

3.1   Aging Trajectory of the Corpus Callosum

The brain is an organ that continuously evolves throughout the lifespan. In particular, numerous markers of aging can be identified using neuroimaging techniques, such as for instance the reduction of the global volume of grey matter in the brain. Numerous brain pathologies – such as Alzheimer's disease – induce some alterations compared to the normal aging process. Establishing the aging trajectory of specific brain features in a healthy population can therefore allow to use a deviation from this normal trajectory as a potential marker of a disease, which opens possibilities to design diagnosis and/or prognosis tools. In a predictive framework, establishing an aging trajectory comes down to obtaining a model that can guess the age of the subject from a given set of brain features.

Amongst the particular parts of the brain that have been identified to display age-related changes, the corpus callosum (CC), the largest commissure of white matter connecting the two hemispheres of the brain, is of particular interest because its integrity is known to be altered in several neurodegenerative diseases, with for instance an abnormally low CC size [25]. However, the age-related changes of the spatial organisation of the CC structural connectivity have not been studied until now. It is well known that the connectivity structure of the CC is topographically organized: among others, the fibers that pass through its most anterior part – called the *genu* – project to the anterior part of the cortex – the frontal lobe, while fibers passing through the posterior CC – the *splenium* – project to the back of the brain – the occipital lobe. In the present paper, we use DWI data to investigate whether the spatial organization of the CC connectivity changes with age.

3.2   Data and Experiments

Our data comes from the enhanced Nathan Kline Institute-Rockland Sample[1]. A small sub-sample of the participants enrolled in this initiative have been scanned using structural and diffusion-weighted MRI, allowing us to work with data from 65 participants aged 36 to 77 year-old. We analyzed the structural MRI (MPRAGE sequence, voxel size: 1 mm, volume size: $256 \times 256 \times 176$, acquisition time: 6 mn, see Fig. 1(A) using the *freesurfer* software suite[2] to identify the CC (Fig. 1(B) and define a set of 1000 target regions for the tractography

---

[1] http://fcon_1000.projects.nitrc.org/indi/enhanced/.
[2] https://surfer.nmr.mgh.harvard.edu/.

(Fig. 1(C). The DWI data (EPI sequence, 137 directions, voxel size: 2 mm, volume size: $106 \times 90 \times 64$, acquisition time: 6 mn) was pre-processed with the FSL software suite[3]. The rest of the experiments were conducted using in-house software (see Fig. 2 for an illustration of the resulting parcellations and graphs).

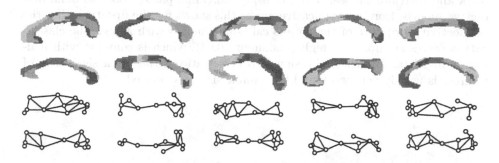

Fig. 2. Top: 10 examples of parcellations of the corpus callosum (with $R = 12$ parcels). Bottom: the resulting graphs, displayed in a local rectangular coordinate system.

We seperated the 65 participants into two groups, the 32 oldest and the 33 youngest, to set up a binary classification problem. Our experiments therefore asked whether it is possible to predict the age group of an individual from the graph G of his/her CC, which would imply, if successfull, that the spatial organization of the CC connectivity does indeed change with aging. We used a Support Vector Machine to perform this classification task, using five kernels amongst the ones described above: (i) K_{random} (using $\sigma_1 = \sigma_2 = 0.5$, as with all the following walk-based kernels); (ii) K_{mean} with $\mathcal{P} \in \{2\}$ (i.e. walks with 2 nodes), that we will denote K_{mean_2}; (iii) K_{mean} with $\mathcal{P} \in \{3\}$, hereafter K_{mean_3}; (iv) K_{mean} with $\mathcal{P} \in \{2, 3\}$, hereafter K_{mean_23}; (v) K_{edit} (with $\gamma = 1000$ and $\sigma_E = 1$). We assessed the quality of the predictions with the mean classification accuracy obtained using a cross-validation scheme that included 500 randomly drawn balanced data splits, each with 56 and 9 participants respectively in the training and test set. Given the small size of the dataset, this procedure ensures obtaining an unbiased estimate of the mean accuracy.

We conducted two sets of experiments. First, in order to study the influence of the number of graph nodes R, we computed the classification accuracy when R is kept constant – taken in $\{4, 6, 8, \cdots, 68, 70\}$ – for all splits of the cross-validation. Secondly, we performed a model order selection, choosing R within the same range, separately for each split, using an inner cross-validation within the training set. The selected value was used to fit the model on the full training set and test it on the left-out data.

---

[3] http://fsl.fmrib.ox.ac.uk.

3.3   Results

The results of the experiments are shown on Fig. 3. All the classification results are above chance level (0.5), establishing the two main outcomes of the paper: from a methodological point of view, this demonstrates the validity of our framework and therefore the feasability of using individual parcellations to build predictive models; from an applicative angle, this shows for the first time that the connectivity structure of the corpus callosum changes with aging. The classification scores are not very high (maximum: 0.68), which is common with neuroimaging data, for which the signal to noise ratio of the brain signatures of interest is usually very weak and the sample size very limited.

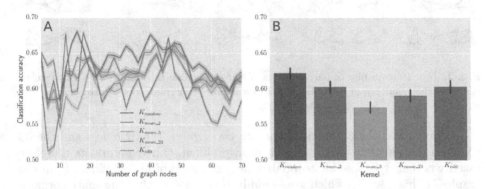

Fig. 3. Mean classification accuracy (± standard error) for five kernels. A: in function of the number R of graph nodes (kept constant for all folds of the cross-validation). Right: when R is selected through an inner cross-validation.

Figure 3(A) shows that the performance of the model strongly depends on the number R of graph nodes. Interestingly, the peak performances are not obtained for the same value of R for all kernels: the accuracy peaks for the random walk and edit distance kernels between $R = 12$ and 18, while the mean kernels perform best around $R = 44$. Figure 3(B) shows that when performing model order selection, K_{random} outperforms the other kernels, followed by K_{mean_2} and K_{edit}.

4   Discussion and Future Work

In this paper, we have introduced a new framework that allows designing predictive models from individual DWI-based parcellations. To the best of our knowledge, this is the first of its kind.

The behavior of the different kernels we used is interesting since they are sensitive to various types of information. The fact that K_{mean_2} outperforms K_{mean_3} and K_{mean_23} indicates that most of the information lies at the local level, and more precisely at the level of pairs of adjacent nodes. Adding an extra node in the paths seems to introduce noise rather than information, which could be expected from the noisy character of the parcellation process illustrated by the variability of the graphs on Fig. 2. But the higher performances of K_{random}, which can catch both local and global features, suggests that some extra information might lie at a more global scale. Also note that K_{edit} is more unstable than the other kernels, which can be caused by its non positive-definiteness [19] or by a failure of the Munkres-based assignment on our noisy parcellations.

This opens several directions for future work. First, the graph construction should benefit from concatenating different parcellations into multi-scale hierarchical graphs. Indeed this will provide a more robust representation of the data and render more explicit the combination of local and global information. Furthermore, because these graphs have a geometric embedding, the use of combinatorial maps and pyramids [3] could also be beneficial. Secondly, improvements should come from the graph kernel itself. We believe that the level of expressivity of the kernel on this data should increase by incorporating ideas from recent work such as [5] or using more structured base elements such as graphlets [22].

From an applicative perspective, we have demonstrated that the connectivity structure of the corpus callosum changes with age. However, additional work is needed to understand the nature of these modifications. Furthermore, in order to apply this framework on clinical data and identify biomarkers of a given neurological disorder, two main lines of work lie ahead of us. First, we will have to assess more finely the aging trajectory – of the corpus callosum or any other brain region – using a regression model that would offer a direct prediction of the age of the subject (similarly to [16]). Second, we will have to improve the performances far beyond the classification accuracies obtained here. This will require working with much larger datasets and therefore using more scalable kernels such as introduced in [15].

References

1. Bai, L., Rossi, L., Zhang, Z., Hancock, E.R.: An aligned subtree kernel for weighted graphs. In: ICML, pp. 30–39 (2015)
2. Behrens, T., Berg, H.J., Jbabdi, S., Rushworth, M., Woolrich, M.: Probabilistic diffusion tractography with multiple fibre orientations: what can we gain? NeuroImage **34**(1), 144–155 (2007)
3. Brun, L., Kropatsch, W.: Introduction to combinatorial pyramids. In: Bertrand, G., Imiya, A., Klette, R. (eds.) Digital and Image Geometry. LNCS, vol. 2243, pp. 108–128. Springer, Heidelberg (2001). doi:10.1007/3-540-45576-0_7
4. Dupé, F.-X., Brun, L.: Tree covering within a graph kernel framework for shape classification. In: Foggia, P., Sansone, C., Vento, M. (eds.) ICIAP 2009. LNCS, vol. 5716, pp. 278–287. Springer, Heidelberg (2009). doi:10.1007/978-3-642-04146-4_31
5. Feragen, A., Kasenburg, N., Petersen, J., de Bruijne, M., Borgwardt, K.: Scalable kernels for graphs with continuous attributes. In: NIPS, pp. 216–224 (2013)
6. Fornito, A., Zalesky, A., Breakspear, M.: The connectomics of brain disorders. Nat. Rev. Neurosci. **16**(3), 159–172 (2015)
7. Ghanbari, Y., Smith, A.R., Schultz, R.T., Verma, R.: Identifying group discriminative and age regressive sub-networks from DTI-based connectivity via a unified framework of non-negative matrix factorization and graph embedding. Med. Image Anal. **18**(8), 1337–1348 (2014)

8. Haussler, D.: Convolution kernels on discrete structures. Technical report UCSC-CRL-99-10, Department of Computer Science, Univ. of California at Santa Cruz (1999)

9. Kamiya, K., Amemiya, S., Suzuki, Y., Kunii, N., Kawai, K., Mori, H., Kunimatsu, A., Saito, N., Aoki, S., Ohtomo, K.: Machine learning of DTI structural brain connectomes for lateralization of temporal lobe epilepsy. Magn. Reson. Med. Sci. **15**(1), 121–129 (2016)

10. Kashima, H., Tsuda, K., Inokuchi, A.: Marginalized kernel between labeled graphs. In: ICML, pp. 321–328 (2003)

11. Kawahara, J., Brown, C.J., Miller, S.P., Booth, B.G., Chau, V., Grunau, R.E., Zwicker, J.G., Hamarneh, G.: BrainNetCNN: convolutional neural networks for brain networks; towards predicting neurodevelopment. NeuroImage, September 2016

12. Kriege, N., Mutzel, P.: Subgraph matching kernels for attributed graphs. In: ICML, pp. 1015–1022 (2012)

13. Mahé, P., Vert, J.P.: Graph kernels based on tree patterns for molecules. Mach. Learn. **75**(1), 3–35 (2009)

14. Messmer, B.T., Bunke, H.: A new algorithm for error-tolerant subgraph isomorphism detection. IEEE PAMI **20**(5), 493–504 (1998)

15. Morris, C., Kriege, N.M., Kersting, K., Mutzel, P.: Faster kernels for graphs with continuous attributes via hashing. In: ICDM (2016)

16. Mwangi, B., Hasan, K.M., Soares, J.C.: Prediction of individual subject's age across the human lifespan using diffusion tensor imaging: a machine learning approach. NeuroImage **75**, 58–67 (2013)

17. Mwangi, B., Wu, M.J., Bauer, I.E., Modi, H., Zeni, C.P., Zunta-Soares, G.B., Hasan, K.M., Soares, J.C.: Predictive classification of pediatric bipolar disorder using atlas-based diffusion weighted imaging and support vector machines. Psychiatry Res. Neuroimaging **234**(2), 265–271 (2015)

18. Neuhaus, M., Bunke, H.: Bridging the gap between graph edit distance and kernel machines. World Scientific (2007)

19. Pękalska, E., Haasdonk, B.: Kernel discriminant analysis for positive definite and indefinite kernels. IEEE PAMI **31**(6), 1017–1032 (2009)

20. Riesen, K., Bunke, H.: Approximate graph edit distance computation by means of bipartite graph matching. Image Vis. Comput. **27**(7), 950–959 (2009)

21. Scholkopf, B., Smola, A.J.: Learning with Kernels: Support Vector Machines, Regularization, Optimization, and Beyond. MIT Press, Cambridge (2001)

22. Shervashidze, N., Vishwanathan, S., Petri, T., Mehlhorn, K., Borgwardt, K.M.: Efficient graphlet kernels for large graph comparison. AISTATS **5**, 488–495 (2009)

23. Siuly, S., Zhang, Y.: Medical big data: neurological diseases diagnosis through medical data analysis. Data Sci. Eng. **1**(2), 54–64 (2016)

24. Suard, F., Rakotomamonjy, A., Bensrhair, A.: Kernel on bag of paths for measuring similarity of shapes. In: ESANN, pp. 355–360 (2007)

25. Van Schependom, J., Jain, S., Cambron, M., Vanbinst, A.M., De Mey, J., Smeets, D., Nagels, G.: Reliability of measuring regional callosal atrophy in neurodegenerative diseases. NeuroImage Clin. **12**, 825–831 (2016)

26. Varoquaux, G., Thirion, B.: How machine learning is shaping cognitive neuroimaging. GigaScience **3**(1), 28 (2014)

27. Vishwanathan, S.V.N., Schraudolph, N.N., Kondor, R., Borgwardt, K.M.: Graph kernels. J. Mach. Learn. Res. **11**(Apr), 1201–1242 (2010)

28. Wilson, R.C., Zhu, P.: A study of graph spectra for comparing graphs and trees. Pattern Recogn. **41**(9), 2833–2841 (2008)

Learning Graph Matching with a Graph-Based Perceptron in a Classification Context

Romain Raveaux[(⊠)], Chloé Martineau, Donatello Conte,
and Gilles Venturini

Laboratoire d'Informatique (EA 6300),
Université de Tours, 64 avenue Jean Portalis, Tours, France
romain.raveaux@univ-tours.fr, chloémartineau99@gmail.com

Abstract. Many tasks in computer vision and pattern recognition are formulated as graph matching problems. Despite the NP-hard nature of the problem, fast and accurate approximations have led to significant progress in a wide range of applications. Learning graph matching functions from observed data, however, still remains a challenging issue. This paper presents an effective scheme to parametrize a graph model for object matching in a classification context. For this, we propose a representation based on a parametrized model graph, and optimize it to increase a classification rate. Experimental evaluations on real datasets demonstrate the effectiveness (in terms of accuracy and speed) of our approach against graph classification with hand-crafted cost functions.

Keywords: Learning graph matching · Graph classification · Graph edit distance

1 Introduction

Graphs are frequently used in various fields of computer science since they constitute a universal modeling tool which allows the description of structured data. The handled objects and their relations are described in a single and human-readable formalism. Hence, tools for graphs supervised classification and graph mining are required in many applications such as pattern recognition [1], chemical components analysis [2], structured data retrieval [3]. Different approaches have been proposed during the last decade to tackle the problem of graph classification. A first one consists in transforming the initial problem in a common statistical pattern recognition one by describing the graphs with vectors in a Euclidean space [2]. Another family of approaches also consists in using kernel-based machine learning algorithms. Contrary to the approaches mentioned above, the graphs are not explicitly but implicitly projected in a Euclidean space, through the use of a graph kernel which computes inner products in the graph space. Many kernels have been proposed in the literature [4]. Another possible approach also consists in projecting the graphs in a Euclidean space of a given dimension but using a distance matrix between each of the graphs. In such cases, a dissimilarity measure between graphs has to be designed. Kernels can be derived

© Springer International Publishing AG 2017
P. Foggia et al. (Eds.): GbRPR 2017, LNCS 10310, pp. 49–58, 2017.
DOI: 10.1007/978-3-319-58961-9_5

from the distance matrix. It is the case for multidimensional scaling methods proposed in [5].

All the aforementioned approaches aim at projecting the graphs into a vector or kernel space however this process may impact the interpretability of the results. This paper deals with paradigms that operate directly on the graph space and can thus capture more structural distortions.

Graph space $(d : G \times G \to \mathbb{R})$. To classify unknown objects using the K-Nearest Neighbor paradigm, one needs to define a metric that measures the distance between the unknown object and the elements in the learning database. The similarity or dissimilarity between two graphs requires the computation and the evaluation of the "best" matching between them. Since exact isomorphism rarely occurs in pattern analysis applications, the matching process must be error-tolerant, i.e., it must tolerate differences on the topology and/or its labeling. For instance, in the Graph Edit Distance (GED) [1], the graph matching process and the dissimilarity computation are linked through the introduction of a set of graph edit operations. Each edit operation is characterized by a cost, and GED is the total cost of the least expensive set of operations that transform one graph into another one. Since graph matching is NP-hard most research has long focused on developing accurate and efficient approximate algorithms.

Recent studies have revealed that simple graphs with hand-crafted structures and dissimilarity functions, typically used in graph matching, are insufficient to capture the inherent structure underlying the problem at hand. As a consequence, a better optimization of the graph matching objective does not guarantee better correspondence accuracy [6,7] and neither better classification rate. To tackle this issue a set of parameters in the graph matching problem has to be learned. Such a learned matching function would better model the inherent structure of the classification problem without losing the interpretability of the results.

2   Problem Statement

In this section, we formally define the problem of learning discriminative graph matching.

Attributed graph is considered as a triple (V, E, L) such that: V is a set of vertices. E is a set of edges such as $E \subseteq V \times V$. L is a set of attributes of the nodes and edges. For the sake of clarity, we abuse the set notation such that L_i is a label associated to vertex v_i and L_{ij} is a label associated to an edge (v_i, v_j).

Graph Matching Problem. Let $G^1 = (N^1, E^1, L^1)$ and $G^2 = (N^2, E^2, L^2)$ be two graphs, with $N^1 = \{1, \cdots, n\}$ and $N^2 = \{1, \cdots, m\}$. In order to apply deletion or insertion operation on nodes, node sets are augmented by dummy elements. The deletion of each node $v_i \in N^1$ is modeled as a mapping $v_i \to \epsilon_i^2$ where ϵ_i^2 is the dummy element associated with v_i. As a consequence, the set N^2 is increased by $\max(0, n-m)$ dummy elements ϵ^2 to form a new set $V^2 = N^2 \cup \epsilon^2$. The node set N^1 is increased similarly by $\max(0, m - n)$ dummy elements ϵ^2

to form $V^1 = N^1 \cup \epsilon^1$. Note that V^1 and V^2 have the same cardinality ($n1 = n2 = \max(n, m)$). A solution of graph matching is defined as a subset of possible correspondences $y \subset V^1 \times V^2$, represented by a binary assignment matrix $Y \in \{0, 1\}^{n_1 \times n_2}$, where n_1 and n_2 denote the size of V^1 and V^2, respectively. If $v_i^1 \in V^1$ matches $v_a^2 \in V^2$, then $Y_{i,a} = 1$, and $Y_{i,a} = 0$ otherwise. We denote by $y \in \{0, 1\}^{n_1 n_2}$, a column-wise vectorized replica of Y. With this notation, graph matching problems can be expressed as finding the assignment vector y^* that minimizes a score function $d(G^1, G^2, y)$ as follows:

Definition 1. *Graph Matching formulation*

$$y^* = \operatorname*{argmin}_{y} \quad d(G^1, G^2, y), \tag{1a}$$

$$\text{subject to} \quad y \in \{0, 1\}^{n_1 n_2} \tag{1b}$$

$$\sum_{i=1}^{n_1} y_{i,a} = 1 \quad \forall a \in [1, \cdots, n_2] \tag{1c}$$

$$\sum_{a=1}^{n_2} y_{i,a} = 1 \quad \forall i \in [1, \cdots, n_1] \tag{1d}$$

The function $d(G^1, G^2, y)$ measures the dissimilarity of graph attributes, and is typically decomposed into a first order dissimilarity function $d_V(L_i^1, L_a^2)$ for a node pair $v_i^1 \in V^1$ and $v_a^2 \in V^2$, and a second-order similarity function $d_E(L_{ij}^1 L_{ab}^2)$ for an edge pair $e_{ij}^1 \in E^1$ and $e_{ab}^2 \in E^2$. Dissimilarity functions are usually represented by a symmetric dissimilarity matrix D, where a non-diagonal element $D_{ia;jb} = d_E(L_{ij}^1, L_{ab}^2)$ contains the edge dissimilarity of two correspondences (v_i^1, v_a^2) and (v_j^1, v_b^2) and a diagonal term $D_{ia;ia} = d_V(L_i^1, L_a^2)$ represents the node dissimilarity of a correspondence (v_i^1, v_a^2).

Thus, the matching function of graph matching is defined as:

$$d(G^1, G^2, y) = \sum_{y_{ia}=1} d_V(L_i^1, L_a^2) + \sum_{y_{ia}=1} \sum_{y_{jb}=1} d_E(L_{ij}^1, L_{ab}^2) = y^T D y \tag{2}$$

In essence, the score accumulates all the dissimilarity values relevant to the assignment. The Definition 1 is referred to as an integer quadratic programming. More precisely, it is the quadratic assignment problem, which is known to be NP-hard. Many efficient approximate algorithms have been proposed for this problem [8–11].

Parametrized Graph Matching. In the context of scoring functions defined in Eq. 2, an interesting question is what can be learned to improve graph matching. To address this, we parameterize Eq. 2 as follows. Let $\pi(a) = i$ denote an assignment of node v_a^2 in G^2 to node v_i^1 in G^1, i.e. $y_{ia} = 1$. A joint feature map $\Phi(G^1, G^2, y)$ is defined by aligning the relevant dissimilarity values of Eq. 2 into a vectorial form as: $\Phi(G^1, G^2, y) = [\cdots, d_V(L_{\pi(a)}^1, L_a^2), \cdots, d_E(L_{\pi(a)\pi(b)}^1, L_{ab}^2), \cdots]$

By introducing weights on all elements of this feature map, we obtain a discriminative score function:

$$d(G^1, G^2, y, \beta) = \beta\Phi(G^1, G^2, y) \tag{3a}$$

$$= [\cdots, d_V(L^1_{\pi(a)}, L^2_a)\beta_a, \cdots, d_E(L^1_{\pi(a)\pi(b)}, L^2_{ab})\beta_{ab}, \cdots] \tag{3b}$$

where β is a weight vector encoding the importance of node and edge dissimilarity values. In the case of uniform weights, i.e. $\beta = 1 \ \forall\beta$, Eq. 3 it reduces to the conventional graph matching score function of Eq. 2: $d(G^1, G^2, y) = d(G^1, G^2, y; 1)$.

The discriminative weight formulation is general in the sense that it can assign different parameters for individual nodes and edges. However, it does not learn a graph model underlying the feature map, and requires a reference graph G^2 at query time, whose attributes cannot be modified in the learning phase.

Graph Classification Problem. For sake of clarity, the rest of the paper is focused on a 2-class problem but the paradigm can be extended to a multi-class problem. A linear classifier is a function that maps its input $x \in \mathbb{R}^q$ (a real-valued vector) to an output value $f(x) \in \{0, 1\}$ (a single binary value).

$$f(x) = \begin{cases} 1 & \text{if } \beta \cdot x + b > 0 \\ 0 & \text{otherwise} \end{cases}$$

where β is a vector of real-valued weights, $w \cdot x$ is the dot product $\sum_{i=1}^{q} \beta_i x_i$, where q is the number of inputs to the classifier and b is the bias. The bias shifts the decision boundary away from the origin and does not depend on any input value. The value of $f(x)$ (0 or 1) is used to classify x as either a positive or a negative instance, in the case of a binary classification problem. If b is negative, then the weighted combination of inputs must produce a positive value greater than $|b|$ in order to push the classifier over the 0 threshold.

To extend this paradigm to graph, let $\mathcal{D}$ be the set of graphs. Given a graph training set $TrS = \{<G_i, c_i>\}_{i=1}^{M}$, where $G_i \in \mathcal{D}$ is a graph and $c_i \in \mathcal{C}$ is the class of the graph among the two classes. The learning of a graph classifier consists in inducing from TrS a mapping function $f(G) \rightarrow \mathcal{C}$ which assigns a class to an unknown graph.

$$f(G) = \begin{cases} 1 & \text{if } \beta \cdot \Phi(G, G^m, y) + b > 0 \\ 0 & \text{otherwise} \end{cases} \quad \text{With } G^m \text{ a model graph}$$

Let $\Delta(TrS, f)$ be a function computing the error rate obtained by a classifier f. We represent the error for the pth training sample by $error_p = \delta(C_p, f(G_p))$, where C_p is the target value, $f(G_p)$ is the value produced by the classifier and $\delta(a, b)$ is the Kronecker Delta function. The error rate (Δ) is the mean of errors

$error_p$ over the set TrS between the ground-truth values and values produced by the classifier. Straightforwardly, we define $\eta = 1 - \Delta$ as the classification rate.

To address the problem of learning graphs matching, we start with the discriminative weight formulation of Eq. 3. We learn the weights β from labelled examples from TrS minimizing the function Δ. The objective function is the error rate function with extra β weights.

3   State of the Art

The literature on learning similarity/dissimilarity matching functions can be roughly categorized into two parts whether the objective is to minimize an error rate on the number of matched graph components (matching level) or an error rate on a classification task (classification level).

Matching level. In this category, the purpose is to minimize the average Hamming distance between a ground-truth's correspondence and the automatically deducted correspondence. Caetano et al. [6] use a 60-dimensional node similarity function for appearance similarity and a simple binary edge similarity for edges. Leordeanu et al. [12] do not use d_V, and instead employ a multidimensional function d_E for dissimilarity of appearance, angle, and distance. The work of Torresani et al. [13] can be viewed as adopting 2-dimensional d_V and d_E functions for measuring appearance dissimilarity, geometric compatibility, and occlusion likelihood. In [14] a method to learn the real numbers for the insertion $d_V(\epsilon \rightarrow v)$ and deletion $d_V(v \rightarrow \epsilon)$ costs on nodes and edges is proposed. An extension to substitution costs is presented in [15]. While the optimization methods for learning these functions are different, all of them are essentially aimed at learning common weights for all the edge and node dissimilarity functions in a matching context. The discriminative weight formulation Eq. 3 is more general in the sense that it can assign different parameters for individual nodes and edges. In [7], the discriminative weight formulation is also employed. The learning problem is turned into a regression problem and a structured support vector machine (SSVM) is used to minimize it.

Classification level. Learning graph matching in a classification context is more challenging since the ground truth is given at the class level and not at the node/edge level. In [8], a grid search on a validation set is used to determine the values of the parameters $\beta_{del}^n = \beta_{ins}^n$, which corresponds to the cost of a node deletion or insertion, and $\beta_{del}^e = \beta_{ins}^e$, which corresponds to the costs of an edge deletion or insertion. Neuhaus and Bunke [16] address the issue of learning dissimilarity functions for numerically labeled graphs from a corpus of sample graphs. A system of self-organizing maps (SOMs) that represent the distance measuring spaces of node and edge labels was proposed. The learning process is based on the concept of self-organization. It adapts the edit costs in such a way that the similarity of graphs from the same class is increased, whereas the similarity of graphs from different classes decreases. Two limitations can be put forward (i) attributes must be numeric vectors and (ii) the method aimed at learning common weights for all the edges and nodes ($\beta_{del}, \beta_{ins}, \beta_{sub}$). From the same authors, in [17], the graph matching process is formulated in a stochastic context and perform a maximum likelihood parameter estimation of the distribution of matching operations. The underlying distortion model is learned using an Expectation Maximization algorithm. The matching costs are adapted so as to decrease the distance between graphs from the same class, leading to compact graph clusters.

Adapting methods that operate at the matching level is not trivial since node correspondences must be inferred from the class label. The neural methods proposed in [16] works at the classification level but it is limited to vector attributes and common weights shared to all nodes and edges. The former limitation is leveraged in [17] thanks to a probabilistic framework but the Expectation Maximization algorithm is not robust as the neural-based minimizer. In this paper we propose to merge both ideas, a neural-based algorithm and the discriminative weight formulation to learn graph matching dissimilarity functions in a classification context.

4   Proposal: A Graph-Based Perceptron

The perceptron is an algorithm for learning a binary classifier $\mathcal{C} = \{0, 1\}$. In the context of neural networks, a perceptron is an artificial neuron using the Heaviside step function as the activation function. A global picture of the graph-based perceptron is depicted in Fig. 1. The conventional perceptron is adapted to graphs thanks to three main features: (a) The learning rule to update the weight vector β. (b) The graph matching algorithm to find y^*. (c) The graph model G^m.

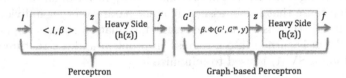

Perceptron   Graph-based Perceptron

Fig. 1. Overview of the perceptron and our proposal a modified perceptron for graph

Learning rule. The learning rule aims at modifying β. The weights should be updated in cases of wrong classifications. The correction must take into account the amount and the sign of the committed error.

$$\text{Learning rule: } \beta(t+1) = \beta(t) + \alpha(c_i - c_k)\Phi(G^i, G^m, y) \tag{4}$$

To show the time-dependence of β, we use $\beta_i(t)$ as the weight at time t. The parameter α is the learning rate, where $0 < \alpha \leq 1$. $(c_i - c_k)$ is the error function. This error is positive if $(c_i > c_k)$ or negative if $(c_i < c_k)$. The learning rule is the steepest gradient descent. It tries to reduce the error in the direction of the error descending along the gradient. If we consider the $\Phi(G^i, G^m, y)$ entries associated with weight β respectively.

Graph matching solver. Many efficient approximate algorithms have been proposed to solve the graph matching problem defined in Definition 1. In [8], Riesen et al. have reformulated the Quadratic Assignment Problem of Definition 1 to a Linear Sum Assignment Problem (LSAP). Nodes of both graphs are involved in the assignment problem. A cost matrix is computed to enumerate pair-wise node distances. The LSAP can be solved in polynomial time $O(n^3)$ which makes this approach very fast.

Graph model. The graph matching is computed between an input graph G^i and a model graph G^m. The choice of a model graph among a set of graphs is of first interest. The model graph should represent the diversity of attributes and topologies which can be found in the graph set TrS. The graph model selection rule is defined as follows: $G^m = arg \max_{G \in TrS} |G|$. With $|G| = |V| + |E|$. Accordingly G^m is the largest graph of the set. In such a way that G^m may gather a large diversity of attributes along with different structures. Other definition could hold such as the median graph definition but this is beyond the scope of the paper.

Learning algorithm. We design the learning algorithm of the graph-based perceptron. Algorithm 1 is an $O(\#iter.|TrS|)$ deterministic algorithm. Solving the parametrized graph matching problem is indicated in Line 8. Line 9 is the classification step while lines 10 to 12 are the application of the learning rule defined Eq. 4 when the classification is wrong. Finally, it is worth mentioning that classifying an entire test set (TeS) is done by only $|TeS|$ call to the graph matching algorithm involved in the function Φ. This low complexity makes it a fast classifier.

Data: $TrS = \{< G^i, c_i >\}_{i=1}^M$
Data: $\#iter$ is the maximum number of iterations
Data: α learning rate
Result: Learned β. A weight vector
1 $\beta \leftarrow 0$ and $t \leftarrow 0$
2 **while** $error > 0$ and $iter < \#iter$ **do**
3   error $\leftarrow 0$ and iter $\leftarrow 0$
4   **for** $G_i \in TrS$ **do**
5          $y^* \leftarrow \underset{y}{argmin} \quad \beta(t) \cdot \Phi(G^i, G^m, y)$ // Solve problem in Definition 1
6          $c_i \leftarrow heavyside(\beta(t) \cdot \Phi(G^i, G^m, y^*))$
7          $c_k \leftarrow getLabel(G^i)$
8   **if** $c_i - c_k \mathrel{!=} 0$ **then**
9              $\beta(t+1) \leftarrow \beta(t) + \alpha(c_i - c_k)\Phi(G^i, G^m, y^*)$
10   error$\leftarrow$ error $+1$
11   **end**
12         $t \leftarrow t + 1$
13   **end**
14   error $\leftarrow$ error$/|TrS|$
15   iter $\leftarrow$ iter $+1$
16 **end**

Algorithm 1. Learning graph-based perceptron scheme

5   Experiments

Two graph databases LETTER HIGH (LETTER for short) and GREC were chosen from from the IAM repository [18]. Each database consists on a set of different graph instances divided in different classes where each class is composed of a training set and a test set. Datasets are described in Table 1. Matching functions d_v and d_e were taken from [8].

Table 1. Summary of graph data set characteristics.

Database	size (TrS, TeS)	♯classes	Node labels	Edge labels	$\overline{|V|}$	$\overline{|E|}$	max $|V|$	max $|E|$	Balanced
LETTER (high)	(750, 750)	15	x, y	None	4.7	4.5	9	9	Y
GREC	(286, 528)	22	x, y	Line types	11.5	12.2	25	30	Y

A commonly used approach in pattern classification is based on nearest-neighbor classification. That is, an unknown object is assigned the class or identity of its closest known element, or nearest neighbor (1-NN). Two versions were involved in the tests. A 1-NN with no weights ($\beta = 1$) called NW-1-NN and a tuned 1-NN (T-1-NN) where $\beta^e_{del} = \beta^e_{ins}$ and $\beta^n_{del} = \beta^n_{ins}$ values are taken from [8]. To assess the performance of our learning scheme and our new classifier, two experiments were performed. First, the impact of the learning rate α was studied on 2 classes of the GREC dataset (class 0 and 1) and second, pair-wise binary classifications were carried out among all classes of two datasets GREC and LETTER. To sum-up all theses experiments, the mean classification rate ($\overline{\eta}$) during the training and the test phases are reported along with the standard deviation ($std(\eta)$). The time in milliseconds to classify all instances is also considered. In Fig. 2, the impact of the learning rate is depicted. A high learning rate leads to fast convergence with many oscillations around 80% of classification rate, while at the opposite a low learning rate implies a slow but stable convergence. A trade-off can be found with an intermediate value ($\alpha = 0.01$). This value was chosen to perform the rest of the experiments with a number of iterations set to 100. To continue on the learning capability of the Algorithm 1, in Table 2, the classification rate obtained during the learning phase are tabulated (column η_{TrS}). A first comment leads to say that with the highest classification rate GREC was easier to learn than LETTER. A second comment is the clear capability of learning of our method. In fact, a dummy classifier with "bad" weights $\beta = 1$ would produce a random classification and a classification rate of 0.5. Finally, binary classifications results on $(22 - 1)^2 = 441$ and 196 pairs of classes for GREC and LETTER, respectively, are synthesized in Table 2. First, on the speed side, our classifier is by far the fastest with a speed gain of about 350 (350 times faster). In fact, time complexity of our graph-based perceptron is linear in function of the test set size ($|TeS|$) whereas the complexity of the 1-NN

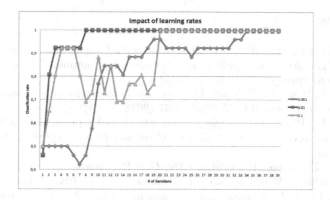

Fig. 2. Learning rate as a function of the number of iterations

Table 2. Classification results on GREC and LETTER. The best results are marked in bold style.

	GREC					LETTER				
	$\overline{\eta_{TrS}}$	$\overline{\eta}$	$std(\eta)$	$\overline{time}$	$std(time)$	$\overline{\eta_{TrS}}$	$\overline{\eta}$	$std(\eta)$	$\overline{time}$	$std(time)$
Proposal	0.9733	0.9488	0.1054	**87.31**	24.49	0.8610	0.8262	0.1279	**31.09**	6.42
NW-1-NN ($\beta = 1$)	NA	0.5235	0.0561	1588.83	870.46	NA	**0.9735**	0.0294	1584.15	510.37
T-1-NN [8]	NA	**0.9992**	0.0096	1789.52	990.08	NA	**0.9735**	0.0295	1573.96	490.51

grows quadratically in function of $|TrS|.|TeS|$. On the classification rate side, on GREC, our proposal clearly outperformed the NW-1-NN classifier with no-weights while obtaining similar results than the T-1-NN classifier. On LETTER, the situation is different, the NW-1-NN classifier provides astonished results as good as the T-1-NN. We can conclude that dissimilarity functions d_V and d_E are well suited on their own for the problem and that performances come from the good graph prototypes of TrS. With a single model graph our approach does not succeed to capture the whole variability of the problem. However, the 15% loss of accuracy is counter balanced by a large speed-up.

6   Conclusion

In this paper, a graph classifier operating in the graph space was presented. A graph-based perceptron was proposed to learn discriminative graph matching in a classification context. Graph matching was parametrized to build a weighted formulation. This weighted formulation is used to define a perceptron classifier. Weights are learned thanks to the gradient descent algorithm. Classification results on two publicly available datasets demonstrated a large speed-up in classification (350 times faster in average) with a loss of accuracy of 4% in average. As the conventional perceptron, the graph-based perceptron will be extended to multi-class problems. Another perceptive is to extend our work to multiple layers and consequently to learn mid-level graph-based representations.

References

1. Riesen, K.: Structural Pattern Recognition with Graph Edit Distance: Approximation Algorithms and Applications. Advances in Computer Vision and Pattern Recognition. Springer, Heidelberg (2015)
2. Gaüzère, B., Brun, L., Villemin, D.: Two new graphs kernels in chemoinformatics. Pattern Recogn. Lett. **33**(15), 2038–2047 (2012)
3. Raveaux, R., Burie, J.-C., Ogier, J.-M.: Structured representations in a content based image retrieval context. J. Vis. Commun. Image Represent. **24**(8), 1252–1268 (2013)
4. Neuhaus, M., Bunke, H.: Bridging the Gap Between Graph Edit Distance and Kernel Machines. World Scientific Publishing Co., Inc., River Edge (2007)
5. Roth, V., Laub, J., Kawanabe, M., Buhmann, J.M.: Optimal cluster preserving embedding of nonmetric proximity data. IEEE Trans. Pattern Anal. Mach. Intell. **25**(12), 1540–1551 (2003)
6. Caetano, T.S., McAuley, J.J., Cheng, L., Le, Q.V., Smola, A.J.: Learning graph matching. IEEE Trans. Pattern Anal. Mach. Intell. **31**(6), 1048–1058 (2009)
7. Cho, M., Alahari, K., Ponce, J.: Learning graphs to match. In: IEEE International Conference on Computer Vision (ICCV 2013), Sydney, Australia, 1–8 December 2013, pp. 25–32. IEEE Computer Society (2013)
8. Riesen, K., Bunke, H.: Approximate graph edit distance computation by means of bipartite graph matching. Image Vis. Comput. **27**(7), 950–959 (2009)
9. Neuhaus, M., Riesen, K., Bunke, H.: Fast suboptimal algorithms for the computation of graph edit distance. In: Yeung, D.-Y., Kwok, J.T., Fred, A., Roli, F., Ridder, D. (eds.) SSPR /SPR 2006. LNCS, vol. 4109, pp. 163–172. Springer, Heidelberg (2006). doi:10.1007/11815921_17
10. Raveaux, R., Burie, J.-C., Ogier, J.-M.: A graph matching method and a graph matching distance based on subgraph assignments. Pattern Recogn. Lett. **31**(5), 394–406 (2010)
11. Serratosa, F.: Fast computation of bipartite graph matching. Pattern Recogn. Lett. **45**, 244–250 (2014)
12. Leordeanu, M., Sukthankar, R., Hebert, M.: Unsupervised learning for graph matching. Int. J. Comput. Vis. **96**(1), 28–45 (2012)
13. Torresani, L., Kolmogorov, V., Rother, C.: Feature correspondence via graph matching: models and global optimization. In: Forsyth, D., Torr, P., Zisserman, A. (eds.) ECCV 2008. LNCS, vol. 5303, pp. 596–609. Springer, Heidelberg (2008). doi:10.1007/978-3-540-88688-4_44
14. Cortés, X., Serratosa, F.: Learning graph-matching edit-costs based on the optimality of the oracle's node correspondences. Pattern Recogn. Lett. **56**, 22–29 (2015)
15. Cortés, X., Serratosa, F.: Learning graph matching substitution weights based on the ground truth node correspondence. IJPRAI **30**(2), 1650005 (2016)
16. Neuhaus, M., Bunke, H.: Self-organizing maps for learning the edit costs in graph matching. IEEE Trans. Syst. Man Cybern. Part B **35**(3), 503–514 (2005)
17. Neuhaus, M., Bunke, H.: Automatic learning of cost functions for graph edit distance. Inf. Sci. **177**(1), 239–247 (2007)
18. Riesen, K., Bunke, H.: IAM graph database repository for graph based pattern recognition and machine learning. In: da Vitoria Lobo, N., Kasparis, T., Roli, F., Kwok, J.T., Georgiopoulos, M., Anagnostopoulos, G.C., Loog, M. (eds.) Structural, Syntactic, and Statistical Pattern Recognition. LNCS, vol. 5342, pp. 287–297. Springer, Heidelberg (2008)

A Nested Alignment Graph Kernel Through the Dynamic Time Warping Framework

Lu Bai[1], Luca Rossi[2(✉)], Lixin Cui[1(✉)], and Edwin R. Hancock[3]

[1] Central University of Finance and Economics, Beijing, China
cuilixin@cufe.edu.cn
[2] Aston University, Birmingham, UK
l.rossi@aston.ac.uk
[3] University of York, York, UK

Abstract. In this paper, we propose a novel nested alignment graph kernel drawing on depth-based complexity traces and the dynamic time warping framework. Specifically, for a pair of graphs, we commence by computing the depth-based complexity traces rooted at the centroid vertices. The resulting kernel for the graphs is defined by measuring the global alignment kernel, which is developed through the dynamic time warping framework, between the complexity traces. We show that the proposed kernel simultaneously considers the local and global graph characteristics in terms of the complexity traces, but also provides richer statistic measures by incorporating the whole spectrum of alignment costs between these traces. Our experiments demonstrate the effectiveness and efficiency of the proposed kernel.

1 Introduction

In pattern recognition, graph kernels are powerful tools for applying standard machine learning techniques to graph datasets [24]. These kernels are typically used in conjuction with kernel methods such as Support Vector Machines (SVM) and kernel Principle Component Analysis (kPCA) for the purposes of classification or clustering [4,21].

The idea underpinning most existing graph kernels is that of decomposing graphs into substructures and comparing pairs of specific isomorphic substructures. Some examples are graph kernels based on counting pairs of isomorphic (a) walks [27], (b) paths [1], and (c) restricted subgraph or subtree substructures [14]. Other examples include the work of Bach [2], who proposed a family of kernels for comparing point clouds. These kernels are based on a local tree-walk kernel between subtrees, which is defined by a factorization on suitably defined graphical models of the subtrees. Wang and Sahbi [28], on the other hand, defined a graph kernel for action recognition. They first describe actions in the videos using directed acyclic graphs (DAGs). The resulting kernel is defined as an extending random walk kernel by counting the number of isomorphic walks of DAGs. Harchaoui and Bath [18] proposed a segmentation graph kernel for images by counting the inexact isomorphic subtree patterns between image segmentation

© Springer International Publishing AG 2017
P. Foggia et al. (Eds.): GbRPR 2017, LNCS 10310, pp. 59–69, 2017.
DOI: 10.1007/978-3-319-58961-9_6

graphs. Other state-of-the-art graph kernels include the subtree-based hypergraph kernel [7], the Lovász graph kernel [19], the aligned subgraph kernel [10], the subgraph matching kernel [21], the fast depth-based subgraph kernel [6], the optimal assignment kernel [22], and the aligned Jensen-Shannon subgraph kernel [11].

Unfortunately, all the aforementioned graph kernels tend to capture only local characteristics of graphs, since they usually use substructures of limited sizes. As a result, these kernels may fail to reflect global graph characteristics. To overcome this shortcoming, Johansson et al. [19] developed a family of global graph kernels using geometric embeddings. Specifically, they use the Lovász number and its associated orthonormal representation to capture global graph characteristics. Bai et al. and Rossi et al. [4,9,25,26] developed a family of graph kernels based on the classical Jensen-Shannon divergence, as well as its quantum analogue. Specifically, they use either the classical or the quantum walk together with quantum information theoretical measures to probe the global structure of the graph.

The aim of this work is to overcome the gap between local kernels (i.e., kernels based on local substructures of limited sizes) and the global kernels (i.e., global kernels and quantum or classical Jensen-Shannon kernels), by proposing a novel nested alignment kernel for graphs based on their depth-based complexity traces [5] and the dynamic time warping framework [15]. For a pair of graphs, we commence by computing the depth-based complexity traces rooted at the centroid vertices. The resulting kernel is defined by measuring the global alignment kernel [15] between the complexity traces. Recall that the depth-based complexity trace of a graph is based on a family of expansion subgraphs that form a nested sequence which gradually expands from the centroid vertex to the global graph structure. As a consequence, this sequence of subgraphs can reflect both local and global structure information of a graph. Furthermore, we show that the associated global alignment kernel encapsulates the whole spectrum of the alignment cost between the complexity traces. As a result, the proposed kernel can not only simultaneously consider both local and global graph characteristics in terms of the nested depth-based complexity traces, but also provide richer statistic measures by incorporating the whole spectrum of alignment costs between these traces. Experiments demonstrate the effectiveness and efficiency of the proposed kernel.

The remainder of this paper is organized as follows. Section 2 reviews the preliminary concepts that will be used in this work. Specifically, we introduce the global alignment kernel through the dynamic time warping framework and the depth-based complexity trace. Section 3 defines the proposed nested alignment kernel. Section 4 provides the experimental evaluation. Section 6 concludes this work.

2   Preliminary Concepts

In this section, we review some preliminary concepts that will be used in this work. We commence by reviewing the dynamic time warping framework.

Specifically, we introduce the global alignment kernel based on this framework. Finally, we review the concept of depth-based complexity trace of a graph.

2.1 Global Alignment Kernels from the Dynamic Time Warping Framework

In this subsection, we review the global alignment kernel based on the dynamic time warping framework proposed in [15]. Let $\mathbf{T}$ be a set of discrete time series that take values in a space $\mathcal{X}$. For a pair of discrete time series $\mathbf{P} = (p_1, \ldots, p_m) \in \mathbf{T}$ and $\mathbf{Q} = (q_1, \ldots, q_n) \in \mathbf{T}$ with lengths m and n respectively, the alignment π between $\mathbf{P}$ and $\mathbf{Q}$ is defined as a pair of increasing integral vectors (π_p, π_q) of length $l \leq m + n - 1$, where

$$1 = \pi_p(1) \leq \cdots \leq \pi_p(l) = m$$

and

$$1 = \pi_q(1) \leq \cdots \leq \pi_q(l) = n$$

such that (π_p, π_q) is defined to have unitary increments and no simultaneous repetitions. For any index $1 \leq i \leq l - 1$, the increment vector of $\pi = (\pi_p, \pi_q)$ satisfies

$$\begin{pmatrix} \pi_p(i+1) - \pi_p(i) \\ \pi_q(i+1) - \pi_q(i) \end{pmatrix} \in \left\{ \begin{pmatrix} 0 \\ 1 \end{pmatrix}, \begin{pmatrix} 1 \\ 0 \end{pmatrix}, \begin{pmatrix} 1 \\ 1 \end{pmatrix} \right\}. \tag{1}$$

In the dynamic time warping framework [15], the coordinates π_p and π_q of the alignment π define the warping function. Let $\mathcal{A}(m, n)$ be the set of all possible alignments between $\mathbf{P}$ and $\mathbf{Q}$. The dynamic time warping distance between $\mathbf{P}$ and $\mathbf{Q}$ is defined as

$$\mathrm{DTW}(\mathbf{P}, \mathbf{Q}) = \min_{\pi \in \mathcal{A}(m,n)} D_{\mathbf{P},\mathbf{Q}}(\pi), \tag{2}$$

where the cost

$$D_{\mathbf{P},\mathbf{Q}}(\pi) = \sum_{i=1}^{|\pi|} \varphi(p_{\pi_p(i)}, q_{\pi_q(i)}), \tag{3}$$

is defined by a local divergence φ that measures the discrepancy between any pair of elements $p_i \in \mathbf{P}$ and $q_i \in \mathbf{Q}$. Generally, φ can be defined as the squared Euclidean distance, i.e., $\varphi(p, q) = \|p - q\|^2$.

Based on the dynamic time warping distance defined in Eq. (2), a dynamic time warping kernel k_{DTW} [17] between $\mathbf{P}$ and $\mathbf{Q}$ can be defined as

$$k_{\mathrm{DTW}}(\mathbf{P}, \mathbf{Q}) = e^{-\mathrm{DTW}(\mathbf{P},\mathbf{Q})}. \tag{4}$$

Unfortunately, this kernel is not positive definite. This is because the optimal alignment required by the dynamic time warping cannot guarantee transitivity. To overcome the shortcoming, Cuturi [15] considers all possible alignments in

$\mathcal{A}(m,n)$ and proposes another dynamic time warping inspired kernel, i.e., the global alignment kernel, as

$$k_{\text{GA}}(\mathbf{P},\mathbf{Q}) = \sum_{\pi \in \mathcal{A}(m,n)} e^{-D_{\mathbf{P},\mathbf{Q}}(\pi)}, \tag{5}$$

where k_{GA} is positive definite, since it quantifies the quality of both the optimal alignment and all other alignments $\pi \in \mathcal{A}(m,n)$. The kernel k_{GA} elaborates on the dynamic time warping distance by considering the same set of elementary operations [16]. However k_{GA} not only generalizes the dynamic time warping kernel k_{DTW}, but also provides richer statistic measures by incorporating the whole spectrum of alignment costs $\{D_{\mathbf{P},\mathbf{Q}}(\pi), \pi \in \mathcal{A}(m,n)\}$.

Intuitively, the global alignment kernel k_{GA} allows one to define a new graph kernel, by measuring the warping alignment π between any types of graph characteristic sequences (or graph embedding vectors [13]) that have certain element orders with increasing structural variables, e.g., the depth-based complexity traces [5] from expansion subgraphs of increasing sizes, or cycle characteristics with increasing lengths identified from the Ihara zeta function [23].

2.2   Centroid Depth-Based Complexity Traces

We review the concept of the depth-based complexity trace of a graph rooted at the centroid vertex [5]. Let $G(V,E)$ be an undirected graph with vertex set V and edge set E. Based on Dijkstra's algorithm, we commence by computing the shortest path matrix S_G, where each element $S_G(v,u)$ of S_G represents the length of the shortest path between vertices $v \in V$ and $u \in V$. For each vertex $v \in V$, let $S(v)$ be the average length of the shortest paths from v to the remaining vertices, i.e.,

$$S(v) = \frac{1}{|V|} \sum_{u \in V} S_G(v,u). \tag{6}$$

As discussed in [5], the centroid vertex $\hat{v}_C$ of $G(V,E)$ can be identified by selecting the vertex that has the minimum variance of shortest path lengths to the remaining vertices, i.e., the index of $\hat{v}_C$ is

$$\hat{v}_C = \arg\min_v \sum_{u \in V} [S_G(v,u) - S_V(v)]^2. \tag{7}$$

Let $N_{\hat{v}_C}^K$ be a vertex subset of $G(V,E)$ satisfying

$$N_{\hat{v}_C}^K = \{u \in V \mid S_G(\hat{v}_C, u) \leq K\}. \tag{8}$$

For $G(V,E)$ and its centroid vertex $\hat{v}_C$, we construct a family of K-layer expansion subgraphs $\mathcal{G}_K(\mathcal{V}_K; \mathcal{E}_K)$ as

$$\begin{cases} \mathcal{V}_K &= \{u \in N_{\hat{v}_C}^K\}; \\ \mathcal{E}_K &= \{(u,v) \subset N_{\hat{v}_C}^K \times N_{\hat{v}_C}^K \mid (u,v) \in E\}. \end{cases} \tag{9}$$

Note that the number expansion subgraphs is equal to the greatest length L of the shortest paths from the centroid vertex to the remaining vertices of $G(V, E)$. Moreover, the L-layer expansion subgraph is the graph $G(V, E)$ itself. An example of constructing a K-layer subgraph is shown in Fig. 1.

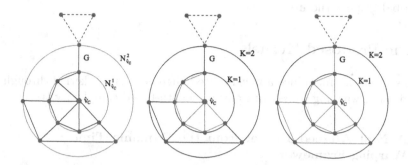

Fig. 1. The left-most figure shows the determination of K-layer centroid expansion subgraphs for a graph $G(V, E)$ which hold $|N_{\hat{v}_C}^1| = 6$ and $|N_{\hat{v}_C}^2| = 10$ vertices. While the middle and the right-most figure show the corresponding 1-layer and 2-layer subgraphs regarding the centroid vertex $\hat{v}_C$, and are depicted by red-colored edges. In this example, the vertices of different K-layer subgraphs regarding the centroid vertex $\hat{v}_C$ are calculated by Eq. (7), and pairwise vertices possess the same connection information in the original graph $G(V, E)$.

Definition (Depth-based complexity traces): For a sample undirected graph $G(V, E)$, let $\{\mathcal{G}_1, \cdots, \mathcal{G}_K, \cdots, \mathcal{G}_L\}$ be the family of K-layer expansion subgraphs rooted at the centroid vertex of $G(V, E)$. Then the depth-based complexity trace $DB(G)$ of $G(V, E)$ is computed by measuring the entropies of the subgraphs [5], i.e.,

$$DB(G) = \{H_S(\mathcal{G}_1), \cdots, H_S(\mathcal{G}_K), \cdots, H_S(\mathcal{G}_L)\}, \qquad (10)$$

where $\cdots, H_S(\mathcal{G}_K)$ is the Shannon entropy associated with the steady state random walk on the K-layer centroid expansion subgraph $\mathcal{G}_K$ [4]. $\square$

The depth-based complexity trace has a number of interesting properties [5]. First, it encapsulates the entropy-based information content flow through the family of K-layer expansion subgraphs rooted at the centroid vertex, and thus reflects rich intrinsic depth topology information of a graph. Second, it can be efficiently computed also on large graphs. This is because it is computed on a small set of expansion subgraphs rooted at the centroid vertex, and the computational complexity is polynomial. Furthermore, based on Eq. (9), we can also observe that the family of K-layer expansion subgraphs rooted at the centroid vertex $\hat{v}_C$ of the graph G constructs a nested sequence. This is because the family of the expansion subgraphs satisfies

$$\hat{v}_C \in \mathcal{G}_1 \cdots \subseteq \mathcal{G}_K \subseteq \cdots \subseteq \mathcal{G}_L \subseteq G.$$

In other words, it represents a sequence of subgraphs that gradually expand from the centroid vertex to the global graph. As a result of it nested nature, the depth-based complexity trace can reflecs both the local and global structure information of a graph. In summary, the depth-based complexity trace provides an elegant way of developing novel fast graph kernels that simultaneously consider local and global graph structures.

3   The Proposed Kernel

In this section, we introduce a novel nested alignment graph kernel through the dynamic time warping framework and the depth-based complexity trace.

3.1   A Nest Aligned Kernel from the Dynamic Time Warping Framework

Let $G_P(V_P, E_P)$ and $G_Q(V_Q, E_Q)$ be a pair of graphs, from a graph set $\mathbf{G}$. We commence by computing the depth-based complexity traces of G_P and G_Q as

$$\mathrm{DB}(G_P) = \{H_S(\mathcal{G}_{P;1}), \cdots, H_S(\mathcal{G}_{P;K}), \cdots, H_S(\mathcal{G}_{P;L^{\max}})\}$$

and

$$\mathrm{DB}(G_Q) = \{H_S(\mathcal{G}_{Q;1}), \cdots, H_S(\mathcal{G}_{Q;K}), \cdots, H_S(\mathcal{G}_{Q;L^{\max}})\},$$

respectively. Here $\mathcal{G}_{P;K}$ and $\mathcal{G}_{Q;K}$ are the K-layer expansion subgraphs rooted at the centroid vertices of G_P and G_Q, and $L^{\max}$ is the greatest length of the shortest paths rooted at the centroid vertices over all graphs in $\mathbf{G}$. Note that, for G_P and G_Q and the greatest lengths M and N of the shortest paths rooted at their centroid vertices, if $K \geq M$ and $K \geq M$ their K-layer expansion subgraphs are themselves, i.e., their global structures. Based on the global alignment kernel defined in Sect. 2.1, we develop a new nested alignment graph kernel k_{NA} between G_P and G_Q as

$$k_{\mathrm{NA}}(G_P, G_Q) = k_{\mathrm{GA}}(\mathrm{DB}(G_P), \mathrm{DB}(G_Q))$$
$$= \sum_{\pi \in \mathcal{A}(L^{\max}, L^{\max})} e^{-D_{\mathbf{P},\mathbf{Q}}(\pi)}, \qquad (11)$$

where π denotes the warping alignment between $\mathrm{DB}(G_P)$ and $\mathrm{DB}(G_Q)$, $\mathcal{A}(L^{\max}, L^{\max})$ denotes all possible alignments, and $D_{\mathbf{P},\mathbf{Q}}(\pi)$ is the alignment cost defined in Eq. (3). Note that we cannot prove that the proposed kernel k_{NA} is positive definite. Although our kernel is based on the global alignment kernel k_{GA}, which is a positive definite kernel, the time series compared by k_{NA} are not defined over the same underlying space but on two different graphs. Future work will explore the possibility of creating a positive definite kernel by computing the depth-based complexity traces over a common structure obtained by combining the input graphs.

As we have observed, the depth-based complexity trace reflects the nested entropy-based information and thus simultaneously considers the local and global graph structures. Furthermore, the proposed kernel $k_{NA}(G_P, G_Q)$ is based on all possible warping alignments between depth-based complexity traces of the input graphs. As a result, $k_{NA}(G_P, G_Q)$ can simultaneously capture richer local and global graph characteristics in terms of all possible alignments between the nested depth-based complexity traces.

3.2 Computational Analysis

For a pair of graphs both having n vertices, computing the nested alignment kernel k_{GA} has time complexity $O(n^3)$. This is because computing the depth-based complexity trace of a graph relies on the computation of the shortest path matrix and thus has time complexity $O(n^3)$. Furthermore, computing all possible alignments between the depth-based complexity traces has time complexity $O((L^{max})^2)$, where L^{max} is the greatest length of the shortest paths rooted at the centroid vertices of the two graphs and is lower than the vertex number n. As a result, the proposed kernel k_{GA} has polynomial time complexity $O(n^3)$.

4   Experimental Evaluations

4.1   Graph Datasets

We evaluate our kernels on standard graph datasets. These datasets include: MUTAG, PTC, COIL5, Shock and CATH2. Details of these datasets are shown in Table 1.

MUTAG: The MUTAG dataset consists of graphs representing 188 chemical compounds labeled according to whether or not they affect the frequency of genetic mutations in the bacterium *Salmonella typhimuriums* and aims to predict whether each compound is associated with mutagenicity.

PTC: The PTC (The Predictive Toxicology Challenge) dataset records the carcinogenicity of several hundred chemical compounds for male rats (MR), female rats (FR), male mice (MM) and female mice (FM). These graphs are very small, i.e., 20–30 vertices, and sparsem, i.e., 25–40 edges. We select the graphs of male rats (MR) for evaluation. There are 344 test graphs in the MR class.

COIL5: The COIL5 dataset is abstracted from the COIL image database. The COIL database consists of images of 100 3D objects. In our experiments, we use the images for the first five objects. For each of these objects we employ 72 images captured from different viewpoints. For each image we first extract corner points using the Harris detector, and then establish Delaunay graphs based on the corner points as vertices. Each vertex is used as the seed of a Voronoi region, which expands radially with a constant speed. The linear collision fronts of the

Table 1. Information on the selected graph based bioninformatics datasets

Datasets	MUTAG	PTC	COIL	Shock	CATH2
Max # vertices	28	109	241	33	568
Min # vertices	10	2	72	4	143
Mean # vertices	17.93	25.60	144.90	109.63	308.03
# graphs	188	344	360	150	190
# classes	2	2	5	5	2

regions delineate the image plane into polygons, and the Delaunay graph is the region adjacency graph for the Voronoi polygons.

Shock: The Shock dataset consists of graphs from the Shock 2D shape database. Each graph is a skeletal-based representation of the differential structure of the boundary of a 2D shape. There are 150 graphs divided into 10 classes.

CATH2: The CATH2 dataset is harder to classify, since the proteins in the same topology class are structurally similar. The protein graphs are 10 times larger in size than chemical compounds, with 200 . 300 vertices. There is 190 testing graphs in the dataset.

5   Experiments on Standard Graph Datasets

We evaluate the performance of the nested alignment graph kernel (NAGK) on a number of graph classification tasks. Furthermore, we also compare our kernel with three state-of-the-art kernels, including (1) the Jensen-Shannon graph kernel (JSGK) [4], (2) the random walk graph kernel (RWGK) [20], (3) the unaligned quantum Jensen-Shannon graph kernel (QJSK) [9], and (4) the Lovász graph kernel (LGK) [19].

We compute the kernel matrix associated with each kernel on each dataset. We perform 10-fold cross-validation using a C-Support Vector Machine (C-SVM) to compute the classification accuracies, using LIBSVM software library [12]. We use nine samples for training and one for testing. The parameters of the C-SVMs are optimized on each training set using cross-validation. We report the average classification accuracy and the runtime for each kernel in Table 2 and Table 3. The runtime is measured under Matlab R2015a running on a 2.5 GHz Intel 2-Core processor (i.e., i5-3210 m).

In terms of classification accuracy, Table 2 indicates that the proposed NAGK kernel can significantly outperform the alternative state-of-the-art graph kernels, excluding the QJSK kernel on the COIL5 and Shock datasets. However, the proposed NAGK kernel is still competitive to the QJSK kernel on the COIL5 dataset and outperforms the QJSK kernel on the MUTAG, PTC and CATH2 datasets. The reasons for this effectiveness are twofold. First, as we have stated, the depth-based complexity traces used by the proposed NAGK kernel encapsulate nested entropy-based information that extend from the centroid vertex to the global

Table 2. Classification accuracy (In % ± Standard error) runtime in second.

Datasets	MUTAG	PTC	COIL5	Shock	ATH2
NAGK	**84.22** ± .50	**58.00** ± .64	69.75 ± .65	37.60 ± .62	**74.00** ± .83
JSGK	83.11 ± .80	57.29 ± .41	69.13 ± .79	21.73 ± .76	72.26 ± .76
RWGK	80.77 ± .75	53.97 ± .31	14.21 ± .65	0.33 ± .37	–
QJSK	82.72 ± .44	56.70 ± .49	**70.11** ± .61	**40.60** ± .92	71.11 ± .88
LGK	80.83 ± .43	56.29 ± .47	–	31.80 ± .89	–

Table 3. Runtime for various kernels.

Datasets	MUTAG	PTC	COIL5	Shock	CATH2
NAGK	$8.6 \cdot 10^2$	$2.3 \cdot 10^3$	$3.3 \cdot 10^3$	$3.8 \cdot 10^2$	$9.4 \cdot 10^2$
JSGK	$1.0 \cdot 10^0$	$1.0 \cdot 10^0$	$1.0 \cdot 10^0$	$1.0 \cdot 10^0$	$1.0 \cdot 10^0$
RWGK	$4.6 \cdot 10^1$	$6.7 \cdot 10^1$	$1.1 \cdot 10^3$	$2.3 \cdot 10^1$	–
QJSK	$2.0 \cdot 10^1$	$1.0 \cdot 10^2$	$1.0 \cdot 10^3$	$1.4 \cdot 10^1$	$4.4 \cdot 10^3$
LGK	$1.0 \cdot 10^3$	$7.4 \cdot 10^3$	–	$1.0 \cdot 10^3$	–

graph structure. As a consequence, the proposed NAGK kernel can simultaneously consider the local and global graph characteristics. By contrast, the QJSK and JSGK kernels can only reflect global graph characteristics, whereas the LGK and RWGK can only reflect local graph characteristics. Second, the proposed NAGK kernel is based on all possible alignments between the complexity traces, and thus reflects rich statistic measures by incorporating the whole spectrum of alignment costs. On the other hand, we observe that the QJSK kernel based on the global von Neumann entropy from the continuous-time quantum walk is the most competitive kernel to the proposed NAGK kernel, though the QJSK kernel can only reflect global characteristics. This is because the entropy measure from the quantum walk can reflect richer intrinsic topology information than that from the classical steady state random walk (for the proposed NAGK kernel). This in turn suggest the possibility of further extending the NAGK kernel using quantum walks to extract an analogous of the depth-based complexity trace used in this study.

In terms of runtime, the proposed the NAGK kernel is not the fastest kernel, when compared to the other graph kernels. However, we can observe that the proposed NAGK kernel can always complete the computation of the kernel matrices, unlike some alternative graph kernels (e.g., the LGK and RWGK kernels), which failed complete the computation in a reasonable time.

6 Conclusion

In this paper, we have proposed a novel nested alignment graph kernel. The kernel is an adaptation of the dynamic time warping framework based kernel (i.e., the global alignment kernel) to graphs. To this end, we made use of the depth-based complexity traces of graphs, a powerful and fast to compute graph descriptor. Unlike most existing graph kernels that only probe local or global graph characteristics, the proposed kernel simultaneously considers local and global graph characteristics and thus reflects the presence of richer structural patterns. The experiments have demonstrated the effectiveness and efficiency of the proposed kernel.

Our future work is to extend the proposed kernel to attributed graphs that encapsulate vertex and edge labels. Moreover, we would also like to further develop novel graph kernels through the dynamic time warping framework associated with other types of (hyper)graph characteristic sequences, e.g., the cycle numbers identified by the Ihara zeta function, the time-varying entropies computed from the continuous-time or discrete-time quantum walk [8,9], and the depth-based hypergraph complexity traces [3]. Finally, we are also interested in developing novel graph kernels for time-varying financial market networks [29], using the dynamic time warping framework.

Acknowledgments. This work is supported by the National Natural Science Foundation of China (Grant no. 61503422 and 61602535), the Open Projects Program of National Laboratory of Pattern Recognition, the Young Scholar Development Fund of Central University of Finance and Economics (No. QJJ1540), and the program for innovation research in Central University of Finance and Economics.

References

1. Alvarez, M.A., Qi, X., Yan, C.: A shortest-path graph kernel for estimating gene product semantic similarity. J. Biomed. Semant. **2**, 3 (2011)
2. Bach, F.R.: Graph kernels between point clouds. In: Proceedings of ICML, pp. 25–32 (2008)
3. Bai, L., Escolano, F., Hancock, E.R.: Depth-based hypergraph complexity traces from directed line graphs. Pattern Recogn. **54**, 229–240 (2016)
4. Bai, L., Hancock, E.R.: Graph kernels from the jensen-shannon divergence. J. Math. Imaging Vis. **47**(1–2), 60–69 (2013)
5. Bai, L., Hancock, E.R.: Depth-based complexity traces of graphs. Pattern Recogn. **47**(3), 1172–1186 (2014)
6. Bai, L., Hancock, E.R.: Fast depth-based subgraph kernels for unattributed graphs. Pattern Recogn. **50**, 233–245 (2016)
7. Bai, L., Ren, P., Hancock, E.R.: A hypergraph kernel from isomorphism tests. In: Proceddings of ICPR, pp. 3880–3885 (2014)
8. Bai, L., Rossi, L., Cui, L., Zhang, Z., Ren, P., Bai, X., Hancock, E.R.: Quantum kernels for unattributed graphs using discrete-time quantum walks. Pattern Recogn. Lett. **87**, 96–103 (2017)
9. Bai, L., Rossi, L., Torsello, A., Hancock, E.R.: A quantum jensen-shannon graph kernel for unattributed graphs. Pattern Recogn. **48**(2), 344–355 (2015)
10. Bai, L., Rossi, L., Zhang, Z., Hancock, E.R.: An aligned subtree kernel for weighted graphs. In: Proceedings of ICML, pp. 30–39 (2015)

11. Bai, L., Zhang, Z., Wang, C., Bai, X., Hancock, E.R.: A graph kernel based on the Jensen-Shannon representation alignment. In: Proceedings of IJCAI, pp. 3322–3328 (2015)
12. Chang, C.C., Lin, C.J.: Libsvm: a library for support vector machines. ACM Trans. Intell. Syst. Technol. **2**(3), 27 (2011)
13. Conte, D., Ramel, J., Sidere, N., Luqman, M.M., Gaüzère, B., Gibert, J., Brun, L., Vento, M.: A comparison of explicit and implicit graph embedding methods for pattern recognition. In: Proceedings of GbRPR, pp. 81–90 (2013)
14. Costa, F., De Grave, K.: Fast neighborhood subgraph pairwise distance kernel. In: Proceedings ICML, pp. 255–262 (2010)
15. Cuturi, M.: Fast global alignment kernels. In: Proceedings of ICML, pp. 929–936 (2011)
16. Cuturi, M., Vert, J., Birkenes, Ø., Matsui, T.: A kernel for time series based on global alignments. In: Proceedings of ICASSP, pp. 413–416 (2007)
17. Haasdonk, B., Bahlmann, C.: Learning with distance substitution kernels. In: Rasmussen, C.E., Bülthoff, H.H., Schölkopf, B., Giese, M.A. (eds.) DAGM 2004. LNCS, vol. 3175, pp. 220–227. Springer, Heidelberg (2004). doi:10.1007/978-3-540-28649-3_27
18. Harchaoui, Z., Bach, F.: Image classification with segmentation graph kernels. In: Proceedings of CVPR, pp. 1–8 (2007)
19. Johansson, F.D., Jethava, V., Dubhashi, D.P., Bhattacharyya, C.: Global graph kernels using geometric embeddings. In: Proceedings of ICML, pp. 694–702 (2014)
20. Kashima, H., Tsuda, K., Inokuchi, A.: Marginalized kernels between labeled graphs. In: Proceedings of ICML, pp. 321–328 (2003)
21. Kriege, N., Mutzel, P.: Subgraph matching kernels for attributed graphs. In: Proceedings of ICML (2012)
22. Kriege, N.M., Giscard, P., Wilson, R.C.: On valid optimal assignment kernels and applications to graph classification. In: Proceedings of NIPS, pp. 1615–1623 (2016)
23. Ren, P., Aleksić, T., Wilson, R.C., Hancock, E.R.: A polynomial characterization of hypergraphs using the Ihara zeta function. Pattern Recogn. **44**(9), 1941–1957 (2011)
24. Riesen, K., Bunke, H.: Graph classification and clustering based on vector space embedding. World Scientific Publishing Co., Inc., River Edge (2010)
25. Rossi, L., Torsello, A., Hancock, E.R.: Measuring graph similarity through continuous-time quantum walks and the quantum jensen-shannon divergence. Phys. Rev. E **91**(2), 022815 (2015)
26. Rossi, L., Torsello, A., Hancock, E.R., Wilson, R.C.: Characterizing graph symmetries through quantum jensen-shannon divergence. Phys. Rev. E **88**(3), 032806 (2013)
27. Urry, M., Sollich, P.: Random walk kernels and learning curves for gaussian process regression on random graphs. J. Mach. Learn. Res. **14**(1), 1801–1835 (2013)
28. Wang, L., Sahbi, H.: Directed acyclic graph kernels for action recognition. In: Proceedings of ICCV, pp. 3168–3175 (2013)
29. Ye, C., Comin, C.H., Peron, T.K.D., Silva, F.N., Rodrigues, F.A., da Costa, F., Torsello, A., Hancock, E.R.: Thermodynamic characterization of networks using graph polynomials. Phys. Rev. E **92**(3), 032810 (2015)

Graph Applications

GERoMe – A Novel Graph Extraction Robustness Measure

Dominik Drees$^{(\boxtimes)}$, Aaron Scherzinger, and Xiaoyi Jiang

Faculty of Mathematics and Computer Science, University of Münster,
Münster, Germany
{dominik.drees,scherzinger,xjiang}@uni-muenster.de

Abstract. The extraction of graph structures in Euclidean vector space
is a topic of interest with applications in many fields, e.g., the biomedical domain. While a number of different approaches have been presented,
a quantitative evaluation of those algorithms remains a challenging task:
Manual generation of ground truth for real-world data is often time-
consuming and error-prone, and while tools for generating synthetic
datasets with corresponding ground truth exist, this data often does
not reflect the complexity in morphology and topology that real-world
scenarios show. As a complementary or even alternative approach, we
propose GERoMe, a novel graph extraction robustness measure, which
quantifies the stability of algorithms that extract multigraphs with associated node positions from non-graph structures. Our method takes edge-
associated properties into consideration and does not necessarily require
ground truth data. Moreover, available ground truth information can be
incorporated to additionally evaluate the correctness of the graph extraction algorithm. We demonstrate the usefulness and applicability of our
approach in an exemplary study on synthetic and real-world data.

Keywords: Graph extraction · Evaluation · Robustness · Stability

1 Introduction

Extracting graphs which are embedded in Euclidean space from non-graph like
structures has been a topic of interest in various areas of research, especially with
regard to biomedical applications. Here, researchers may be interested in the
general structure and topology of the graph, the position of branching points, or
specific (e.g., morphologic) properties of individual edges, e.g., in the analysis of
hepatic blood vasculature [10], airway trees [12], neural systems [11] or lymphatic
vessel systems [6]. There exist several publications which focus on the extraction
of embedded graphs from 2D or 3D images [1,7]. Moreover, there is an interest
in the simultaneous extraction of geometrical and morphological edge-associated
properties from the original dataset, e.g., [2,3,9].

While a number of algorithms exist which produce plausible results, providing an objective evaluation of the quality of the extracted graph remains a
challenging task. Although manual ground truth generation is conceivable for the

© Springer International Publishing AG 2017
P. Foggia et al. (Eds.): GbRPR 2017, LNCS 10310, pp. 73–82, 2017.
DOI: 10.1007/978-3-319-58961-9_7

topological structure of the graph and node positions, it is time-consuming and error-prone, especially for 3D structures such as complex vessel networks in medical imaging. Even more so, an accurate manual annotation of edge-associated properties such as volume or average radius appears almost impossible in 3D. Although tools for producing synthetic datasets have been presented, they only include a limited number of edge-associated properties. Moreover, the complexity of the generated datasets does not compare to real-world data (see Sect. 2).

As a complement or even an alternative to using synthetic data, we propose GERoMe – a novel graph extraction robustness measure, which is able to quantify the stability of extraction algorithms on arbitrary (e.g., real-world) input data, without requiring ground truth information. Moreover, we introduce a graph similarity measure which can be used to evaluate the accuracy of a graph extraction algorithm if ground truth information is available.

For a given input, a set of transformations, and any edge-associated property, our method generates a scalar robustness index. This is achieved by applying one of the transformations to the input data, and using the result to extract a graph, which is then retransformed into the original space. This graph is matched with a template graph directly extracted from the input. For each transformation, a similarity measure is computed based on the difference in features of matched edges and the quality of the matching itself. The similarities for all transformations are then combined to form the robustness index GERoMe. Our method does not require ground truth data for evaluating the robustness of an algorithm. However, if it is available for the desired properties, ground truth data can be used as the template graph. In this case the resulting GERoMe value quantifies the accuracy of the examined algorithm in conjunction with its robustness.

The extracted graphs can be of arbitrary structure, may include multiple edges connecting two nodes (i.e., they may be multigraphs), and can be evaluated for arbitrary real-valued edge-associated properties. The input data for the considered graph extraction algorithm can be of arbitrary nature, as long as a geometric transformation can be applied to it. We demonstrate the applicability of our approach in an exemplary study using a preliminary version of the algorithm proposed in [3] on artificial and real-world datasets.

The remainder of this paper is structured as follows. In the following section we give an overview of related publications. Afterwards we provide an in-depth description of our proposed method. Finally, we exemplarily apply the proposed graph robustness measure to an existing algorithm and discuss the results.

2   Related Work

Drechsler et al. [2] have proposed a graph extraction method for hepatic blood vasculature. In order to evaluate their algorithm, they rotate and resample the original volume and plot the number of nodes and edges in the generated graph for various rotation angles. They observe that their algorithm is not rotation-invariant, but note that an ideal algorithm should fulfill this requirement.

A possible validation strategy for graph extraction methods is the use of synthetic data for which ground truth information is available. VascuSynth [5] is a tool for the simulation of 3D medical images of blood vasculature. In addition to the raw image data it provides ground truth data which includes a segmentation, the generated graph (i.e., node positions and edges), as well as radius, length, and flow for each edge of the graph. However, the generated vessel networks always have a tree-like topology and thus do not include cycles or multiple edges connecting the same pair of nodes. Moreover, the approach only simulates images of blood vasculature where the generated vessels are of relatively simple morphology. The resulting data sets thus do not heavily challenge graph extraction algorithms in that regard.

One important aspect of this paper is matching edges of two (multi-)graphs. Traditional graph matching, which aims to find a mapping between the nodes of two graphs, is a current and popular research topic [4]. Frameworks for matching multigraphs with additional attributes exist (e.g., [13]) and may in principle be applied to the matching problem in this paper. However, although these approaches may incorporate geometrical information, traditional graph matching algorithms heavily rely on the second order (i.e., topological) information present in the graph. As it turns out, when matching two graphs for the purpose of this paper, geometrical information can be expected to be fairly reliable, while the topology of the generated graphs may differ (depending on the examined algorithm and the input data). We therefore employ a direct edge matching approach using geometric and additional edge associated information (see Sect. 3).

3   Method

An embedded multigraph shall be defined as a tuple $G = (N, E)$ of a set of nodes $N \in \mathbb{R}^n$ (we assume nodes with the same spatial position to be identical) and a set of edges $E \subset (N \times N \times \mathbb{N})$. Edges (n_1, n_2, i) are defined by two nodes $n_1, n_2 \in N$ and a unique identifier i. Additionally, edges $e \in E$ have m associated real-valued properties $P_i \geq 0, i \in \{1, \dots m\}$.

3.1   The Graph Extraction Robustness Measure

The graph extraction robustness measure (GERoMe), which will be denoted $\mathcal{G}$ for the remainder of this paper, provides a stability measure for multigraph extraction algorithms. Conceptually, it describes a process which compares the results of the extraction algorithm $\mathcal{A}$ on a transformed version of the input s to a template graph. The template graph can either be given as ground truth G_{GT}, or – e.g., if ground truth information for the property of interest is not available – extracted from the input dataset without applying any transformation, i.e., $G_{tpl} = \mathcal{A}(s)$. The input dataset is then transformed by T, and the result is used as input to the examined graph extraction algorithm. For a robust algorithm, the result can be expected to be similar to the template graph (after retransforming one of the results into the original space using T^{-1}) for *any* T. Therefore, the measure $\mathcal{G}$ is defined as the minimum similarity S_P (see Subsect. 3.2) over all elements of a

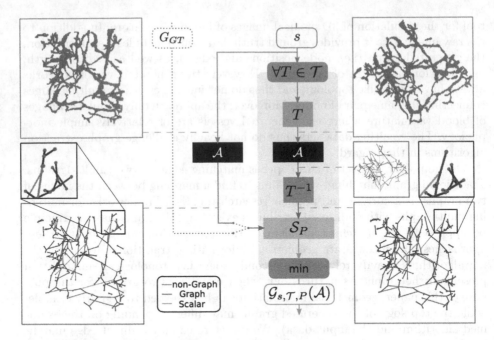

Fig. 1. A schematic overview of the proposed method. $\mathcal{T}$ is a set of transformations, P is an edge-associated property, $\mathcal{A}$ is a graph extraction algorithm, s is a non-graph structure. Annotated images on the sides show intermediate results of the approach when applied to a preliminary version of the algorithm described in [3] and a lymphatic vessel foreground segmentation dataset [6].

set of transformations $\mathcal{T}$ for a given dataset. This process is illustrated in Fig. 1. Hence, T must be an automorphism that can be applied to both the input dataset s and an extracted graph. Moreover, for a perfect extraction algorithm $\mathcal{A}^*$ for the corresponding edges e and e^* in $\mathcal{A}^*(s)$ and $(T^{-1} \circ \mathcal{A}^* \circ T)(s)$ one should have $P(e) \approx P(e^*)$ for any property P. As an example, if T includes a scaling operation (on s), and the information extracted via $\mathcal{A}^*$ includes the distance between two nodes $P_{distance}$ for all edges, T^{-1} subsequently must scale $P_{distance}$ accordingly. For many properties in real world applications, this is the case if T is a rigid-body transformation. More formally, given the parameters mentioned above, this procedure can be defined as follows:

$$\mathcal{G}_{s,\mathcal{T},P}(\mathcal{A}) = \min_{T \in \mathcal{T}} \mathcal{S}_P(G_{tpl}, (T^{-1} \circ \mathcal{A} \circ T)(s)) \tag{1}$$

It should be noted that $\mathcal{G}_{s,\mathcal{T},P} \in [0,1]$. A robust extraction algorithm $\mathcal{A}$ will produce similar graphs regardless of any transformation $T \in \mathcal{T}$, yielding a GERoMe-value near the optimal value 1. If ground truth information for P in form of a ground truth graph G_{GT} is available, $\mathcal{G}$ also includes information about the accuracy of $\mathcal{A}$ for $G_{tpl} = G_{GT}$. Otherwise, we set $G_{tpl} = \mathcal{A}(s)$ and only quantify the robustness of the algorithm.

3.2   Graph Similarity

In order to compare two embedded multigraphs, they need to be matched. Since we are interested in differences in edge properties, and since nodes have an associated position, it is sufficient to find a *matching* $M_{G_1,G_2} \subset E_1 \times E_2$ for two graphs $G_1 = (N_1, E_1), G_2 = (N_2, E_2)$ which matches edges in G_1 to edges in G_2. Note that not all edges in E_1 or E_2 have to be part of the matching, but any edge in E_1 or E_2 can only be part of *one* pair in M_{G_1,G_2}.

$$M_{G_1,G_2} \subset E_1 \times E_2 \Rightarrow \forall e_1 \in E_1 : |\{(e_1, e) \in M_{G_1,G_2}\}| \leq 1$$
$$\wedge \forall e_2 \in E_2 : |\{(e, e_2) \in M_{G_1,G_2}\}| \leq 1 \qquad (2)$$

Moreover, given a property P we define the relative error E_P of two edges e_1, e_2:

$$E_P(e_1, e_2) = \frac{|P(e_1) - P(e_2)|}{\max(P(e_1), P(e_2))} \in [0, 1] \qquad (3)$$

Then, given a graph matching M_{G_1,G_2} and a property P, the relative error of a graph matching can be defined using (3):

$$E_P(M_{G_1,G_2}) = \frac{1}{|M_{G_1,G_2}|} \sum_{(e_1,e_2)\in M} E_P(e_1, e_2) \qquad (4)$$

However, $E_P(M_{G_1,G_2})$ ignores edges in the original graphs that have not been matched. Therefore, for two graphs G_1 and G_2 we define the similarity (in terms of the property P) as follows:

$$S_P(G_1, G_2) = (1 - E_P(M_{G_1,G_2})) \cdot \frac{2|M_{G_1,G_2}|}{|E_1| + |E_2|} \qquad (5)$$

The term $\frac{2|M_{G_1,G_2}|}{|E_1|+|E_2|}$, i.e., the edge match ratio, can be understood as the DICE index for $E_1 \cap E_2 := M_{G_1,G_2}$. For $|E_1| = |E_2|$ the term simulates (arbitrarily) pairing all leftover (i.e., non-matched) edges while setting the relative error of all of these fake matches to 1.

3.3   Matching

In order to compute a matching, we utilize the Hungarian method [8] using a distance d between two edges. As a basis for d we first define d' which only relies on the spatial positions and Euclidean distances between the node positions of two edges. The distance d' is calculated by concatenating the nodes for both edges to form a $2n$-dimensional vector, and computing the Euclidean distance. This punishes edges that share one node but not the other harder than the sum of node distances. Since the order of nodes is arbitrary, the minimum distance of both unique node pairing permutations is denoted d'.

$$d'((n_1, n_2, \ldots), (n'_1, n'_2, \ldots)) = \min(||(n_1 \circ n_2) - (n'_1 \circ n'_2)||_2,$$
$$||(n_1 \circ n_2) - (n'_2 \circ n'_1)||_2) \qquad (6)$$

The distance d is then defined by increasing the base distance given by d' if the average of relative property errors (3) is large:

$$d(e_1, e_2) = \frac{d'(e_1, e_2)}{1 - \frac{1}{m} \sum\limits_{i \in [1,m]} E_{P_i}(e_1, e_2)} \tag{7}$$

In order to omit false positive matches produced by the Hungarian method, we set all distances above a certain threshold t to the same value d_{max}. The threshold is chosen to be equal to the $\frac{2 \cdot \min(|E_1|, |E_2|)}{|E_1| \cdot |E_2|}$-quantile (i.e., the $2 \cdot \min(|E_1|, |E_2|)$'th smallest value) of the set of all $|E_1| \cdot |E_2|$ edge distances. Finally, matches reported by the Hungarian method with a distance greater than t are ignored. In this way, obvious matches can still be found by the matching algorithm, while edges that do not have a correspondence in the other graph stay unmatched and do not skew the overall result by interfering with other matches in the search for a global minimum. In total, no more than $\min(|E_1|, |E_2|)$ matches can be found. However, two edges connecting the same nodes will have similar distances to corresponding edges in the other graph. Therefore, $\min(|E_1|, |E_2|)$ cannot be a hard cutoff point. In order to include all likely match candidates the $2 \cdot \min(|E_1|, |E_2|)$'th smallest value is chosen. It should be noted that the threshold is designed for real-world applications such as the extraction of blood or lymphatic vasculature. Extreme cases where a large percentage of nodes are connected by multiple edges may thus require a larger threshold.

4   Exemplary Study

In order to demonstrate the applicability and usefulness of our proposed method and measure, a preliminary version of the graph creation and feature extraction algorithm proposed in [3] will be evaluated in terms of its robustness as an exemplary study. The algorithm first creates a voxel skeleton from a binary volumetric input dataset, from which it then extracts a graph embedded in 3D space and calculates both geometric as well as morphologic edge-associated properties using the skeleton and the original input volume. For the purpose of this study we restrict the set of examined edge-properties to *length* (the length of a branch when following the medial line), *distance* (the Euclidean distance of the connected nodes), *straightness* $= \frac{distance}{length}$, *avgRadius* (i.e., the average distance of a centerline to the surface of the branch) and *volume* (the total volume occupied by a branch in the original binary volume). For the set of applied transformations $\mathcal{T}$, 4 rotation axes (the three coordinate axes as well as $(1, 1, 1)$) are taken into account. For each axis, 36 equally distributed rotations (from 0 to 2π) are generated, resulting in $|\mathcal{T}| = 4 \cdot 36 = 144$. Using these parameters, $\mathcal{G}_{s, \mathcal{T}, P}$ is applied to the 3D ground truth segmentation of an artificial blood vessel tree structure generated by VascuSynth [5] as well as a segmentation of an ultramicroscopy image of human lymphatic vessel tissue [6] (the latter depicted in Fig. 1). The transformation T is applied to the 3D image data by resampling the volume. Since a rotation T (or T^{-1}) does not change the values of the edge-associated properties, an extracted graph can be transformed by applying T merely to the node positions. Since both datasets contain vessels of low *avgRadius*, the transformed volume (i.e., $T(s)$) is generated by doubling the resolution in each dimension in order to reduce the error introduced in the resampling step. As this may allow for a better accuracy in the generation of the intermediate voxel skeleton and the extraction of edge-associated properties, the resolution of original dataset is also octupled prior to starting the graph extraction process. Since there is no ground truth information available for the real-world dataset, and the ground truth for the synthetic dataset does not include all properties of interest, we use graphs extracted from the input dataset as template graphs and thus only consider the robustness of the algorithm.

4.1   Synthetic Data

The similarity of the original graph extracted from a synthetic blood vasculature dataset generated using VascuSynth and a transformed version is illustrated in Fig. 2a for 5 selected properties. The set of applied transformations comprises 36 rotations around the x-axis of the coordinate system. As can be seen, the plot shows 4 peaks for all properties which correspond to the angles in which the transformed volume is aligned with the grid of the original volume (i.e., all angles that are multiples of $\frac{\pi}{2}$). This illustrates that the observed error can partially be attributed to the resampling process rather than the graph extraction algorithm itself. Moreover, the relative error of some properties seems to be affected more by the transformation process than others: Both *avgRadius* and *straightness* are less affected than *distance, length,* and especially *volume.* The relative errors of *length* and *distance* are probably caused by small variations in node positions, while this does not have such a strong effect on $straightness = \frac{distance}{length}$.

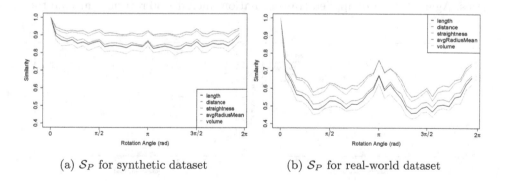

(a) $\mathcal{S}_P$ for synthetic dataset   (b) $\mathcal{S}_P$ for real-world dataset

Fig. 2. Intermediate results of the procedure depicted in Fig. 1 for a synthetic (a) and a real-world dataset (b) without using ground truth information. Graphs have been extracted using a preliminary version of the algorithm described in [3]. For each dataset, the similarity $\mathcal{S}_P$ of the graph extracted from the original dataset and the graphs extracted after applying 36 rotations around the x-axis are shown for 5 selected properties.

Table 1. The robustness measure GERoMe $\mathcal{G}$ applied to a preliminary version of the algorithm proposed in [3] for 5 selected properties, using the synthetic dataset generated by VascuSynth and a real-world lymphatic vessel dataset. The sets of transformations comprise 36 rotations around each of the coordinate axes $(\mathcal{T}_x, \mathcal{T}_y, \mathcal{T}_z)$ as well as the axis $(1, 1, 1)$ $(\mathcal{T}_{xyz})$.

$\mathcal{G}_{s,\mathcal{T},P}(\mathcal{A})$		*length*	*distance*	*straightness*	*avgRadius*	*volume*
Synthetic dataset	$\mathcal{T}_x$	0.822	0.835	0.891	0.900	0.787
	$\mathcal{T}_y$	0.816	0.828	0.874	0.892	0.789
	$\mathcal{T}_z$	0.830	0.837	0.880	0.893	0.801
	$\mathcal{T}_{xyz}$	**0.730**	**0.735**	**0.790**	**0.799**	**0.675**
Real-world dataset	$\mathcal{T}_x$	0.460	0.486	0.528	0.553	0.429
	$\mathcal{T}_y$	0.460	0.484	0.556	0.563	0.429
	$\mathcal{T}_y$	0.559	0.575	0.651	0.656	0.528
	$\mathcal{T}_{xyz}$	**0.298**	**0.312**	**0.341**	**0.371**	**0.272**

The *volume* is heavily affected by errors in the resampling process. The relative error of *avgRadius* is likely caused by errors in the resampling process as well, but to a lesser extent, since the property is averaged along the run of a branch. GERoMe values for sets of rotations for the 4 considered rotation axes are shown in Table 1.

4.2 Real-World Lymphatic Vessel Data

The similarity of the original graph extracted from a real-world lymphatic vasculature dataset and a transformed version is shown in Fig. 2b for 5 selected properties. Again, the set of applied transformations includes 36 rotations around the

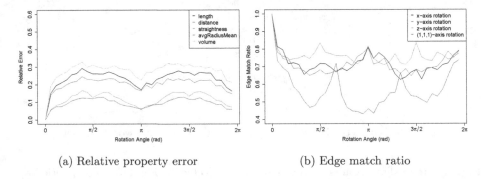

(a) Relative property error   (b) Edge match ratio

Fig. 3. This figure shows intermediate measures generated from a real-world dataset for a preliminary version of the algorithm described in [3]. In (a), the relative errors of 5 selected properties are plotted for 36 rotations $\mathcal{T}$ around the x-axis. In (b) the edge match ratios for 36 rotations around the x-, y-, and z-axis, as well as (1,1,1) are shown.

x-axis of the coordinate system. In comparison to the similarities extracted from the synthetic dataset, the property similarities S_P are much lower. These relatively low similarity values originate from both the relative property error (see Fig. 3a) as well as the edge match ratio (see Fig. 3b). As observed for the synthetic dataset, the similarity, the edge match ratio, and (to a lesser extent) the relative error seem to peak whenever the voxel grids of the original volume and the transformed volume are aligned, i.e., for rotation angles that are multiples of $\frac{\pi}{2}$ for coordinate axes as rotational axes, and for multiples of $\frac{2\pi}{3}$ for $(1, 1, 1)$. Again, this suggests that at least part of the dissimilarity originates from resampling errors (even more so for rotational axes other than the coordinate axes, since resampling errors are introduced in 3, and not only in 2 dimensions in this case). However, this does not imply a weakness of the proposed method itself, as the parameter T as well as optional upsampling can and should always be kept constant when comparing methods and specified along with the results. The fact that a rotational axis of $(1, 1, 1)$ produces larger resampling errors also becomes apparent in the final GERoMe values (see Table 1): For both datasets the minimum similarity for all properties was reached for a transformation around this (non-aligned) axis. Moreover, it can be observed that the amount of the relative error introduced by the transformation and resampling process seems to be relatively independent of the dataset: Just like it is the case for the synthetic dataset, *avgRadius* and *straightness* seem to be less affected than *distance*, *length*, and *volume*.

Another aspect to note is that at least the examined algorithm does not produce outliers in terms of the similarity between two graphs for any transformation. This is an important property of robust graph extraction algorithms. Any potentially generated outliers would immediately become visible in $\mathcal{G}$, as it is defined as the minimum of all similarities.

These results also show that the examined extraction algorithm produces much more stable results for the synthetic dataset than for the real-world dataset. This indicates that evaluating graph extraction algorithms solely on the basis of synthetic datasets is a highly problematic strategy. In combination with difficulties in obtaining ground truth annotations for real-world data this underlines the usefulness of our method.

5   Conclusion

We have proposed GERoMe, a novel robustness measure for graph extraction algorithms. Our approach does not necessarily require ground truth data, and can be applied to any algorithm which extracts (multi-)graphs that are embedded in Euclidean space from non-graph structures for which an edge property-preserving transformation is defined. If ground truth data is available, the method and the introduced similarity measure can be used to quantify the accuracy of graph extraction algorithms in conjunction with the robustness. In addition to the node positions, we use edge-associated properties to distinguish edges, which is useful for matching true multigraphs. We have demonstrated the applicability and usefulness of our method in an exemplary study on synthetic and real-world medical 3D image data. We are convinced that GERoMe may prove useful for evaluating graph extraction algorithms, especially in cases where ground truth data is not available. In the future, we plan to use GERoMe to study and compare the performance of state-of-the-art graph extraction algorithms. Moreover, we would like to augment and generalize GERoMe to utilize information from node-associated properties in both the matching process and the similarity measure itself. Additionally, it may be worthwhile to consider and compare alternative matching approaches which utilize the expected spatial proximity of matched edges if the runtime of the proposed method (which is dominated by the Hungarian algorithm) is problematic.

References

1. Chen, Y., Laura, C.O., Drechsler, K.: Generation of a graph representation from three-dimensional skeletons of the liver vasculature. In: Proceeding of 2nd International Conference on BioMedical Engineering and Informatics, pp. 1–5 (2009)
2. Drechsler, K., Laura, C.O.: Hierarchical decomposition of vessel skeletons for graph creation and feature extraction. In: IEEE International Conference on Bioinformatics and Biomedicine, pp. 456–461 (2010)
3. Drees, D., Scherzinger, A., Hägerling, R., Kiefer, F., Jiang, X.: Graph creation and feature extraction for vessel networks of arbitrary topology in large volumetric datasets (in preparation)
4. Foggia, P., Percannella, G., Vento, M.: Graph matching and learning in pattern recognition in the last 10 years. Int. J. Pattern Recogn. Artif. Intell. **28**(1), 1450001 (2014)
5. Hamarneh, G., Jassi, P.: Vascusynth: Simulating vascular trees for generating volumetric image data with ground truth segmentation and tree analysis. Comput. Med. Imaging Graph. **34**(8), 605–616 (2010)
6. Hägerling, R., et al.: Novel diagnostic approach identifies the vascular pathology causing lymphedema in WILD sydrome (in preparation)
7. Klette, G.: Branch voxels and junctions in 3D skeletons. In: Proceeding 11th International Workshop on Combinatorial Image Analysis, pp. 34–44 (2006)
8. Kuhn, H.W.: The hungarian method for the assignment problem. Naval Res. Logistics Q. **2**(1–2), 83–97 (1955)
9. Rodriguez, A., Ehlenberger, D.B., Hof, P.R., Wearne, S.L.: Rayburst sampling, an algorithm for automated three-dimensional shape analysis from laser scanning microscopy images. Nat. Protoc. **1**(4), 2152–2161 (2006)
10. Selle, D., Preim, B., Schenk, A., Peitgen, H.O.: Analysis of vasculature for liver surgical planning. IEEE Trans. Med. Imaging **21**(11), 1344–1357 (2002)
11. Wearne, S., Rodriguez, A., Ehlenberger, D., Rocher, A., Henderson, S., Hof, P.: New techniques for imaging, digitization and analysis of three-dimensional neural morphology on multiple scales. Neuroscience **136**(3), 661–680 (2005)
12. Wood, S.A., Zerhouni, E.A., Hoford, J.D., Hoffman, E.A., Mitzner, W.: Measurement of three-dimensional lung tree structures by using computed tomography. J. Appl. Physiol. **79**(5), 1687–1697 (1995)
13. Yan, J., Wang, J., Zha, H., Yang, X., Chu, S.M.: Consistency-driven alternating optimization for multigraph matching: a unified approach. IEEE Trans. Image Process. **24**(3), 994–1009 (2015)

Speeding-Up Graph-Based Keyword Spotting in Historical Handwritten Documents

. Michael Stauffer[1,4]([✉]), Andreas Fischer[2,3], and Kaspar Riesen[1]

[1] Institute for Information Systems, University of Applied Sciences
and Arts Northwestern Switzerland, Riggenbachstr. 16, 4600 Olten, Switzerland
{michael.stauffer,kaspar.riesen}@fhnw.ch
[2] Department of Informatics, University of Fribourg, 1700 Fribourg, Switzerland
andreas.fischer@unifr.ch
[3] Institute for Complex Systems, University of Applied Sciences
and Arts Western Switzerland, 1705 Fribourg, Switzerland
[4] Department of Informatics, University of Pretoria, Pretoria, South Africa

Abstract. The present paper is concerned with a graph-based system for Keyword Spotting (KWS) in historical documents. This particular system operates on segmented words that are in turn represented as graphs. The basic KWS process employs the cubic-time bipartite matching algorithm (BP). Yet, even though this graph matching procedure is relatively efficient, the computation time is a limiting factor for processing large volumes of historical manuscripts. In order to speed up our framework, we propose a novel fast rejection heuristic. This heuristic compares the node distribution of the query graph and the document graph in a polar coordinate system. This comparison can be accomplished in linear time. If the node distributions are similar enough, the BP matching is actually carried out (otherwise the document graph is rejected). In an experimental evaluation on two benchmark datasets we show that about 50% or more of the matchings can be omitted with this procedure while the KWS accuracy is not negatively affected.

Keywords: Handwritten keyword spotting · Bipartite graph matching · Fast rejection · Filtering graph matching

1 Introduction

An automatic full transcriptions of historical handwritten documents is often negatively affected by both the degenerative conservation state of scanned documents and different writing styles. Thus, *Keyword Spotting (KWS)* as a more error-tolerant, flexible, and suitable approach has been proposed [1–4]. KWS refers to the task of retrieving any instance of a given query word in a document. This task is of high relevance due to a global trend towards digitalisation of paper-based archives and libraries. Similar to handwriting recognition, textual KWS can be divided into two different approaches *online* and *offline* KWS, respectively. The former has access to temporal information, while the latter is

© Springer International Publishing AG 2017
P. Foggia et al. (Eds.): GbRPR 2017, LNCS 10310, pp. 83–93, 2017.
DOI: 10.1007/978-3-319-58961-9_8

limited to spatial information only. The focus of this paper is on historical documents, and thus, offline KWS, referred to as KWS from now on, can be applied only.

KWS approaches can be divided into *template-based* or *learning-based* algorithms. Template-based matching algorithms such as for example *Dynamic Time Warping (DTW)* [2,5,6] directly match sample images of the keyword with document images. Learning-based algorithms [3,4,7], on the other hand, derive character or word models from learning samples. The latter typically achieve higher accuracies than template-based approaches but are limited by the need for a considerable amount of learning samples. Template-based approaches, in contrast, require only one or a few keyword instances and are thus more flexible. In this paper, we focus on template-based KWS using different graph representations of handwritten words.

Even though graphs gained noticeable attention in diverse applications [8,9], we observe only limited attempts where graphs are used to represent handwriting for KWS [10–14]. This is particularly interesting as graphs are, in contrast with feature vectors, flexible enough to adapt their size to the size and complexity of the underlying handwriting. Moreover, graphs are capable to represent binary relationships in the handwriting (e.g. strokes between two keypoints). Overall, graphs offer a more natural and comprehensive way to represent handwritten characters or words when compared to feature vectors. Additionally, various procedure for efficiently evaluating the dissimilarity of graphs, commonly known as *graph matching*, have been proposed in the last decade [9].

Yet, in the case of searching n keywords in a certain document (represented by a set of graphs G), we need to match $n \times |G|$ pairs of graphs. Even when a fast graph matching procedure is employed, this large amount of matchings can substantially slow down the complete KWS process. To speed up the KWS procedure the number of graph matchings actually carried out, can be reduced by efficiently filtering graphs from G with a low similarity to the current query graph q. This approach is known as *fast rejection* [3,5,7] and the focus of the present paper. That is, we introduce a novel heuristic for fast and accurate filtering of irrelevant document graphs given a certain query graph.

The remainder of this paper is organised as follows. In Sect. 2, the proposed fast rejection approach to speed up graph-based KWS is introduced. The datasets employed as well as the different graph representations are reviewed in Sect. 3. An experimental evaluation and comparison with the original framework is given in Sect. 4. Finally, Sect. 5 concludes the paper and outlines possible future research activities.

2   Fast Rejection of Document Graphs

Given a set of document graphs $G = \{g_1, \ldots, g_N\}$ as well as query graph q (used to represent a certain keyword), the process of KWS performs a matching of q with all graphs from G. We employ the *Bipartite Graph Edit Distance (BP)* [15], and thus observe cubic time complexity for these pairwise dissimilarity computations. The present paper introduces a fast rejection approach in order to substantially reduce the number of document graphs needed to be matched with q. The motivation is to filter document graphs without relevance to the given keyword and thus speeding up the KWS procedure without negatively affecting the retrieval accuracy.

The basic idea of our approach is as follows. Before actually carrying out the graph matching, we first measure the dissimilarity between histograms based on a specific segmentation of the graphs by means of a polar coordinate system. We denote this fast graph dissimilarity computation by *Polar Graph Dissimilarity (PGD)* from now on. If the PGD is below a certain threshold D for a pair of graphs (q, g_i), we carry out the computationally more expensive BP matching procedure [13]. Otherwise, we define the distance between q and g_i to be ∞. Formally,

$$d(q, g_i) \begin{cases} \infty, & \text{if } PGD(q, g_i) > D \\ BP(q, g_i), & \text{otherwise} \end{cases}, \tag{1}$$

where q and g_i denotes the query and document graph, respectively. Increasing the threshold D generally reduces the number of filtered document graphs. Likewise, the number of filtered graphs is increased when D is decreased. The overall aim is to find a good tradeoff between low matching time (due to many filterings) and high KWS accuracy.

Our novel dissimilarity model PGD has been inspired by the scale-invariant shape descriptor *Contour Points Distribution Histogram (CPDH)* for 2D-shape matching [16]. The basic idea behind this shape descriptor is to segment equidistant contour points by a specific polar coordinate system. A given shape image is formally described by a histogram CPDH $= \{h_1, \ldots, h_i, \ldots, h_n\}$ where h_i basically consists of the number of contour points n_i in the corresponding segment.

We adopt this procedure in order to measure the dissimilarity between graphs in linear time. Rather than contour points, however, we make use of nodes as shown in Fig. 1. For all of our graphs that represent segmented words, nodes are labelled with two-dimensional numerical labels, while edges remain unlabelled (see Sect. 3 for details).

To create a histogram for a given graph g, we first calculate the centre of mass (x_m, y_m) of g and then transform the (x, y)-coordinates of each node label $\mu(v) = (x, y) \in \mathbb{R}^2$ into polar coordinates (see Fig. 1a)[1]. Formally,

$$\rho = \sqrt{(x - x_m)^2 + (y - y_m)^2} \text{ and } \theta_i = \text{atan2}((y - y_m)/(x - x_m)),$$

where ρ denotes the radius from the centre of g to the node position and $-\pi \leq \theta_i < \pi$ refers to the angle from the x-axis to the node position (computed via arctangent function with two arguments in order to return the correct quadrant). Next, we define a bounding circle C given by the maximum radius $\rho_{\max}$ that surrounds all nodes of graph g. We segment C based on the number of different

---

[1] Node coordinates are *a priori* denormalised by the standard deviation of all node coordinates, for further details we refer to [13].

(a) Circle C with Centre of Mass (x_m, y_m) and Radius ρ_{max}

(b) Segmentation of C into Bins

(c) Number of Nodes per Bin

Fig. 1. Construction of the polar graph dissimilarity.

radii u_{max} and angles v_{max} into $u_{max} \times v_{max}$ bins (in Fig. 1b $u_{max} = 3$ and $v_{max} = 8$ resulting in 24 bins). Every bin b_i is defined by two radii $\rho_{i_{min}}$ and $\rho_{i_{max}}$, and two angles $\theta_{i_{min}}$ and $\theta_{i_{max}}$, and thus every node $v \in V$ with coordinates (ρ, θ) can be assigned to the corresponding bin b_i with $\rho_{i_{min}} \leq \rho < \rho_{i_{max}}$ and $\theta_{i_{min}} \leq \theta < \theta_{i_{max}}$. Finally, we count the number of nodes of g in each bin and build a corresponding histogram $H = \{h_1, \dots, h_n\}$ for graph g (see Fig. 1c). To measure the dissimilarity between two histograms H_1 and H_2, an arsenal of different distance measures have been proposed [17]. In the present paper, we make use of the χ^2 distance.

We further refine the computation of our fast graph dissimilarity computation by implementing a recursive *quadtree* segmentation. The idea is formalised in Algorithm 1. First, the procedure is initialised by an external call with $l = 1$ (i.e. $PGD(1, g_1, g_2)$). On the basis of two graphs g_1 and g_2, the histograms H_1 and H_2 are created with respect to u_{max} and v_{max} (see line 2 of Algorithm 1)[2]. Next, the χ^2-distance between the two histograms is measured (see line 2). If the current recursion level l is equal to the maximal recursion depth r, the distance is returned (see lines 4 and 5). Otherwise, both graphs g_1 and g_2 are segmented into four independent subgraphs. Each of these subgraphs represent the nodes and edges in one of the four quadrants in circle C (see line 6). Eventually, for each subgraph pair, the PGD is measured by means of a recursive function call (see line 7). This procedure is repeated until the current recursion level l is equal to the user-defined maximum depth r.

3   Handwriting Graphs

Our novel algorithm for fast rejection is evaluated in the context of KWS on two different manuscripts. First, the *George Washington (GW)* letters that are

---

[2] Note that u_{max} and v_{max} can be defined for every recursion level separately.

Algorithm 1. Polar Graph Dissimilarity (PGD)

Input: Graphs g_1 and g_2, number of radii and segments u_{max} and v_{max}, recursion depth r
Output: Polar graph dissimilarity between graph g_1 and g_2
1: **function** PGD(l, g_1, g_2)
2:     Create H_1 based on g_1, u_{max}, v_{max}, and H_2 based on g_2, u_{max}, v_{max}
3:     Calculate χ^2-distance $d(H_1, H_2)$
4:     **if** l equal r **then**
5:         **return** d
6:     Segment g_1 and g_2 based on quadtree to g_{1_1}, g_{1_2}, g_{1_3}, g_{1_4} and g_{2_1}, g_{2_2}, g_{2_3}, g_{2_4}
7:     **return** $(\sum_{i=1}^{4} \text{PGD}(l+1, g_{1_i}, g_{2_i})) + d$

---

written in English and consist of twenty pages with a total of 4,894 words stemming from handwritten letters with only minor writing variations and signs of degradation[3]. Second, the *Parzival (PAR)* manuscript that is written in Middle High German and consists of 45 pages with a total of 23,478 words stemming from handwritten letters with low writing variations but markable signs of degradation[4].

We extract graphs from segmented words of both documents by means of the following four graph extraction algorithms (originally presented in [14]).

- **Keypoint:** The first graph extraction algorithm makes use of keypoints in the word images such as start, end, and junction points. These keypoints are represented as nodes that are labelled with the corresponding (x, y)-coordinates. Between pairs of keypoints further intermediate points are converted to nodes and added to the graph in equidistant intervals. Finally, undirected edges are inserted into the graph for each pair of nodes that is directly connected by a stroke.
- **Grid:** The second graph extraction algorithm is based on a grid-wise segmentation of the word images. For every segment, a node is inserted into the graph and labelled by the (x, y)-coordinates of the centre of mass of this segment. Undirected edges are inserted between two neighbouring segments that are actually represented by a node. Eventually, the inserted edges are reduced by means of a *Minimal Spanning Tree* algorithm.
- **Projection:** The next graph extraction algorithm works very similar to **Grid**. However, this methods is based on an adaptive segmentation of word images by means of projection profiles (using horizontal and vertical projection profiles).
- **Split:** The last graph extraction algorithm is based on an iterative segmentation of word images. Segments are iteratively split into smaller subsegments until the width and height of all segments is below a certain threshold.

---

[3] George Washington Papers at the Library of Congress, 1741–1799: Series 2, Letterbook 1, pp. 270–279 & 300–309, http://memory.loc.gov/ammem/gwhtml/gwseries2.html.

[4] Parzival at IAM historical document database, http://www.fki.inf.unibe.ch/databases/iam-historical-document-database/parzival-database.

The dynamic range of the (x, y)-coordinates of each node label $\mu(v)$ is normalised with a z-score. Formally,

$$\hat{x} = \frac{x - \mu_x}{\sigma_x} \text{ and } \hat{y} = \frac{y - \mu_y}{\sigma_y}, \qquad (2)$$

where (μ_x, μ_y) and (σ_x, σ_y) represent the mean and standard deviation of all (x, y)-coordinates in the graph under consideration.

On the resulting sets of word graphs, ten different keywords are manually selected on both datasets to optimise several system parameters (see Sect. 4.2). For validation these keywords are matched against a validation set that consists of 1,000 different random words including at least 10 instances of all 10 keywords. The optimised systems are eventually evaluated on the same training and test sets as used in [4]. All templates of a keyword present in the training set are used for KWS. In Table 1 a summary of the datasets is given.

Table 1. The number of keywords as well as the size of the training and test sets for both documents.

Dataset	Keywords	Train	Test
GW	105	2,447	1,224
PAR	1,217	11,468	6,869

4 Experimental Evaluation

4.1 Basic KWS Systems

For evaluating our proposed fast rejection heuristic, we consider the graph-based KWS system introduced in [13] and the four types of handwriting graphs described in Sect. 3. The original KWS system [13] is termed BP from now on, while our extended model with fast rejection is termed BP-FR.

To evaluate the KWS performance, two different metrics are used for global and local thresholds. In the case of global thresholds, the *Average Precision (AP)* is measured, which is the area under the Recall-Precision curve for all keywords given a single (global) threshold. In the case of local thresholds, we compute the *Mean Average Precision (MAP)*, that is the mean of all APs for each individual keyword query. To measure the effects of our fast rejection filter, we compute the relative amount of pairwise matchings that is filtered by BP-FR (termed *Filter Rate (FR)* from now on).

4.2 Optimisation of the Parameters

For the basic KWS system BP and the four graph representations, we adopt parameters from previous work [13,14]. For our extension BP-FR the following parameters are additionally optimised on the validation set.

First, the parameters of PGD are optimised with respect to MAP. That is, we employ PGD (rather than BP) as basic matching procedure in our KWS framework. On the validation set different polar segmentations (defined via u_{max} and v_{max}) are tested for two recursion levels (i.e. we define the maximal recursion depth to $r = 2$). For $l = 1$, the parameter combinations $u_{max} = \{1, 2, 3, 4, 5, 6\} \times v_{max} = \{4, 8, 12, 16, 20, 24, 28, 32, 36, 40\}$ are evaluated, while for $l = 2$ the parameter combinations $u_{max} = \{1, 2, 3, 4\} \times v_{max} = \{2, 4, 6, 8, 10\}$ are tested. Hence, we evaluate $6 \times 10 \times 4 \times 5 = 1,200$ parameter combinations for every graph extraction method. In Table 2 the best performing parameters are presented for every graph extraction method and both datasets.

Table 2. Optimal u_{max} and v_{max} for PGD on both recursion levels l.

Method	GW				PAR			
	$l = 1$		$l = 2$		$l = 1$		$l = 2$	
	u_{max}	v_{max}	u_{max}	v_{max}	u_{max}	v_{max}	u_{max}	v_{max}
Keypoint	4	12	1	6	3	20	2	6
Grid	5	24	1	4	4	20	1	6
Projection	5	16	1	4	3	36	3	4
Split	4	20	1	4	3	40	2	6

For fast rejection in our extension BP-FR we evaluate different thresholds $D = \{5, 10, \ldots, 195, 200\}$. In Fig. 2, the MAP and FR are shown for every tested threshold D. By increasing D we observe that the KWS performance is improved in general. Simultaneously, the number of filtered graphs is decreasing (making the KWS process slower in general). Threshold D is finally determined such that the MAP is maximal (or not further improved, when D is increased). In Table 3

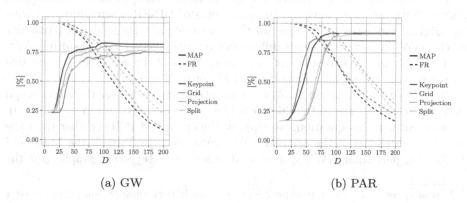

(a) GW   (b) PAR

Fig. 2. Mean average precision (MAP) and filter rate (FR) as function of the threshold D.

Table 3. Optimal D for BP-FR and corresponding filter rate (FR).

Method	GW			PAR		
	D	MAP	FR	D	MAP	FR
Keypoint	100	82.8	61.1	95	91.7	71.5
Grid	165	75.6	46.0	70	86.5	85.6
Projection	115	80.7	56.9	130	92.2	70.9
Split	155	76.4	44.6	145	90.9	57.5

the selected threshold D is given for each graph extraction method and both datasets.

4.3   Results and Discussion

We compare the optimised system BP-FR on the independent test sets with the original KWS framework BP [13] (without fast rejection). In Table 4 the mean average precision (MAP) for local thresholds, the average precision (AP) for global thresholds, as well as the filter rate (FR) is given for both BP and BP-FR. On the GW dataset we observe a filter rate between 50% and 70% (i.e. only 50% to 30% of all comparisons have to be carried out by the bipartite matching algorithm). Due to this filtering, we decrease the computation time of the complete KWS experiment by about 80 to 150 h on the different graph representations. Similar (or even better) filter rates can be observed on the second dataset[5].

Regarding the effects of our fast filtering on the KWS performance, we observe that the MAP is not negatively affected on both datasets. On the contrary, the filtering of irrelevant documents via PGD actually improves the MAP by about 5% and 10% on the GW and PAR dataset, respectively.

Regarding the AP (employed for global rather than local thresholds), we observe both deteriorations and improvements of BP-FR when compared with the original framework. Yet, most of the deviations are negligible. In particular on the GW dataset only small differences are observed on the resulting APs. On PAR we observe two substantial deteriorations of the AP. Yet, in these two cases we observe very high filter rates of about 60% and 70%.

Regarding the results in Table 4 the question arises whether the novel graph dissimilarity PGD would be able to achieve a competitive KWS accuracy. In order to answer this question, we employ the optimised PGD (rather than the bipartite matching) in the original KWS framework. In Table 5, the MAP and AP of this particular KWS system is shown on the Keypoint graphs (for the

---

[5] Actually, we carry out our experiment on a high performance computing cluster with dozens of CPU nodes. Hence, these readings are approximated by means of the average matching time per keyword measured on the validation set in a sequential scenario.

Table 4. Mean average precision (MAP) using local thresholds, average precision (AP) using a global threshold, and filter rate (FR) for KWS using the original bipartite graph matching without rejection (BP) and with the proposed fast rejection (BP-FR). With ± we indicate the relative percental gain or loss in the accuracy of BP-FR when compared with BP.

	Method	GW					PAR				
		MAP	±	AP	±	FR	MAP	±	AP	±	FR
BP	Keypoint	66.08		54.99		0.00	62.04		60.74		0.00
	Grid	60.02		46.44		0.00	56.50		44.08		0.00
	Projection	61.43		48.69		0.00	66.23		60.61		0.00
	Split	60.23		47.96		0.00	59.44		55.46		0.00
BP-FR	Keypoint	68.81	+4.12	55.68	+1.25	69.04	67.70	+9.12	58.03	−4.46	58.72
	Grid	62.59	+4.27	47.48	+2.23	54.65	63.41	+12.23	38.59	−12.45	78.71
	Projection	64.65	+5.25	50.41	+3.53	61.04	72.02	+8.74	55.83	−7.89	58.10
	Split	63.49	+5.41	46.95	−2.11	47.70	65.65	+10.45	56.97	+2.72	39.24

Table 5. Mean average precision (MAP) using local thresholds, average precision (AP) using a global threshold for KWS using the original bipartite graph matching (BP), and the polar graph dissimilarity (PGD) on the `Keypoint` graphs.

	GW		PAR	
	MAP	AP	MAP	AP
BP	66.08	54.99	62.04	60.74
PGD	58.54	44.77	42.65	31.63

other graphs similar results are obtained). We observe that this system achieves worse results than BP on both datasets (regarding both MAP and AP). Hence, we conclude that PGD itself is not powerful enough to serve as basic dissimilarity model for graph-based KWS. Yet, as seen in the previous evaluation in Table 4, the PGD as fast rejection criterion in conjunction with BP is clearly beneficial.

5 Conclusion and Outlook

In the present paper a fast rejection approach for graph-based KWS is introduced. The rejection is based on a novel graph dissimilarity model, which compares the histograms of the node distributions in a polar coordinate system.

We compare our extended model with the original KWS framework without rejection ability on two benchmark datasets. We observe that our novel rejection approach reduces the amount of graph matchings by 50% or more on both datasets (in fact, filter rates of up to 80% are observed). Our rejection criterion is computed in linear time, while the actual graph matching needs cubic time. Hence, a dramatic speed up of the complete KWS process is achieved. Moreover, we can conclude that our novel extension for speeding up the existing KWS framework does not negatively influence the spotting accuracy.

In future work we aim at extending our novel graph dissimilarity model. For instance, we could consider not only nodes but also edges in the histograms.

Acknowledgments. This work has been supported by the Hasler Foundation Switzerland.

References

1. Manmatha, R., Han, C., Riseman, E.: Word spotting: a new approach to indexing handwriting. In: Computer Vision and Pattern Recognition, pp. 631–637 (1996)
2. Rath, T., Manmatha, R.: Word image matching using dynamic time warping. In: Computer Vision and Pattern Recognition, vol. 2, pp. II-521–II-527 (2003)
3. Rodríguez-Serrano, J.A., Perronnin, F.: Handwritten word-spotting using hidden Markov models and universal vocabularies. Pattern Recogn. **42**(9), 2106–2116 (2009)
4. Fischer, A., Keller, A., Frinken, V., Bunke, H.: Lexicon-free handwritten word spotting using character HMMs. Pattern Recogn. Lett. **33**(7), 934–942 (2012)
5. Rodriguez, J.A., Perronnin, F.: Local gradient histogram features for word spotting in unconstrained handwritten documents. In: International Conference on Frontiers in Handwriting Recognition, pp. 7–12 (2008)
6. Rodríguez-Serrano, J.A., Perronnin, F.: A model-based sequence similarity with application to handwritten word spotting. IEEE Trans. Pattern Anal. Mach. Intell. **34**(11), 2108–20 (2012)
7. Perronnin, F., Rodriguez-Serrano, J.A.: Fisher kernels for handwritten word-spotting. In: International Conference on Document Analysis and Recognition, pp. 106–110 (2009)
8. Conte, D., Foggia, P., Sansone, C., Vento, M.: Thirty years of graph matching in pattern recognition. Int. J. Pattern Recogn. Artif. Intell. **18**(03), 265–298 (2004)
9. Riesen, K.: Structural pattern recognition with graph edit distance. Advances in Computer Vision and Pattern Recognition, Cham (2015)
10. Wang, P., Eglin, V., Garcia, C., Largeron, C., Llados, J., Fornes, A.: A novel learning-free word spotting approach based on graph representation. In: International Workshop on Document Analysis Systems, pp. 207–211 (2014)
11. Bui, Q.A., Visani, M., Mullot, R.: Unsupervised word spotting using a graph representation based on invariants. In: International Conference on Document Analysis and Recognition, pp. 616–620 (2015)
12. Riba, P., Llados, J., Fornes, A.: Handwritten word spotting by inexact matching of grapheme graphs. In: International Conference on Document Analysis and Recognition, pp. 781–785 (2015)
13. Stauffer, M., Fischer, A., Riesen, K.: Graph-based keyword spotting in historical handwritten documents. In: International Workshop on Structural, Syntactic, and Statistical Pattern Recognition (2016)
14. Stauffer, M., Fischer, A., Riesen, K.: A novel graph database for handwritten word images. In: Robles-Kelly, A., Loog, M., Biggio, B., Escolano, F., Wilson, R. (eds.) S+SSPR 2016. LNCS, vol. 10029, pp. 553–563. Springer, Cham (2016). doi:10.1007/978-3-319-49055-7_49

15. Riesen, K., Bunke, H.: Approximate graph edit distance computation by means of bipartite graph matching. Image Vis. Comput. **27**(7), 950–959 (2009)
16. Shu, X., Wu, X.J.: A novel contour descriptor for 2D shape matching and its application to image retrieval. Image Vis. Comput. **29**(4), 286–294 (2011)
17. Serratosa, F., Sanfeliu, A.: Signatures versus histograms: definitions, distances and algorithms. Pattern Recogn. **39**(5), 921–934 (2006)

Detecting Alzheimer's Disease
Using Directed Graphs

Jianjia Wang$^{(\boxtimes)}$, Richard C. Wilson, and Edwin R. Hancock

Department of Computer Science, University of York, York YO10 5DD, UK
jw1157@york.ac.uk

Abstract. The neurobiology of Alzheimer's disease (AD) has been extensively studied by applying network analysis techniques to activation patterns in fMRI images. However, the structure of the directed networks representing the activation patterns, and their differences in healthy and Alzheimer's people remain poorly understood. In this paper, we aim to identify the differences in fMRI activation network structure for patients with AD, late mild cognitive impairment (LMCI) and early mild cognitive impairment (EMCI). We use a directed graph theoretical approach combined with entropic measurements to distinguish subjects falling into these three categories and the normal healthy control (HC) group. We explore three methods. The first is based on applying linear discriminant analysis to vectors representing the in and out degree statistics of different anatomical regions. The second uses an entropic measure of node assortativity to gauge the asymmetries in the node with in and out degree. The final approach selects the most salient anatomical brain regions and uses the degree statistics of the connecting directed edges.

Keywords: fMRI Networks · Directed graphs entropy · Alzheimer's disease (AD)

1 Introduction

Functional magnetic resonance imaging (fMRI) provides a sophisticated means of studying the neuropathophysiology associated with Alzheimer's disease (AD) [1]. Specifically, the blood oxygen level-dependent (BOLD) signal in fMRI indicates the activation potential of different brain regions, and neuronal activity between the various brain regions can be determined by measuring the correlation between activation signals. The resulting network representation of region activity has proved useful in understanding the functional working of the brain [2]. Functional neuroimaging has also proved useful in understanding Alzheimer's disease (AD) via the analysis of intrinsic brain connectivity [3]. Abnormal brain function in AD is characterized by progressive impairment of episodic memory and other cognitive domains, resulting in dementia and, ultimately, death [5]. Although there is converging evidence about the identity of the affected regions in fMRI, it is not clear how this abnormality affects the functional organization of the whole brain.

© Springer International Publishing AG 2017
P. Foggia et al. (Eds.): GbRPR 2017, LNCS 10310, pp. 94–104, 2017.
DOI: 10.1007/978-3-319-58961-9_9

Tools from complex network analysis provide a convenient approach for understanding the functional association of different regions in the brain [3]. The approach is to characterize the topological structures present in the brain and to quantify the functional interaction between brain regions, using the mathematical study of networks and graph theory. Graph theory offers an attractive route since it provides effective tools for characterizing network structures together with their intrinsic complexity. This approach has led to the design of several practical methods for characterizing the global and local structure of undirected graphs [4]. Features based on the global and local measures of connectivity are widely used in functional brain analysis [6]. By comparing the structural and functional network topologies between different populations of subjects, graph theory provides meaningful and easily computable measurements to reveal connectivity abnormalities in both neurological and psychiatric disorders [5].

Unfortunately, there is relatively little literature aimed at studying structural network features using directed graphs. The reason for that is the vast majority of techniques suggested by graph theory pertain to undirected rather than directed graphs. However, directed graphs are a more natural representation for brain structure, since they allow the temporal causality of activation signals for different anatomical structures in the brain. Moreover, Granger causality provides a powerful tool that can be used to investigate the direction of information flow between different brain regions [6]. When combined with machine learning algorithms, classification exhibited from directed graphs provides an effective way of detecting functional regions associated with Alzheimer's disease [6]. By explicitly defining anatomical and functional connections in a directed manner between brain regions, fMRI data may be analyzed in a more detailed way and used to identify the different stages of neurodegenerative diseases [5,6].

This paper is motivated by the need to fill this important gap in the literature, and to establish effective methods for measuring the structural properties of directed graphs representing inter-regional casual networks extracted from fMRI brain data. In particular, in order to characterize the functional organization of the brain, our approach uses as its starting point the von Neumann entropy for directed graphs. In a recent paper, Ye et al. [4] have derived an approximation of the Neumann entropy of a directed graph that depends on the in and out degrees of nodes in a directed graph. Thus it provides a natural way of capturing the flow of information across a directed network, based on the asymmetry of edges entering and exiting its nodes. We aim to use the directed network entropy to develop graph analytical methods to measure the degree of functional connectivity in brain networks.

We demonstrate that the resulting techniques can be used to distinguish the fMRI data from healthy controls and AD objects. The AD subjects exhibit significantly lower regional connectivity and exhibit disrupted the global functional organization when compared to healthy controls. Moreover, we apply linear discriminant analysis to brain network data from two groups of subjects with early mild cognitive impairment (EMCI) and late mild cognitive impairment (LMCI). Our results indicate that the node in and out degree statistics together with their associated von Neumann entropy may be useful as a graph-based indicator to distinguish Alzheimer's disease subjects from normal healthy control population.

2 Directed Graphs in fMRI Networks

2.1 Preliminaries

Let $G(V, E)$ be a directed graph with node set V and directed edge set $E \subseteq V \times V$. Each edge $e = (u, v) \in E$, has a start vertex u and end-vertex v. The adjacency matrix A of the directed graph is defined as

$$A = \begin{cases} 1 & \text{if } (u, v) \in E \\ 0 & \text{otherwise.} \end{cases} \tag{1}$$

For the node u the in-degree and out-degree of node are

$$d_u^{in} = \sum_{v \in V} A_{vu} \quad d_u^{out} = \sum_{v \in V} A_{uv} \tag{2}$$

and the total degree of node in the directed graph is $d_u = d_u^{in} + d_u^{out}$. An edge is said to be unidirectional if $A_{uv} = 1$ and $A_{vu} = 0$, and bidirectional if $A_{uv} = 1$ and $A_{vu} = 1$.

2.2 Von Neumann Entropy for Directed Graphs

For an undirected graph the von Neumann entropy [7] computed from the normalised Laplacian spectrum has been shown to be effective for network characterization. In fact, Han et al. [8] have shown how to approximate the calculation of von Neumann entropy in terms of simple degree statistics. Their approximation allows the cubic complexity of computing the von Neumann entropy from the Laplacian spectrum, to be reduced to one of quadratic complexity using simple edge degree statistics, i.e.

$$S_U = 1 - \frac{1}{|V|} - \frac{1}{|V|^2} \sum_{(u,v) \in E} \frac{1}{d_u d_v} \tag{3}$$

This expression for the von Neumann entropy has been shown to be an effective tool for characterizing structural properties of networks. Moreover, it has extremal values for cycles and fully connected graphs. Ye et al. [4] have extended this result to directed graphs by distinguishing between the in-degree and out-degree of nodes, giving the following expression for the entropy

$$S_D = 1 - \frac{1}{|V|} - \frac{1}{2|V|^2} \sum_{(u,v) \in E_1} \frac{d_u^{in}}{d_v^{in} d_u^{out2}} + \sum_{(u,v) \in E_2} \frac{1}{d_u^{out} d_v^{out}} \tag{4}$$

where the edge set E is partitioned into two disjoint subsets E_1 and E_2, which respectively contain the unidirectional and directional edges.

The two subsets E_1 and E_2 satisfy the conditions that $E_1 = \{(u,v)|(u,v) \in E \cap (v,u) \notin E\}$, $E_2 = \{(u,v)|(u,v) \in E \cap (v,u) \in E\}$. $E_1 \cup E_2 = E$, $E_1 \cap E_2 = \emptyset$. If most of the edges in the graph are unidirectional, i.e., $|E_1| \gg |E_2|$, then the graph is said to be strongly directed. In this case we can ignore the entropy associated with the summation over E_2, giving the approximate entropy for strongly directed graphs as

$$S_{SD} = 1 - \frac{1}{|V|} - \frac{1}{2|V|^2} \sum_{(u,v)\in E} \frac{d_u^{in}}{d_u^{out}} \cdot \frac{1}{d_v^{in} d_u^{out}} \tag{5}$$

There are thus two factors determining the entropy. The first is the ratio of the in to out degree of the start node u of the directed edge, i.e. $\rho_u = \frac{d_u^{in}}{d_u^{out}}$, while the second is the directed version of the edge entropy, i.e. $\frac{1}{d_u^{out} d_v^{in}}$. The former weights the contributions of the entropy associated with the directed edges exiting node u. The contributions to the entropy are thus large if the ratio ρ_u is small, and directed edge connects nodes with large both out and in degree.

2.3   Entropic Edge Assortativity for Directed Graphs

For undirected graphs, the assortativity is the tendency of nodes to connect to those of similar degree. This concept can be extended to directed graphs if we measure the tendency of nodes to connect to those nodes of similar in and out degree. Foster et al. [11] define the directed assortativity as

$$r(\alpha, \beta) = \frac{1}{|E|} \frac{\sum_{(u,v)\in E}[(d_u^\alpha - \bar{d}_u^\alpha)(d_v^\beta - \bar{d}_v^\beta)]}{\sigma^\alpha \sigma^\beta} \tag{6}$$

where $\alpha, \beta \in \{in, out\}$ is the incoming and outgoing direction for a directed edge. $\bar{d}_u^\alpha = |E|^{-1} \sum_{(u,v)\in E} d_u^\alpha$ and $\sigma^\alpha = \sqrt{|E|^{-1} \sum_{(u,v)\in E}(d_u^\alpha - \bar{d}_u^\alpha)^2}$. The similar definitions are for $\bar{d}_v^\beta$ and σ^β.

Ye [10] adopts a different approach to defining degree assortativity for directed graphs based on von Neumann entropy decomposition. The method is based on the observation that edges associated with high degree nodes have large entropy and preferentially attach to clusters in a graph. The entropic assortativity measurement provides a novel way to analyze the graph structure. For instance, with the approximation for the von Neumann entropy for directed graph S_D, the coefficient of directed edge assortativity is given by [10]

$$R = \frac{\sum_{(u,v)\in E}[(S_{uv}^u - S_{uv}^{\bar{u}})(S_{uv}^v - S_{uv}^{\bar{v}})]}{\sigma_u^S \sigma_v^S} \tag{7}$$

where S_{uv}^u associate the entropy of all the outgoing edges from vertex u, and S_{uv}^v are all the incoming edges of vertex v.

3   Experiments and Evaluations

In this section, we describe the application of the above methods to the analysis of interregional connectivity structure for fMRI activation networks for normal and Alzheimer subjects. We first examine the differences in degree distribution for the four groups of subjects. Then we apply the entropy-based analysis to distinguish Early Mild Cognitive Impairment (EMCI) and Late Mild Cognitive Impairment (LMCI).

3.1   fMRI Data Set

The fMRI data comes from the ADNI initiative [9]. Each image volume is acquired every two seconds with Blood-Oxygenation-Level-Dependent (BOLD) signals. The fMRI voxels here have been aggregated into larger regions of interest (ROIs). The different ROI's correspond to different anatomical regions of the brain and are assigned anatomical labels to distinguish them. There are 96 anatomical regions in each fMRI image. The correlation between the average time series in different ROIs represents the degree of functional connectivity between regions which are driven by neural activities [12].

A directed graph with 96 nodes is constructed for each patient based on the magnitude of the correlation and the sign of the time-lag between the time-series for different anatomical regions. To model causal interaction among ROIs, the directed graph uses the time lagged cross-correlation coefficients for the average time series for pairs of ROIs. We detect directed edges by finding the time-lag that results in the maximum value of the cross-correlation coefficient. The direction of the edge depends on whether the time lag is positive or negative. We then apply a threshold to the maximum values to retain directed edges with the top 40% of correlation coefficients. This yields a binary directed adjacency matrix for each subject, where the diagonal elements are set to zero. Those ROIs which have missing time series data are discarded.

Subjects fall into four categories according to their degree of disease severity. The classes are full Alzheimer's (AD), Late Mild Cognitive Impairment (LMCI), Early Mild Cognitive Impairment (EMCI) and Normal Healthy Controls (HC). The LMCI subjects are more severely affected and close to full Alzheimer's, while the EMCI subjects are closer to the healthy control group (Normal). We have fMRI data for 30 AD subjects, 34 LMC subjects, 47 EMCI subjects, and 38 normal healthy control subjects.

3.2   Alzheimer's Classification

We first investigate the in and out degree distribution of the data by showing a scatter plot in-degree versus out-degree for each directed edge in the data. In order to extract potential structural difference, the distribution of points in the scatter plot is analyzed using a general linear model. Figure 1 shows the scatter plots of in-degree versus out-degree, comparing the first AD vs. Normal and

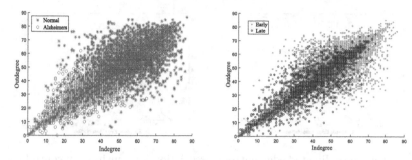

Fig. 1. The in-degree/out-degree distribution for edges in the directed graphs in Normal Healthy Control and Alzheimer's groups (left), Early Mild Cognitive Impairment(EMCI) and Late Mild Cognitive Impairment (LMCI) (right). The blue stars represent the edges in normal patients' graphs which occupy the high degree region with large variance. The red cycles show the AD patients' graphs with narrow and low degree occupation. (Color figure online)

secondly EMCI vs. LMCI respectively. The obvious difference is that normal subjects exhibit a high degree of interregional connection compared to Alzheimer's subjects. A similar effect is shown by Early and Late detection groups. Table 1 shows the coefficients of a linear model with 95% confidence bounds and root mean square error.

The results of fitting the linear model show that the in and out degree distributions for the nodes in the AD and LMCI groups of subjects have a greater slope than those of the Normal and Early groups. This implies that there is a greater imbalance in in-degree and out-degree in the Alzheimers and late detection groups. In other words, the nodes in the fMRI inter-regional connectivity graphs for these two groups tend to have larger in-degree than out-degree. Moreover, the small value of RMSE in these two groups reveals that for Alzheimer's subjects the scatter about the regression lines is smallest. By contrast, for the normal and early control subjects the scatter is significantly higher. This underlines the imbalance in in-degree for the subjects belonging to the diseased groups.

We can explore this asymmetry of in and out degree in more detail using Ye's entropy assortativity measure. This gauges the extent to which nodes to connect to others with similar in-degree or out-degree [6]. To represent the structural

Table 1. Liner polynomial model to fit the edge in-degree/out-degree distribution

Group of subjects	Coef (α)	BSC (α)	Coef (β)	BSC (β)	R^2	RMSE
AD	0.8582	[0.8406, 0.8758]	5.445	[4.719, 6.171]	0.7604	7.2444
Normal	0.6103	[0.5848, 0.6357]	22.45	[20.94, 23.96]	0.3771	11.3445
EMCI	0.7235	[0.7034, 0.7436]	14.6	[13.5, 15.7]	0.5253	10.3959
LMCI	0.9236	[0.9098, 0.9375]	2.933	[2.356, 3.509]	0.8395	6.4426

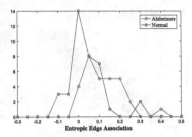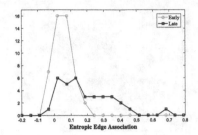

Fig. 2. Histogram of directed edge entropy association for four healthy control groups. The normal and early patients exhibit low entropy association for each edge compared to the late and AD groups which the distributions shift to high entropy region.

difference regarding the entropy associated with degree of each node, we plot the histogram of edge entropy assortativity in Fig. 2. It shows the difference in entropy of the directed edges for subjects in AD vs. Normal, and EMCI vs. LMCI. By comparing the directed edges in the AD and normal groups, we conclude that the edges in the directed graphs for Alzheimer's subjects tend to have a higher value of entropy, and this reveals the structure is weakly connected with a lower average in out to in degree ratio. A similar effect is shown in the EMCI and LMCI subject groups. For late Alzheimer's subjects, the shift in entropy to the right represents the weak degree connection in the nodes. This clearly reveals the loss of interregional connection for directed edges in Alzheimer's.

Finally, the in-degree and out-degree of nodes are used as the features to distinguish the different group of subjects. For each edge, we construct four dimensional feature vectors with two nodes and in and out degree measurements on each node. So the graph can be represented by these directed edges associated with four-dimensional feature vectors. We perform the linear discriminant analysis (LDA) on the Alzheimer's (AD) and Normal healthy control groups as the training process to find the decision boundary. Then the LDA model is applied on the EMCI and LMCI groups to classify patients. We compare the results and the labels to get classification accuracy.

Table 2 shows the classification accuracy of linear discriminant analysis(LDA). The directed graphs for the AD and Normal subjects are used as the training data to find the decision boundary. The performance of the resulting LDA classier is high with an accuracy of 87.87% when computed using 10-fold cross-validation. We randomly divide the AD and Normal subjects into 10 disjoint subsets of equal

Table 2. The classification accuracy with linear discriminant analysis(LDA) for training data (AD/Normal) and testing data (EMCI/LMCI) (in %)

LDA	Accuracy	Sensitivity	Specificity	Positive predictivity
AD/Normal	87.87 ± 0.58	88.59	87.10	88.00
EMCI/LMCI	80.47 ± 0.41	75.85	86.18	87.14

size. Remove one subset, train the LDA model using the other nine subsets. This process is repeated by removing each of the ten subsets once at a time and then average the classification accuracy. In order to evaluate the performance of classification, we provide results for sensitivity and specificity for LDA classifier. The sensitivity indicates the percentage of Alzheimer's people who are correctly identified. It reaches 88.59% which represents the high percentage of correctly classified. In addition, the specificity shows the true negative that is the healthy people correctly identified as healthy. It is 87.10% revealing most normal healthy people are correctly identified in the Normal group. Similarly to the LDA in AD and Normal classier, for the discrimination of subjects belonging to the EMCI and LMCI groups, we obtain a classification accuracy of 80.47%. Although this result is acceptable, the sensitivity is reduced to 75.85% indicating some percentage of patients are not correctly classified in LMCI groups.

3.3   Identifying Salient Nodes for Disease Classification

Identifying diseased regions in the brain is also important in the study in Alzheimer's analysis. Several studies have shown that in anatomical structures the corresponding ROIs are important for understanding brain disorders [1,3]. Here we compute the difference of out-degree and in-degree in our study and investigate the method for identification of the disease nodes in patients with Alzheimer's.

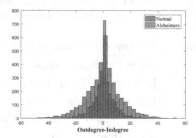 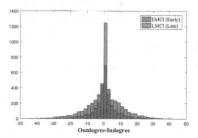

Fig. 3. Histogram of degree difference between Alzheimer's (AD) and Normal Healthy Controls (HC) groups. The normal and early patients exhibit wide bound range compared to the late and AD groups which the distributions narrows around zero.

We first compute the histogram of degree imbalance, i.e. out-degree minus in-degree for each node. Figure 3 compares histograms obtained for AD and HC, and for EMCI and LMCI. The obvious feature is that the directed graphs for HC (normal) and EMCI (early development) groups give a much broader range of degree difference compared to that for the AD (fully developed disease) and LMCI (late development) groups. In other words for subjects with fully developed AD, there is a loss of connection between brain regions and gives rise to a narrowing of the distribution of degree difference.

We now plot the difference in directed edge entropy between corresponding regions (nodes) in the directed graphs for the AD and HC groups. We find a

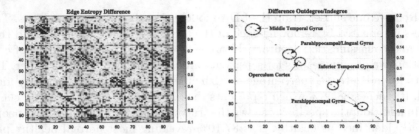

Fig. 4. Directed edge entropy difference between Alzheimer's (AD) and Normal Healthy Controls (HC) groups (left). The ratio of out-degree and in-degree difference corresponding to each ROI in two groups of AD and Normal patients (right). The significant changes of degree ratio in each nodes associate to the similar pattern in edge entropy plot, which illustrates the disease area in the brain.

similar feature pattern of the degree difference in both plots as shown in Fig. 4. The entropic measurements associated with degree difference in the brain areas, such as the Temporal Gyrus, Parahippocampal Gyrus, Operculum Cortex and Lingual Gyrus, suggest that subjects with AD experience loss of interconnection in their brain network during the progression of the disease.

Table 3. Top 10 ROIs with the significant difference between groups of AD and Normal. These ROIs are extracted from the absolute value of out-degree to in-degree ratio.

Graph measure	ROI number	Corresponding area in brain
Out-degree/In-degree ratio difference	83	Right Parahippocampal Gyrus
	14	Left Inferior Temporal Gyrus
	27	Left Paracingulate Gyrus
	65	Right Temporal Fusiform Cortex
	93	Right Heschl's Gyrus
	43	Left Parietal Operculum Cortex
	75	Right Paracingulate Gyrus
	38	Left Temporal Fusiform Cortex
	42	Left Central Opercular Cortex
	5	Left Inferior Frontal Gyrus

As listed in Table 3, the ten anatomical regions with the largest entropy differences for subjects with full AD are right Parahippocampal Gyrus, left Inferior Temporal Gyrus, left Paracingulate Gyrus, right Temporal Fusiform Cortex, right Heschl's Gyrus, left Parietal Operculum Cortex, right Paracingulate Gyrus, left Temporal Fusiform Cortex, left Central Opercular Cortex and left Inferior Frontal Gyrus. This result is consistent with the previous study [5,6], which suggested that the middle temporal gyrus is an important region in AD pathology [3].

Table 4. The LDA classification accuracy with top 20 selected ROIs to distinguish AD/Normal and EMCI/LMCL (in %)

LDA	Accuracy	Sensitivity	Specificity	Positive Predictivity
AD/Normal	90.52 ± 0.67	91.36	89.61	91.20
EMCI/LMCI	86.20 ± 0.81	83.90	90.12	89.26

The parahippocampal gyrus has consistently been reported as being an affected region in EMCI and AD [11]. The loss of connection between these brain regions results in significant functional impairment between healthy subjects and patients with AD.

We now repeat our LDA analysis using just the salient regions listed in Table 3, since it is the impairment of connections to these anatomical structures that appears to determine the onset of AD. We perform LDA on the 4 vectors representing the pairs of listed anatomical regions. The classification accuracy is shown in Table 4. In comparison to the previous results in Table 2, the accuracy increases by about 3% in AD/Normal groups and 6% in the EMCI/LMCL groups. All other performances are also improved with these selected degree features.

4   Conclusions

In conclusion, this paper is motivated by filling the gap in the literature of analyzing fMRI regional brain interaction networks using directed graphs. We commence from the recently developed simplified approximations to the von Neumann entropy of directed graphs, which are dependent on the graph size and the in and out degree statistics of vertices. In order to characterize the functional organization of the brain, assortativity of nodes in directed graphs provides insights into the neuropathology of Alzheimer's disease. Entropic measurements associated with node degree identifies the edge connection features which offer high discrimination between subjects suffering from AD and normal subjects.

References

1. van den Heuvel, M., Pol, H.E.H.: Exploring the brain network: a review on resting-state fMRI functional connectivity. J. Eur. Neuropsychopharmacol. **20**, 519–534 (2010)
2. Anwar, A.R., et al.: Complex network analysis of resting-state fMRI of the brain. In: 2016 IEEE 38th Annual International Conference of the Engineering in Medicine and Biology Society (EMBC). IEEE (2016)
3. Rubinov, M., Sporns, O.: Complex network measures of brain connectivity: uses and interpretations. Neuroimage **52**(3), 1059–1069 (2010)
4. Ye, C., Wilson, R.C., Comin, C.H., Costa, L.D.F., Hancock, E.R.: Approximate von Neumann entropy for directed graphs. Phys. Rev. E **89**(5), 052804 (2014)

5. Rombouts, S.A., Barkhof, F., Goekoop, R., Stam, C.J., Scheltens, P.: Altered resting state networks in mild cognitive impairment and mild Alzheimer's disease: an fMRI study. Hum. Brain Mapp. **26**(4), 231–239 (2005)
6. Khazaee, A., Ebrahimzadeh, A., Babajani-Ferem, A.: Classification of patients with MCI and AD from healthy controls using directed graph measures of resting-state fMRI. Behav. Brain Res. **322**, 339–350 (2016)
7. Passerini, F., Severini, S.: The von neumann entropy of networks. Int. J. Agent Technol. Syst. **1**, 58–67 (2008)
8. Han, L., Escolano, F., Hancock, E.R., Wilson, R.C.: Graph characterizations from von neumann entropy. Pattern Recogn. Lett. **33**, 1958–1967 (2012)
9. Alzheimer's Disease Neuroimaging Initiative (ADNI). http://adni.loni.usc.edu/
10. Ye, C., Wilson, R.C., Hancock, E.R.: An entropic edge assortativity measure. In: Liu, C.-L., Luo, B., Kropatsch, W.G., Cheng, J. (eds.) GbRPR 2015. LNCS, vol. 9069, pp. 23–33. Springer, Cham (2015). doi:10.1007/978-3-319-18224-7_3
11. Foster, J.G., Foster, D.V., Grassberger, P., Paczuski, M.: Edge direction and the structure of networks. Proc. Natl. Acad. Sci. U.S.A. **107**(24), 10815–10820 (2010)
12. Wang, J., Wilson, R.C., Hancock, E.R.: fMRI activation network analysis using bose-einstein entropy. In: Robles-Kelly, A., Loog, M., Biggio, B., Escolano, F., Wilson, R. (eds.) S+SSPR 2016. LNCS, vol. 10029, pp. 218–228. Springer, Cham (2016). doi:10.1007/978-3-319-49055-7_20

Graph Matching

Error-Tolerant Coarse-to-Fine Matching Model for Hierarchical Graphs

Pau Riba[✉], Josep Lladós, and Alicia Fornés

Computer Science Department, Computer Vision Center,
Universitat Autònoma de Barcelona, Barcelona, Spain
{priba,josep,afornes}@cvc.uab.es
http://www.cvc.uab.es

Abstract. Graph-based representations are effective tools to capture structural information from visual elements. However, retrieving a query graph from a large database of graphs implies a high computational complexity. Moreover, these representations are very sensitive to noise or small changes. In this work, a novel hierarchical graph representation is designed. Using graph clustering techniques adapted from graph-based social media analysis, we propose to generate a hierarchy able to deal with different levels of abstraction while keeping information about the topology. For the proposed representations, a coarse-to-fine matching method is defined. These approaches are validated using real scenarios such as classification of colour images and handwritten word spotting.

Keywords: Graph matching · Hierarchical graph · Graph-based representation · Coarse-to-fine matching

1 Introduction

Graph-based representations play an important role in content-based image retrieval. Using graphs, not only statistical information is codified but also the relations between the compounding parts. The use of graph representations in computer vision has two main requirements. First, the extraction of the structures underlying the visual objects. Second, error-tolerant metrics coping with noise or distortion must be designed. Graph matching is one of the most important challenges of graph processing [6]. Generally speaking, the problem consists in finding the best correspondence between the sets of vertices of two graphs preserving the underlying structures. The intrinsic variability of patterns, noise and errors produced from the graph extraction process, makes mandatory to encode tolerance to errors into graph matching frameworks. Thus error-tolerant graph matching has to be applied.

Graph edit distance [9] is the process of evaluating the similarity of two different graphs computing the minimum edit cost from the source to the target graph in terms of node and edge insertion, deletion and substitution. It is an optimal method and the computational complexity is exponential in the number

© Springer International Publishing AG 2017
P. Foggia et al. (Eds.): GbRPR 2017, LNCS 10310, pp. 107–117, 2017.
DOI: 10.1007/978-3-319-58961-9_10

of nodes. A suboptimal approximation called *bipartite graph matching* was proposed by Riesen *et al.* [16]. It is based on the assignment problem solution using a cost matrix which codifies the edit operations costs. More recently, an efficient approach was proposed by Zhou and De la Torre [20] formulating graph matching as a quadratic assignment. To avoid the computation of the large pairwise affinity matrix, they propose a factorisation into smaller matrices that encode the local structure of each graph and the pairwise affinity between edges.

When dealing with large scale data, indexation strategies are required to prune the number of graph comparisons. Generally, graph indexing is solved by graph factorisation techniques where the dataset of graphs is decomposed in smaller ones representing a codebook of compounding structures. The indexation is formulated in terms of matching the constituent graphs organised in a look-up table structure. Usually, path-based methods are used to split the graphs into small redundant fragments. *GraphGrep* [17] enumerates all the existing paths up to a predefined length. This reduces the search space performing the exact matching using only few graphs. A relevant work was proposed by Yan *et al.* [19]. They propose to use frequent substructures instead of path-based methods as indexing features. Frequent graph substructures are obtained by graph sequentialization, according to a *depth first search* (DFS) traversal of the graph edges. Edge sequences are organised in a prefix tree called the *gIndex* tree. Riba *et al.* [15] proposed a binary embedding for the local context of each node. A vote scheme is used for indexation, so the subgraphs with more votes are more accurately analyzed in a finer matching process.

The above methods rely on local structures rather than global knowledge of the graph. An interesting alternative is to use a scale-space approach where the input data is hierarchically organized, summarizing it in order to avoid complex graph comparisons. Several hierarchical graph approaches have been proposed. Brun and Kropatsch [4] introduces a set of relationships between regions of a partition through irregular graph pyramids. Broelemann *et al.* [3] propose to deal with noise such as spurious nodes and edges through a hierarchical representation of plausible graphs. Ahuja and Todorovic [1] present a region based approach for object recognition based in multi-scale region segmentation. Conte *et al.* [5] propose a similar graph multi-resolution approach in order to improve the object tracking in a video. Mousavi *et al.* [11] use a hierarchical graph representation in order to improve the information codified by graph embedding frameworks. Indexation frameworks have been also proposed.

The main contribution of this work is a hierarchical graph representation and matching able to discard non-promising structures. Our hierarchical information avoids a direct matching at the original graph. The hierarchy is designed to perform a big reduction of the graphs drastically reducing the matching time. The proposed approaches are validated using real scenarios such as classification of colour images and handwritten word spotting. In the next sections we describe the representation, the matching and the results respectively.

2 Hierarchical Attributed Graph Representation

2.1 Hierarchy Construction

A hierarchical graph representing information at different levels of abstraction (contraction) allows to perform the retrieval problem in an abstract manner.

Definition 1 (Hierarchical Graph). *A hierarchical graph* H *is defined as a 6-tuple* $H(V, E_N, E_H, L_V, L_{E_N}, L_{E_H})$ *where* V *is the set of nodes;* $E_N \subseteq V \times V$ *are the neighborhood edges;* $E_H \subseteq V \times V$ *are the hierarchical edges;* L_V, L_{E_N} *and* L_{E_H} *are three labeling functions defined as* $L_V : V \to \Sigma_V \times A_V^k$, $L_{E_N} : E_N \to \Sigma_{E_N} \times A_{E_N}^l$ *and* $L_{E_H} : E_H \to \Sigma_{E_H} \times A_{E_H}^m$, *where* Σ_V, Σ_{E_N} *and* Σ_{E_H} *are three sets of symbolic labels for vertices and edges,* A_V, A_{E_N} *and* A_{E_H} *are three sets of attributes for vertices and edges, respectively, and* $k, l, m \in \mathbb{N}$.

Given a graph G, two functions are needed to construct a hierarchical graph H:

- *Contraction:* $c : G \to H$, defines the groups of nodes that are gathered together. The contraction process can follow different criteria such as topology, features of the nodes or edges, etc. This function follows a clustering process.
- *Embedding:* $\varphi : G \to \mathbb{R}^n$, returns a vectorial representation of the contracted subgraph to be used as an attribute. The embedding function can be seen as a signature of the subgraph that summarizes the information from one level to another (information propagation between levels).

We propose a contraction criterion based on the topology. The embedding function is applied to all contracted groups of nodes propagating the information. This function is application dependent and is specified for each particular case.

2.2 Hierarchy Construction by Community Detection

To determine the group of nodes that are joined into a unique vertex, the Girvan-Newman algorithm [10] is applied. This is a well-known method for community detection in complex systems with complexity $\mathcal{O}(m^2 n)$, where m and n are the number of edges and nodes respectively. It is a global divisive algorithm which removes the appropriate edge at each step until all the edges are deleted. The *betweenness centrality* measure is used as edge selection. The *betweenness centrality* of $e \in E$ is defined as the number of shortest walks between any pair of nodes that cross e. The idea is that the edges with higher centrality are candidates to connect two clusters. After the edge deletion, each connected component is considered as a cluster in the hierarchy. This algorithm consists of 4 steps:

1. Calculate the betweenness centrality (BC) for all edges in the network.
2. Remove the edge with highest BC and generate a cluster for each connected component.

3. Recalculate BCs for all edges affected by the removal.
4. Repeat from step 2 until no edges remain.

The output of this algorithm is a dendrogram providing a hierarchical clustering of the graph nodes. In case of ties (i.e. several edges have the same BC), The edge with more connections in their compounding nodes is deleted. From it, we contract clusters containing at least two nodes. Moreover, it does not allow any node to be a cluster individually. Therefore, the reduction ratio is at least of 2. Afterwards, the corresponding nodes are contracted into only one vertex which is labelled with the embedding function applied to these subgraphs. The idea is that each node of the hierarchy represents a subgraph and provide information about its topology. Finally, connected communities will create connected nodes.

2.3   Splitting of Articulation Points

There are cases where slight deformations in the input graphs can lead to completely different hierarchies. Figure 1 shows a common subgraph that can lead to two possible hierarchies. This ambiguity can result in matching errors. Although *overlapping community detection* techniques have been developed [13], they generate redundant information leading to a bad abstraction. This problem usually comes from a symmetry in the original graph. To tackle with this problem we define *articulation points* as follows:

Definition 2 (Articulation Point). *A node in an undirected graph is an articulation point if and only if removing it the number of connected components of the graph increases.*

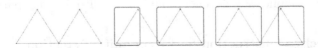

Fig. 1. Ambiguity configuration that can significantly influence in the hierarchy construction, in red two possible clusterings of nodes from the contraction function. (Color figure online)

These nodes are of key importance, if they are classified in an incorrect cluster, they can change significantly the topology. Thus, we propose to split the articulation points of the graphs creating *virtual* nodes and disconnecting them. Hence, the hierarchical representation is stabilised without introducing noise to the data. The articulation points therefore divide and belong to two or more clusters. Introducing this modification to the contraction function, a more stable hierarchy is generated. Figure 2 shows the splitting process in a real scenario where graphs represent skeleton features in handwritten word images.

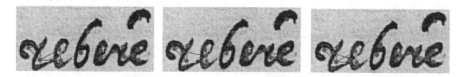

Fig. 2. From left to right: input graph, hierarchy for the proposed contraction function and hierarchy splitting the articulation points. In red, the contracted nodes. (Color figure online)

3   Error Tolerant Hierarchical Matching

As graph matching baseline, we have used the algorithm of *bipartite graph matching* proposed by Riesen and Bunke in [16]. It uses a cost matrix that codifies the edit costs between the source and target nodes. Once the cost matrix is defined, an edit operation is assigned to each node minimising the total cost. We have used the same edit costs as [14]: node substitution cost is based in the distance, the attributes and local structure of incident edges; edge substitution cost is computed in terms of edge attribute, angle and length; predefined costs are defined for node and edge insertion and deletion. Thus, there are 8 parameters: 3 (node substitution), 3 (edge substitution) and 2 (insertion and deletion).

To take advantage of the hierarchical representation, we propose a coarse-to-fine graph matching approach. Let us denote H^i the graph representation at level $i = 1, \ldots, N$. It iteratively refines the matching starting at the coarsest level (i.e. $i = N$). The comparison is performed using bipartite graph matching taking the graph representation at level i without the hierarchical edges. If the distance at level i is small enough, the matching is performed at the next level $(i - 1)$. The threshold to decide whether to advance in the hierarchy or not is application dependent and a threshold is set experimentally. Starting the matching at the abstract level avoids a high number of comparisons at more detailed levels where the graphs are significantly bigger. Ideally, the last level is only used for graphs that are very similar to the input one. The information about the matching level is kept. Figure 3 shows the iterative process to decide whether the graphs match or we can discard the comparisons in any of the abstract levels of the hierarchy.

4   Experiments

4.1   Datasets

Different databases have been used. First, the *Columbia Object Image Library* (COIL-100) [12] and the *Object DataBank* (ODBK) [18] have been used to reproduce the experiments proposed by Mousavi *et al.* [11] in an object classification scenario. Second, the *Barcelona Historical Handwritten Marriages* (BH2M) database has been used in a graph-based word spotting scenario in handwritten documents where graphs are irregular and suffer from high distortions.

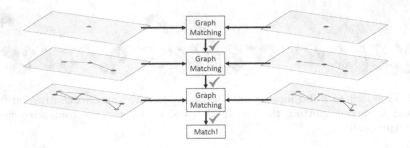

Fig. 3. Coarse-to-fine matching scheme.

The COIL database consists of images of 100 different objects taken at 72 equally spaced poses whereas the ODBK database is formed by 209 3D objects with 14 views. Graph nodes are extracted using the *Harris corner detector*. The edges are generated using the Delaunay triangulation on these nodes. The final graphs are not weighted for the edges and store the coordinates for the nodes. For the experimentation, 15 and 50 classes, with maximum average number of nodes are used. The graphs are divided into three sets, training, validation and test of 360, 75 and 150 for the COIL dataset and 300, 150 and 150 for the ODBK database. Figure 4 shows some examples coming from these databases.

Fig. 4. Example of objects from the COIL-100 and ODBK databases.

The BH2M database [8] corresponds to marriage licenses written between 1617 and 1619. It contains 174 handwritten pages divided into training (100), validation (34) and test (40). The handwritten words are represented by attributed graphs where nodes correspond to basic primitives called *graphemes* [14]. Graphemes defined as convexities are described using the *Blurred Shape Model* (BSM) descriptor [7]. The descriptors extracted from the training set are used to create a *codebook*, from which node labels are set. Edges represent adjacency relations between those primitives. Figure 2 shows an example of the obtained graphs plotted on the image.

4.2   Results

The experiments have been divided into two challenges: *object classification* and *word spotting*. Thresholds have been carefully selected using the validation set.

Object Classification: The use of a richer representation allows us to use a simple classification approach (k-NN) achieving similar results than an scheme with a less expressive representation and complex classifier, with the advantage of reducing the computational cost. The selected embedding function encodes information of the Morgan Index of length 1 and 2 of the previous level. Three approaches have been evaluated for the information selection: averaging the Morgan Index and node position from all contracted nodes, averaging the Morgan Index and selecting the most connected node, and taking the maximum Morgan Index and the most connected node position.

Each level of the hierarchy has been validated alone and combined with the original graph to explore the benefits of the proposed coarse-to-fine matching. All the parameters for the distance computation have been chosen performing a random search in the validation set. Since the graphs are generated using a triangulation, there are not articulation points, therefore, both contraction functions will lead to the same hierarchy.

Table 1 shows the performance for COIL and ODBK databases respectively. For this experiment, the mean of the node positions and the Morgan Index is used as embedding function. The last 3 rows correspond to the performance reported by [11] using their hierarchical representation with the same graphs.

Note that the big loss of performance between the abstract levels is corrected choosing a good trade-off between them. We are able to prune more than the 50% of comparisons at the finest level while achieving good results. For instance, choosing a conservative threshold, the time reduction is half, losing only 2% of accuracy for the COIL database and 1.5 times faster maintaining the same accuracy for the ODBK database. However, relaxing this threshold, we are able to achieve a speed-up of 7× with a loss of 10% in accuracy for the COIL database and 3.3 times faster losing 1% in accuracy for ODBK. In a large scale scenario, this is an acceptable loss to make an application much faster. Compared to [11], our methodology do not achieve as good results as them in the different abstraction levels. One of the main reasons is that our hierarchy is dynamically constructed, not fixing the contraction degree and generating smaller graphs.

Word Spotting: Word spotting is the task of retrieving word images from a document image similar to a given query (text or image). It is formulated as a visual object detection problem. Most word spotting techniques use statistical representations (e.g. HOG, SIFT) of the word images, e.g. [2]. The *embedding function* consists in a vector that counts the number of paths of length up to k from any node to a node with label i. The best configuration has been $k = 0$, i.e. counting the number of nodes with label i (similar to a *bag of words* for the nodes). As a retrieval problem the *mean average precision* (mAP) has been used for the evaluation. Using the same parameters proposed in [14] in order to compute the edit cost operations a mAP of 69.45% is achieved. Reproducing the same experiment using the **first level** of the hierarchy the achieved mAP is 35.67%. By **splitting the articulation points** as proposed in Sect. 2.3 achieves a mAP of 46.37%. Figure 5 shows the interpretation of the hierarchical graph representation in the context of this database. Observe how the graphemes are combined at each level to create more complex shapes like letters, bi-grams and finally words.

Table 1. Performance for Object Classification for COIL (left) and ODBK (right) datasets. Rows are divided in 5 blocks: performance for each level; coarse-to-fine matching using the 1st, 2nd abstract levels and the combination of them; the final row-block correspond the performance reported by Mousavi et al. Columns correspond to the used threshold; accuracy of a k-NN classifier; percentage of avoided comparisons at the base level; time in seconds to perform all the comparisons.

COIL database

	Thresh.	K-NN (%)			AC^a (%)	t (s)
		1	3	5		
Original	–	100.00	100.00	98.00	–	2010
1st abst.	–	72.67	74.67	72.67	–	167
2nd abst.	–	38.00	39.33	44.67	–	13
1st abst.	0.1982	98.00	97.33	93.33	67.37	977
	0.1680	90.00	89.33	82.67	95.41	289
2nd abst.	0.2153	100.00	99.33	96.67	33.68	1444
	0.1895	97.33	94.67	93.33	58.99	937
1st abst.	0.1982	98.67	98.00	92.67	71.63	893
2nd abst.	0.2153					
Original	–	100.00	97.00	90.00	Mousavi et al. [11]	
1st abst.	–	98.17	94.83	88.83		
2nd abst.	–	87.00	81.67	78.17		

ODBK database

	Thresh.	K-NN (%)			AC^a (%)	t (s)
		1	3	5		
Original	–	79.33	76.00	74.00	–	34959
1st abst.	–	58.67	58.00	54.67	–	1954
2nd abst.	–	42.00	41.33	46.00	–	141
1st abst.	0.2396	79.33	76.00	74.00	48.18	22501
	0.2130	78.67	75.33	72.00	79.10	10496
2nd abst.	0.2973	78.67	74.67	72.67	33.76	26111
	0.2573	76.67	71.33	68.67	68.49	12228
1st abst.	0.2130	78.00	74.00	70.67	79.23	10292
2nd abst.	0.2973					
Original	–	66.67	65.33	63.33	Mousavi et al. [11]	
1st abst.	–	66.67	62.67	62.00		
2nd abst.	–	60.00	55.33	53.33		

^a AC stands for Avoided Comparison

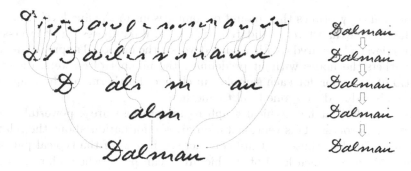

Fig. 5. Hierarchy construction for the word "Dalmau".

Table 2 shows a comparison between: the original graphs, the proposed framework with two thresholds, and graph indexation [15]. Recall (R) and Specificity (SPC) are computed on the selected graphs using the first abstract level as classifier. Notice that the proposed hierarchical framework achieves high specificity whereas keeping a better trade-off with the recall than the indexation approach. Moreover, only losing 8% of mAP which is acceptable for a large scale retrieval we are able to speed up the process almost 5 times.

Table 2. Comparison of the proposed hierarchical framework against an indexation framework [15]. The mean average precision corresponds to the evaluation of the word spotting problem; recall (R) and specificity (SPC) are computed on the selected graphs using the hierarchy or the indexation respectively; finally, time per query is provided.

	mAP (%)	R (%)	SPC (%)	Time/query[a] (s)
Original	69.45	100.00	0.00	19.58
+abst. (t = 0.30)	**68.27**	90.91	69.98	**12.46**
+abst. (t = 0.25)	**61.71**	67.93	97.91	**3.94**
+ [15] (t = 0.20)	66.13	92.54	46.13	16.34
+ [15] (t = 0.30)	61.15	83.55	63.04	12.74

[a]1000 queries selected randomly against 13098 graphs

5   Conclusions

This paper has presented a construction of a hierarchical graph representation by means of contraction and embedding functions. Contraction uses graph clustering techniques to gather nodes and simplify the graph. Moreover, a modification of the contraction function has been proposed to stabilise the hierarchy in certain graphs. The proposed method is able to significantly reduce the graph size allowing a fast graph comparison through a coarse-to-fine matching approach.

This methodology prunes the amount of comparisons in the fine level. The approach has been exhaustively validated using several databases for of large-scale graph retrieval. Compared to other related works, the proposed approach dynamically gathers the nodes without predefining the number of clusters, therefore, the ratio of reduction for each sample can change. Furthermore, the graph size is extremely reduced from one level to another.

We conclude that hierarchical graph representations are a powerful tool in the matching process. This representation gives information about the relation of a group of nodes (those that are contracted) instead of the typical pair-wise relations. Moreover, each level of the hierarchy can be enriched following other indexation methodologies such as [15]. The future work will be focused on the development of matching algorithms using the whole representation at once.

Acknowledgments. This work has been partially supported by the Spanish project TIN2015-70924-C2-2-R, a FPU fellowship FPU15/06264 from the Spanish Ministerio de Educación, Cultura y Deporte, the Ramon y Cajal Fellowship RYC-2014-16831 and the CERCA Programme/Generalitat de Catalunya.

References

1. Ahuja, N., Todorovic, S.: From region based image representation to object discovery and recognition. In: Hancock, E.R., Wilson, R.C., Windeatt, T., Ulusoy, I., Escolano, F. (eds.) SSPR /SPR 2010. LNCS, pp. 1–19. Springer, Heidelberg (2010). doi:10.1007/978-3-642-14980-1_1
2. Almazán, J., Gordo, A., Fornés, A., Valveny, E.: Word spotting and recognition with embedded attributes. IEEE Trans. Pattern Anal. Mach. Intell. **36**(12), 2552–2566 (2014)
3. Broelemann, K., Dutta, A., Jiang, X., Lladós, J.: Hierarchical plausibility-graphs for symbol spotting in graphical documents. In: Lamiroy, B., Ogier, J.-M. (eds.) GREC 2013. LNCS, vol. 8746, pp. 25–37. Springer, Heidelberg (2014). doi:10.1007/978-3-662-44854-0_3
4. Brun, L., Kropatsch, W.: Contains and inside relationships within combinatorial pyramids. Pattern Recognit. **39**(4), 515–526 (2006). Graph-based Representations
5. Conte, D., Foggia, P., Jolion, J.M., Vento, M.: A graph-based, multi-resolution algorithm for tracking objects in presence of occlusions. Pattern Recognit. **39**(4), 562–572 (2006). Graph-based Representations
6. Conte, D., Foggia, P., Sansone, C., Vento, M.: Thirty years of graph matching in pattern recognition. Int. J. Pattern Recognit. Artif. Intell. **18**(03), 265–298 (2004)
7. Escalera, S., Fornés, A., Pujol, O., Radeva, P., Sánchez, G., Lladós, J.: Blurred shape model for binary and grey-level symbol recognition. Pattern Recognit. Lett. **30**(15), 1424–1433 (2009)
8. Fernández-Mota, D., Almazán, J., Cirera, N., Fornés, A., Lladós, J.: BH2M: the Barcelona historical handwritten marriages database. In: International Conference on Pattern Recognition (2014)
9. Gao, X., Xiao, B., Tao, D., Li, X.: A survey of graph edit distance. Pattern Anal. Appl. **13**(1), 113–129 (2010)
10. Girvan, M., Newman, M.E.J.: Community structure in social and biological networks. Natl. Acad. Sci. **99**(12), 7821–7826 (2002)

11. Mousavi, S.F., Safayani, M., Mirzaei, A., Bahonar, H.: Hierarchical graph embedding in vector space by graph pyramid. Pattern Recognit. **61**, 245–254 (2017)
12. Nayar, S., Nene, S., Murase, H.: Columbia object image library (COIL 100). Department of Computer Science, Columbia University, Technical report CUCS-006-96 (1996)
13. Palla, G., Derényi, I., Farkas, I., Vicsek, T.: Uncovering the overlapping community structure of complex networks in nature and society. Nature **435**(7043), 814–818 (2005)
14. Riba, P., Fornés, A., Lladós, J.: Handwritten word spotting by inexact matching of grapheme graphs. In: 13th International Conference on Document Analysis and Recognition, pp. 781–785, August 2015
15. Riba, P., Lladós, J., Fornés, A., Dutta, A.: Large-scale graph indexing using binary embeddings of node contexts. In: Liu, C.-L., Luo, B., Kropatsch, W.G., Cheng, J. (eds.) GbRPR 2015. LNCS, vol. 9069, pp. 208–217. Springer, Cham (2015). doi:10.1007/978-3-319-18224-7_21
16. Riesen, K., Bunke, H.: Approximate graph edit distance computation by means of bipartite graph matching. Image Vis. Comput. **27**(7), 950–959 (2009)
17. Shasha, D., Wang, J.T.L., Giugno, R.: Algorithmics and applications of tree and graph searching. In: Proceedings of the 21st Symposium on Principles of Database Systems, pp. 39–52. ACM, New York (2002)
18. Tarr, M.J.: The object databank (2011)
19. Yan, X., Yu, P.S., Han, J.: Graph indexing: a frequent structure-based approach. In: Proceedings of the International Conference on Management of data, pp. 335–346 (2004)
20. Zhou, F., de la Torre, F.: Factorized graph matching. IEEE Trans. Pattern Anal. Mach. Intell. **38**(9), 1774–1789 (2016)

A Hungarian Algorithm for Error-Correcting Graph Matching

Sébastien Bougleux[1]([⊠]), Benoit Gaüzère[2], and Luc Brun[1]

[1] Normandie Univ, CNRS - ENSICAEN - UNICAEN, Caen, France
bougleux@unicaen.fr
[2] Normandie Univ, INSA de Rouen, Rouen, France

Abstract. Bipartite graph matching algorithms become more and more popular to solve error-correcting graph matching problems and to approximate the graph edit distance of two graphs. However, the memory requirements and execution times of this method are respectively proportional to $(n+m)^2$ and $(n+m)^3$ where n and m are the order of the graphs. Subsequent developments reduced these complexities. However, these improvements are valid only under some constraints on the parameters of the graph edit distance. We propose in this paper a new formulation of the bipartite graph matching algorithm designed to solve efficiently the associated graph edit distance problem. The resulting algorithm requires $\mathcal{O}(nm)$ memory space and $\mathcal{O}(\min(n,m)^2 \max(n,m))$ execution times.

Keywords: Graph edit distance · Bipartite matching · Error-correcting matching · Hungarian algorithm

1 Introduction

Computing an efficient similarity or dissimilarity measure between graphs is a major problem in structural pattern recognition. The graph edit distance (GED), developed in the context of error-correcting graph matching, provides such a measure. It may be understood as the minimal amount of distortion required to transform one graph into another, by a sequence of edit operations applied on nodes and edges, restricted here to substitutions, insertions and removals. Such a sequence is called an edit path. Each possible edit operation is penalized by a non-negative cost, and the integration of these costs over an edit path defines the length (or the cost) of this path. An edit path having a minimal length, among all edit paths transforming one graph into another one defines the GED between these two graphs. Since computing the GED is NP-complete, it is restricted to rather small graphs. So several approaches have been proposed to approximate the GED efficiently and to process larger graphs.

In this paper, graphs are assumed to be simple (no loop nor multiple edge), and each element of the two graphs can be edited only once (no composition of edit operations). Under these hypotheses, each node of a graph G_1 can be

P. Foggia et al. (Eds.): GbRPR 2017, LNCS 10310, pp. 118–127, 2017.
DOI: 10.1007/978-3-319-58961-9_11

either substituted once to a node of another graph G_2, or removed. Similarly, any node of G_2 may be substituted once, or inserted. Since each node of G_1 and G_2 is transformed only once, such operations on nodes can be encoded by a $(n + m) \times (n + m)$ permutation matrix $\mathbf{X}$ [12], where n and m denote the orders of G_1 and G_2. The costs related to these operations can be encoded by a $(n+m) \times (n+m)$ cost matrix $\mathbf{C}$. Using different heuristics [6,12] to design matrix $\mathbf{C}$, an approximation of the GED can be obtained by solving a linear sum assignment problem (LSAP), *i.e.* by computing an optimal permutation matrix $\mathbf{X}$, for instance with the Hungarian algorithm in $O((n + m)^3)$ time complexity.

However, matrix $\mathbf{C}$ contains an important amount of redundant information mainly used to transform the initial graph edit distance problem into a bipartite matching problem (LSAP). The storage of these additional information induces important memory requirements and increases the size of matrix $\mathbf{C}$, which determines the complexity of the algorithm. Moreover, the resulting matrix $\mathbf{X}$ may contain some useless operations. Serratosa [13] proposed to reduce the size of matrix $\mathbf{C}$ in the special case where the graph edit distance fulfills all the axioms of a distance. Such an assumption induces several constraints of the elementary edit costs. Assuming these constraints, Serratosa proposed either to store a $n \times m$ rectangular cost matrix whose optimal solution may be found in $\mathcal{O}(\min(n, m)^2 \max(n, m))$ using the Bourgeois' adaption [4] of the Hungarian algorithm or to store a $\max(n, m) \times \max(n, m)$ cost matrix [14] whose optimal solution may be found by combining the Jonker-Volgenant [8] and Hungarian algorithms. The overall complexity of this last approach is $\mathcal{O}(\max(n, m)^3)$.

Following [12], the approach proposed in this paper approximates the graph edit distance by the Hungarian algorithm. However, our method reformulates the basic problem, hence leading to a $(n + 1) \times (m + 1)$ cost matrix [2]. Note that a similar formulation has been proposed by [7]. However, this formulation is combined with a Jonker-Volegenant matrix reduction and the classical Hungarian algorithm, hence leading to a $\mathcal{O}((n + m)^3)$ overall complexity. In this paper we investigates the basic principles of the Hungarian algorithm in order to adapt it to this new formulation. Such an extension is detailed in Sect. 3 after a short introduction to the Hungarian algorithm in Sect. 2. The resulting algorithm has a worst case complexity of $\mathcal{O}(\min(n, m)^2 \max(n, m))$. Conversely to the methods [13] proposed by Serratosa, our method only assumes that the edit costs are non negative. We also provide in Sect. 4 accuracy and execution times of a previously published quadratic minimizer [2,3] of the GED combined with our new Hungarian algorithm.

2   Bipartite Matching and Hungarian Algorithm

Preliminary Definitions. Given a bipartite graph $(U \cup V, E)$, a *matching* M is a subset of E such that each node in $U \cup V$ is incident to at most one edge of M. It defines a bijective mapping between a subset of U and a subset of V. An edge is *matching edge* if it is in M, else it is an *unmatching edge*. A node incident to an edge of M is *covered* by M, and otherwise *uncovered*. If all nodes of both sets are covered, the two sets have the same size and the matching is called *perfect*. It defines a bijection between U and V, also called an *assignment*.

Consider a matching M with at least two uncovered nodes, one in each set. A path in the bipartite graph is called *alternating* if it alternates between unmatching and matching edges. An alternating path that begins and ends with uncovered nodes is called *augmenting*. If an augmenting path P exists, a new matching is obtained from M by removing the matching edges of P and by inserting the unmatching ones. The new matching augments the number of matching edges by one, and the number of covered nodes by two.

Linear Sum Assignment Problem and Its Dual. Consider two sets U and V with the same size n. Each assignment of an element $i \in U$ to an element $j \in V$ is penalized by a non-negative[1] cost $c_{i,j}$. All costs are encoded through a $n \times n$ matrix $\mathbf{C} = (c_{i,j})_{(i,j) \in U \times V}$, *i.e.* a node-node cost matrix associated with the complete bipartite graph $(U \cup V, U \times V)$. When the assignment of a node i to a node j is forbidden, the cost of the edge (i,j) is commonly set to a large value ω, larger than all costs. The *linear sum assignment problem* (LSAP), or minimal-cost perfect matching problem, consists in finding a perfect matching having a minimal cost L, among all perfect matchings:

$$\underset{\mathbf{X}}{\operatorname{argmin}} \left\{ L(\mathbf{X}, \mathbf{C}) = \sum_{i=1}^{n} \sum_{j=1}^{n} c_{i,j} x_{i,j} \ : \ \mathbf{X} \in \{0,1\}^{n \times n}, \ \mathbf{X1} = 1, \ \mathbf{X}^T 1 = 1 \right\} \quad (1)$$

where $\mathbf{X}$ defines the node-node adjacency matrix of a perfect matching M ($x_{i,j} = 1$ if $(i,j) \in M$ and $x_{i,j} = 0$ else), *i. e.* a *permutation matrix*.

Several algorithms have been developed to find a solution to the LSAP [5]. Among them, the Hungarian algorithm is commonly used to compute approximate GED [2,6,12–14]. When it is properly implemented, it finds a solution in $O(n^3)$ in time and in $O(n^2)$ in space [5,9], in worst-case.

The Hungarian algorithm uses a primal-dual approach to find a solution to the LSAP and its dual problem, known as the maximum labeling problem:

$$\underset{(\mathbf{u}, \mathbf{v})}{\operatorname{argmax}} \left\{ 1^T \mathbf{u} + 1^T \mathbf{v} \ : \ \mathbf{u}, \mathbf{v} \geq 0, \ \mathbf{u1}^T + \mathbf{v1}^T \leq \mathbf{C} \right\} \quad (2)$$

where vectors $\mathbf{u} = (u_i)_{i=1,\ldots,n}$ and $\mathbf{v} = (v_j)_{j=1,\ldots,n}$ associate a label (or capacity) to each node of $U \cup V$. A pair $(\mathbf{u}, \mathbf{v})$ satisfying the constraint $\mathbf{u1}^T + \mathbf{v1}^T \leq \mathbf{C}$ is called a *feasible node labeling*. A pair $(\mathbf{X}, (\mathbf{u}, \mathbf{v}))$ solves the LSAP and its dual iff it verifies the complementary slackness condition:

$$\forall (i,j) \in U \times V, \ ((x_{i,j} = 1) \wedge (u_i + v_j = c_{i,j})) \vee ((x_{i,j} = 0) \wedge (u_i + v_j \leq c_{i,j})) \quad (3)$$

More generally, given a feasible node labeling, let $E^0 = \{(i,j) \in U \times V \ : \ c_{i,j} = u_i + v_j\}$, the graph induced by this set is called the *equality subgraph*. When E^0 contains an optimal perfect matching, it contains also all other ones.

---

[1] If some costs are negative, all costs are shifted by $-\min_{i,j}\{c_{i,j}\}$ [5].

Hungarian Algorithm. Given a cost matrix $\mathbf{C}$, an initial feasible node labeling $(\mathbf{u}, \mathbf{v})$ and an associated matching M (included in the equality subgraph), the Hungarian algorithm proceeds by iteratively updating M and $(\mathbf{u}, \mathbf{v})$ such that two more nodes are covered at each iteration. It is realized by growing a tree of alternating paths in the equality subgraph, called *Hungarian tree*, until an augmenting path is found. At each iteration of the growing process, the tree is augmented by a pair of unmatching and matching edges of the equality subgraph. If this is not possible, because the equality subgraph does not contain enough unmatching edges, the feasible node labeling is revised. We describe the efficient version detailed in [5,9]. The tree is represented by matching edges and by a predecessor array, denoted by pred, which encodes the predecessor (a node of U) of each node of V. Nodes encountered in the tree are encoded by the sets $T_U \subset U$ and $T_V \subset V$. The efficiency of the algorithm relies on maintaining slack variables during the search for an augmenting path: $\forall j \in V \backslash T_V$, $\text{slack}_j = \min\{c_{i,j} - u_i - v_j, \ i \in T_U\}$.

1. If all nodes of U are covered by M, a pair of solutions is found. Else, initialize a Hungarian tree rooted in an uncovered node $i \in U$: $T_U = \{i\}$ and $T_V = \emptyset$. Also, initialize all slack values to $+\infty$.

2. Grow the Hungarian tree in the equality subgraph from a leaf node $i \in T_U$:
   (a) Update neighbors of i to add unmatching edges (i, j) to the tree:

$$\forall j \in V \backslash T_V, \begin{cases} \text{if } c_{i,j} - u_i - v_j < \text{slack}_j \text{ then} \\ \quad \text{slack}_j \leftarrow c_{i,j} - u_i - v_j \\ \quad \text{pred}_j \leftarrow i \\ \text{if } \text{slack}_j = 0 \text{ then } T_V \leftarrow T_V \cup \{j\} \end{cases} \tag{4}$$

   (b) If there is no leaf node in T_V, the tree cannot grow anymore. The dual variables are updated to add at least one unmatching edge in the equality subgraph and in the tree:

$$\delta = \min\{\text{slack}_j, \ j \in V \backslash T_V\} \tag{5}$$
$$\forall i \in T_U, \ u_i \leftarrow u_i + \delta \tag{6}$$
$$\forall j \in T_V, \ v_j \leftarrow v_j - \delta \tag{7}$$
$$\forall j \in V \backslash T_V, \begin{cases} \text{slack}_j \leftarrow \text{slack}_j - \delta \\ \text{if } \text{slack}_j = 0 \text{ then } T_V \leftarrow T_V \cup \{j\} \end{cases} \tag{8}$$

   (c) If there is an uncovered leaf node $j \in T_V$, an augmenting path is found, go to Step 3. Else, the tree is extended with the unmatching edge (i, j) followed by the matching edge (l, j) by inserting l into T_U. Then go to Step 2a with $i \leftarrow l$.

3. Update the matching by backtracking in the tree from the node $j \in V$ found in Step 2c to the root, *i. e.* by traversing an augmenting path. Along this path, each matching edge is removed from the matching and each unmatching edge is inserted. Then go to Step 1.

An initial feasible labeling is usually given by $u_i \leftarrow \min\{c_{i,j}, \forall j \in V\}$ $\forall i \in U$, and $v_j \leftarrow \min\{c_{i,j} - u_i, \forall i \in U\}$ $\forall j \in V$. A matching is then deduced from this labeling by traversing the equality subgraph. More sophisticated methods, such as the one proposed by Jonker and Volgenant [5,8] can also be used.

3   Proposed Adaptation of the Hungarian Algorithm

Error-Correcting Matching and Minimal-Cost Problem. An error-correcting matching from a set U to a set V transforms U into V by editing their elements, together with their attributes. Edit operations are restricted here to substitutions, removals and insertions. Let $U^\epsilon = U \cup \{\epsilon\}$ and $V^\epsilon = V \cup \{\epsilon\}$ be the sets extended by the null element ϵ. Consider the complete bipartite graph $(U^\epsilon \cup V^\epsilon, U^\epsilon \times V^\epsilon)$. An *error-correcting matching* in this graph is a subset of edges connecting each node in U to a unique node of V (substituted by) or to ϵ (removed), and similarly, each node in V to a unique node of U (substituted to) or to ϵ (inserted). Null nodes are unconstrained, they can be connected to zero or more nodes. By considering node-node matrices associated to bipartite graphs, all error-correcting matching are represented by the set of binary matrices:

$$\Pi_{n,m}^\epsilon = \{ \mathbf{X} \in \{0,1\}^{(n+1)\times(m+1)} : x_{n+1,m+1} = 0, \tag{9}$$

$$\forall j = 1,\ldots,m, \ \textstyle\sum_{i=1}^{n+1} x_{i,j} = 1, \ \forall i = 1,\ldots,n, \ \textstyle\sum_{j=1}^{m+1} x_{i,j} = 1 \} \tag{10}$$

Null elements correspond to the last row and the last column. As observed in Eq. 10, they are unconstrained.

Let $\mathbf{C}$ be a $(n+1) \times (m+1)$ cost matrix associated to the complete bipartite graph, *i.e.* a non-negative cost (see Footnote 1) for each substitution, removal and insertion:

$$\mathbf{C} = \begin{pmatrix} c_{1,1} & \cdots & c_{1,m} & c_{1,\epsilon} \\ \vdots & \ddots & \vdots & \vdots \\ c_{n,1} & \cdots & c_{n,m} & c_{n,\epsilon} \\ \hline c_{\epsilon,1} & \cdots & c_{\epsilon,m} & 0 \end{pmatrix} \tag{11}$$

The cost of an error-correcting bipartite matching is then written as

$$L(\mathbf{X},\mathbf{C}) = \sum_{i=1}^{n+1}\sum_{j=1}^{m+1} c_{i,j} x_{i,j} = \sum_{i=1}^{n}\sum_{j=1}^{m} c_{i,j} x_{i,j} + \sum_{i=1}^{n} c_{i,\epsilon} x_{i,m+1} + \sum_{j=1}^{m} c_{\epsilon,j} x_{n+1,j}$$

Transforming U into V, with minimum cost, consists in finding an error-correcting bipartite matching having a minimal cost:

$$\operatorname*{argmin}_{\mathbf{X}} \{ L(\mathbf{X},\mathbf{C}), \ \mathbf{X} \in \Pi_{n,m}^\epsilon \} \tag{12}$$

This is a *linear sum assignment problem with error-correction* (LSAPE). Its dual problem, given by $\max_{(\mathbf{u},\mathbf{v})} \{ \mathbf{1}^T\mathbf{u} + \mathbf{1}^T\mathbf{v} : \mathbf{u}\mathbf{1}^T + \mathbf{v}\mathbf{1}^T \leq \mathbf{C}, \ u_{n+1} = v_{m+1} = 0 \}$, is similar to the labeling problem dual to the LSAP, with two elements constrained to be null (the null elements). Based on these formulations of the LSAPE and its dual, it is not difficult to show that the framework used to analyze and solve the LSAP and its dual problem still apply. The Hungarian algorithm can thus be adapted to find a pair of the primal and dual solutions satisfying Eq. 3. The adaptation concerns the processing of null nodes, since they are unconstrained. While the notion of alternating path and Hungarian tree are unchanged, this modifies the notion of augmenting paths as follows.

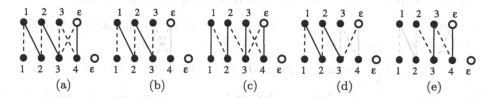

Fig. 1. (a) An incomplete error-correcting matching (solid) and the other edges of the inequality subgraph (dashed). (b) An augmenting path between two uncovered nodes. (c) The new matching obtained by interchanging matching and unmatching edges along this path. (d, e) An augmenting path ending by a null node.

Augmenting Paths. Since null nodes are always unconstrained, any path containing a null node ends by this node. This is equivalent to consider null nodes as never covered. As before (Sect. 2), an augmenting path can end with an uncovered node (Fig. 1(a)), which may thus be a null node (Fig. 1(d)). In this last case, the new matching contains one more covered node and one more matching edge. An augmenting path can also end with a null node incident to a matching edge (Fig. 1(e)). In this case, the new matching augments the number of covered nodes by one while the number of matching edges remains the same. So an augmenting path can be constructed by growing a Hungarian tree until an uncovered node is encountered, including null nodes. Null nodes do not need to be explicitly represented in the tree to find an augmenting path (always leaf nodes). This allows to modify the Hungarian algorithm as follows.

Hungarian Algorithm. Given two sets U and V, and a $(n+1) \times (m+1)$ edit cost matrix (Eq. 11) **C**, consider an initial[2] feasible node labeling $(\mathbf{u}, \mathbf{v})$ and an associated incomplete error-correcting matching M (all nodes are not yet covered). We complete the Hungarian algorithm described in Sect. 2 in order to treat the case of null nodes independently, without altering the global process. To this, the growing of the Hungarian is stopped when a null node is encountered:

---

[2] The Jonker-Volgenant algorithm proposed in [7] can be used to provide a good initialization. Here we adapt the basic one (Sect. 2): $u_i \leftarrow \min\{c_{i,j}, \forall j \in V^\epsilon\}$ $\forall i \in U$, and $v_j \leftarrow \min\{c_{i,j} - u_i, \forall i \in U^\epsilon\}$ $\forall j \in V$, with $u_{n+1} = v_{m+1} = 0$. An error-correcting matching is then deduced as in Sect. 2 by traversing the equality subgraph.

- A null node incident to a matching edge (here an insertion) can be detected in Eqs. 4 and 8 of Step 2 by replacing the instruction $T_V \leftarrow T_V \cup \{j\}$ by:

$$\text{if } (\epsilon, j) \in M \text{ go to Step 3, else } T_V \leftarrow T_V \cup \{j\}. \tag{13}$$

- A null node incident to an unmatching edge (here a removal) can be detected in Step 2c, when there is an edge $(l, \epsilon) \notin M$ in the equality subgraph, *i. e.* if $c_{l,\epsilon} = u_l$. If this is the case, the algorithm goes to Step 3 instead of going to Step 2a. A null node incident to an unmatching edge can also be detected after the update of the dual variables in Step 2b, as detailed below.

Dual variables are updated (Step 2b) such that costs associated to null nodes are also taken into account. Therefore, Eq. 5 is replaced by:

$$\delta = \min\left\{\min\{\text{slack}_j, \ j \in V \backslash T_V\}, \ \min\{c_{i,\epsilon} - u_i, \ i \in T_U\}\right\}. \tag{14}$$

Then, after Eqs. 6 and 7, and just before Eq. 8, if the minimum δ is obtained from an unmatching edges (i, ϵ), an augmenting path is found and the algorithm goes to Step 3.

The proposed modifications allow to cover all nodes of U. Some nodes of V may not be covered, which occurs if $n < m$ or if at least one node in U is assigned to a null node. To find a minimal-cost error-correcting matching, the modified Hungarian algorithm is completed by the following step to cover all nodes of V:

4 When all nodes of U are covered, swap the sets U and V, and go to Step 1 with $\mathbf{C}^T$ and $(\mathbf{v}, \mathbf{u})$ as initial feasible node labeling.

The proposed algorithm finds a minimal-cost error-correcting matching in $O(\min\{n, m\}^2 \max\{n, m\})$ in time and $O(nm)$ in space, see [1] for a proof. These complexities are similar to the ones obtained in [4] for solving the LSAP with rectangular cost matrices.

4   Experiments

Bipartite GED. The other formulations of the LSAPE (Sect. 1), transform the problem into a LSAP with a square cost matrix for BP [12] and SFBP [14], or with a rectangular one for FBP [13]. The Hungarian algorithm used in these works [12], differs from the algorithm presented in Sect. 2 on two aspects: several Hungarian trees are grown at each iteration, and the cost matrix is updated instead of the dual variables. As already discussed [5,9], the version described in this paper has lower execution times. So we have repeated the experiments carried out in [14] on artificially created graphs, with the Hungarian algorithm of Sect. 2 for solving BP and SFBP. Note that our implementation of the Hungarian algorithm is optimized such that forbidden assignments (with a cost equal to ω) are not treated. As already observed in [14], all the methods lead to a similar approximation of the GED. This is also the case of the approach proposed in this paper (denoted by BPE). A more interesting behavior concerns the computational time. Figure 2(a) shows the average run time of 10 computations of FBP, with respect to the order of the graphs. Contrary to what was observed in [14], the shape of the run time surface is symmetric. The run time surface of the other algorithms (BP, SFBP and BPE) have a similar pyramidal shape. As illustrated in Fig. 2(b), BP and SFBP have a similar behavior, with an asymmetry, and are less efficient than FBP and BPE. Observe that these two last approaches have also a similar behavior. Contrary to FBP, BPE does not impose any constraint on the costs.

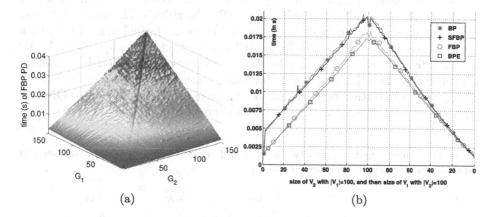

(a)   (b)

Fig. 2. Computational time of the bipartite GED with respect to the graphs' order.

IPFP and GNCCP. As illustrated in [2,3], LSAP methods may also be the core component of different solvers of quadratic programming formulations of the GED. A first method [2] called QAP consists in adapting the IPFP algorithm [10] to the computation of the quadratic formulation of GED. Basically, IPFP iterates over LSAP resolutions to compute a gradient direction leading to an approximate solution of a relaxed version of the quadratic problem. The second proposition [2] uses a convex-concave relaxation of the IPFP approach to tackle drawbacks induced by the influence of initialization and by the final projection step from a stochastic matrix to a mapping one. This approach, denoted GNCCP, iterates over a slightly modified version of IPFP which iterates over LSAP resolutions. Therefore, these two contributions use LSAP as a core component in their respective algorithms. In these experiments, we evaluate the gain obtained by the use of our new algorithm (LSAPE) to resolve LSAP steps in QAP [3] and GNCCP (new in this paper) approaches instead of the classic Hungarian algorithm.

Both algorithms are evaluated on real world chemical datasets[3] composed of different kinds of molecules: Alkane and Acyclic are represented as acyclic graphs of about 8 nodes in average, whereas MAO and PAH are composed of larger graphs, with an average size of 20 nodes. As in [2,6], the cost of substituting nodes and edges has been set to 1, and to 3 for insertions and deletions.

---

[3] Datasets are available at https://iapr-tc15.greyc.fr/links.html.

Table 1 shows average edit distances and computational times obtained by different approaches on the four chemical datasets. A^* approach, on the first line, computes the exact graph edit distance and constitutes a reference for approximation methods. However, due to its high complexity, exact graph edit distances have been only computed for Alkane and Acyclic datasets. The first block of three methods, from line 2 to 4, corresponds to methods based on the bipartite approach. The line denoted as Riesen and Bunke corresponds to the original method proposed in [12], while the two others use a different cost matrix [6] using respectively LSAP and LSAPE algorithms. The next block, lines 5 to 7, corresponds to methods based on the quadratic formulation of the graph edit distance. QAP and QAPE [3] use IPFP algorithm with respectively LSAP and LSAPE algorithms. The line denoted as "Neuhaus" corresponds to another quadratic approach [11] which does not handle insertions and removals of nodes during the optimization process. Finally, the last block corresponds to GNCCP approach [2] using LSAP and LSAPE algorithms.

Table 1. Accuracy and complexity scores. d and t denote respectively the average edit distance and computational time (in seconds).

Algorithm	Alkane		Acyclic		MAO		PAH	
	d	t	d	t	d	t	d	t
A^*	15.47	1.29	17.33	6.02	–	–	–	–
Riesen and Bunke [12]	35.16	0.00135	35.43	0.00109	105	0.00551	138	0.00692
LSAP [6]	34.51	0.00205	32.52	0.00181	56.89	0.02218	123.6	0.03342
LSAPE	34.51	0.00203	32.61	0.00179	56.92	0.02212	123.8	0.03338
QAP [2]	19.28	0.00925	20.51	0.00711	32.97	0.04158	48.5	0.08285
QAPE [3]	19.33	0.00553	20.43	0.00489	32.94	0.03017	48.9	0.04832
Neuhaus [11]	20.5	0.07	25.7	0.0424	59.1	7	52.9	8.2
GNCCP [2]	16.54	0.3474	18.36	0.2481	32.14	4.128	39.2	6.141
GNCCPE	16.83	0.116	19.09	0.07638	32.92	0.4673	38.7	0.8623

As expected, approximations of graph edit distances are not significantly different using either LSAP or LSAPE approaches. Conversely, as previously observed [2,3], methods based on a quadratic formulation obtain better approximations than the ones based on a linear approximation. From a computational point of view, quadratic approaches require more computational time. However, using LSAPE instead of LSAP algorithm leads to a significant improvement on computational times. This gain almost reaches 10 times with MAO dataset. On MAO and PAH datasets, executions times of LSAP and QAPE methods are comparable. Note that we only observe a very tight improvement using LSAPE instead of LSAP within the original bipartite approach (lines 3 and 4). This limited gain can be explained by the fact that most of computational time is spent in computing the cost matrix rather than optimizing the mapping problem.

5   Conclusion

We have presented in this paper a new type of linear sum assignment problem designed to solve efficiently the bipartite graph edit distance. The resulting algorithm only supposes that the basic costs are non negative. It requires the storage of an $(n+1) \times (m+1)$ matrix, n and m being the orders of both graphs and has a time complexity of $\mathcal{O}(\min(n,m)^2 \max(n,m))$. This algorithm may be applied once to obtain a rough estimate of the edit distance or be integrated into more complex iterative quadratic solvers. The speed-up obtained by our algorithm is significant in this last case and opens the way to the computation of the graph edit distance on larger graphs.

References

1. Bougleux, S., Brun, L.: Linear sum assignment with edition. Technical report, Normandie Univ, GREYC UMR 6072, Caen (2016)
2. Bougleux, S., Brun, L., Carletti, V., Foggia, P., Gaüzère, B., Vento, M.: Graph edit distance as a quadratic assignment problem. Pattern Recognit. Lett. **87**, 38–46 (2017)
3. Bougleux, S., Gaüzère, B., Brun, L.: Graph edit distance as a quadratic program. In: International Conference on Pattern Recognition. IEEE (2016)
4. Bourgeois, F., Lassalle, J.: An extension of the Munkres algorithm for the assignment problem to rectangular matrices. Commun. ACM **14**, 802–804 (1971)
5. Burkard, R., Dell'Amico, M., Martello, S.: Assignment Problems. SIAM, Philadelphia (2009)
6. Gaüzère, B., Bougleux, S., Riesen, K., Brun, L.: Approximate graph edit distance guided by bipartite matching of bags of walks. In: Fränti, P., Brown, G., Loog, M., Escolano, F., Pelillo, M. (eds.) S+SSPR 2014. LNCS, vol. 8621, pp. 73–82. Springer, Heidelberg (2014). doi:10.1007/978-3-662-44415-3_8
7. Jones, W., Chawdhary, A., King, A.: Revisiting Volgenant-Jonker for approximating graph edit distance. In: Liu, C.-L., Luo, B., Kropatsch, W.G., Cheng, J. (eds.) GbRPR 2015. LNCS, vol. 9069, pp. 98–107. Springer, Cham (2015). doi:10.1007/978-3-319-18224-7_10
8. Jonker, R., Volgenant, A.: Improving the Hungarian assignment algorithm. Oper. Res. Lett. **5**, 171–175 (1986)
9. Lawler, E.: Combinatorial Optimization: Networks and Matroids. Holt, Rinehart and Winston, New York (1976)
10. Leordeanu, M., Hebert, M., Sukthankar, R.: An integer projected fixed point method for graph matching and map inference. In: Advances in Neural Information Processing Systems, vol. 22, pp. 1114–1122 (2009)
11. Neuhaus, M., Bunke, H.: A quadratic programming approach to the graph edit distance problem. In: Escolano, F., Vento, M. (eds.) GbRPR 2007. LNCS, vol. 4538, pp. 92–102. Springer, Heidelberg (2007). doi:10.1007/978-3-540-72903-7_9
12. Riesen, K., Bunke, H.: Approximate graph edit distance computation by means of bipartite graph matching. Image Vis. Comput. **27**, 950–959 (2009)
13. Serratosa, F.: Fast computation of bipartite graph matching. Pattern Recognit. Lett. **45**, 244–250 (2014)
14. Serratosa, F.: Speeding up fast bipartite graph matching through a new cost matrix. Int. J. Pattern Recognit. **29**(2), 1550010 (2015)

Introducing VF3: A New Algorithm for Subgraph Isomorphism

Vincenzo Carletti$^{(\boxtimes)}$, Pasquale Foggia, Alessia Saggese, and Mario Vento

Department of Information Engineering, Electrical Engineering and Applied
Mathematics, University of Salerno, Salerno, Italy
{vcarletti,pfoggia,asaggese,mvento}@unisa.it
http://mivia.unisa.it

Abstract. Several graph-based applications require to detect and locate occurrences of a *pattern graph* within a larger *target graph*. Subgraph isomorphism is a widely adopted formalization of this problem. While subgraph isomorphism is NP-Complete in the general case, there are algorithms that can solve it in a reasonable time on the average graphs that are encountered in specific real-world applications. In 2015 we introduced one such algorithm, VF2Plus, that was specifically designed for the large graphs encountered in bioinformatics applications. VF2Plus was an evolution of VF2, which had been considered for many years one of the fastest available algorithms. In turn, VF2Plus proved to be significantly faster than its predecessor, and among the fastest algorithms on bioinformatics graphs. In this paper we propose a further evolution, named VF3, that adds new improvements specifically targeted at enhancing the performance on graphs that are at the same time large and dense, that are currently the most problematic case for the state-of-the-art algorithms. The effectiveness of VF3 has been experimentally validated using several publicly available datasets, showing a significant speedup with respect to its predecessor and to the other most advanced state-of-the-art algorithms.

1 Introduction

A graph-based representation is commonly used in several application fields dealing with structured data, i.e. data that can be decomposed into atomic entities and relationships between entities (described using the nodes and the edges of the graph). In the last few years, in several disciplines the trend has been to use larger and larger graph structures, thanks to the increase in the available memory and computational power. Examples are the bioinformatics and chemoinformatics disciplines [2,4,5,11,12], with the obvious application to the structure of molecules or proteins, but also to less obvious information such as the protein or gene interaction networks; Social Network Analysis [19], where graphs are used to model the interactions and relations between people or organizations, in very large social networks like Facebook; semantic technologies, where huge knowledge bases (like DBPedia [13]) are encoded using the Resource Description Framework (RDF), a standardized graph-based representation.

© Springer International Publishing AG 2017
P. Foggia et al. (Eds.): GbRPR 2017, LNCS 10310, pp. 128–139, 2017.
DOI: 10.1007/978-3-319-58961-9_12

A common problem in many graph-based applications is the search for the occurrences of a *pattern graph* within a larger *target graph*. This problem can be formalized as the search for all the *subgraph isomorphisms* between the two graphs [7]. In the general case (i.e. if no restrictive assumptions are made on the graphs) the subgraph isomorphism problem is provably NP-complete. However, many algorithms have been proposed over the years that are fast enough to be practical at least on the actual graphs commonly encountered in some applications [7,9,18]. These algorithms typically use some kind of heuristics, that take advantage of knowledge about the structure of the graphs in the common cases of the addressed applications, although they maintain a worst-case complexity that is exponential. Thus, as new and more complex applications emerge, new algorithms with more suitable heuristics are required to cope efficiently with the new cases at hand.

Most of the recently proposed algorithms follow three different approaches: Tree Search, Constraint Propagation and Graph Indexing. Algorithms based on Tree Search formulate the problem as the exploration of a *search space* (having a tree structure), composed of *states* that represent partial solutions. The search space is visited usually with a depth-first order, using heuristics to avoid exploring useless parts of the space. Two very popular algorithms based on this approach are Ullmann's algorithm [16] and VF2 [8]; this latter emerged in several benchmarks as the fastest algorithm at the time of its introduction. Other more recent algorithms in this family are RI/RI-DS [3] and VF2Plus [6], that were expressly designed to be efficient on large bioinformatics graphs.

Algorithms based on Constraint Propagation view the search for subgraph isomorphisms as a Constraint Satisfaction Problem, where the goal is to find an assignment of values to a set of variables that satisfies a set of mutual constraints. In particular, for each node of the pattern a domain of compatibility is mantained, containing the potential matching nodes in target. Local constraints (e.g. node or edge consistency) are propagated to different parts of the graphs reducing the domains, until only few candidate matchings remain, that can be explored to find the solutions. An early algorithm following this approach is McGregor's [14]; more recent proposals are by Zampelli et al. [20], Solnon et al. [15] and Ullmann [17].

The last approach, Graph Indexing, originates from graph database applications, where the goal is to retrieve, from a large set of graphs, only the ones containing the desired pattern. To this aim, an index structure is built that makes possible to quickly verify if the pattern is present or not in a target graph, usually without even requiring to load the whole target in memory, thus filtering out unfruitful targets. In general, after the index verification is passed, a more costly refinement phase is needed to actually determine if and where the pattern graph is present. GADDI [21] and TurboISO [10] are recent algorithms based on this approach.

In this paper we present a novel subgraph isomorphism algorithm called VF3. VF3 can be considered an evolution of VF2Plus [6], introduced in 2015 specifically for addressing the very large graphs that occur in several bioinformatics applications. Like its predecessor, VF3 is based on the Tree Search approach, and uses several heuristics to prune the search space. While VF2Plus is particularly effective on graphs that are large but sparse, the improvements introduced in VF3 significantly increase the performance when the graphs become more dense, without compromising the performance on sparse graphs. Thus, the new algorithm has a much broader field of applicability; in particular, it becomes the fastest algorithm on a class of graphs (simultaneously large and dense) that present serious problems for the other state-of-the-art algorithms. The effectiveness of the new algorithm has been verified experimentally with a thorough testing in comparison with VF2Plus and with other state of the art algorithms, using different publicly available databases.

2   The Base of VF3: The VF2Plus Algorithm

In this section we provide a brief introduction to VF2Plus [6], upon which VF3 is based, while next section will be devoted to the novel parts introduced in VF3.

2.1   Graph Matching and State Space Representation

Given two graphs $G_1 = (V_1, E_1)$ and $G_2 = (V_2, E_2)$, graph matching is the problem of finding a mapping function $M : V_1 \rightarrow V_2$ that satisfies a given set of structural constraints. In the case of the subgraph isomorphism, as detailed in [7,9,18], the function M must be injective and *structure preserving*, i.e. it must preserve both the *presence* and the *absence* of the edges between corresponding pairs of nodes.

The problem of finding a matching between two graphs is addressed by VF3, similarly to its predecessors, using a *State Space Representation* (SSR). Each *state* s of the SSR represents a partial mapping $\widetilde{M}(s) \subseteq M$ that is consistent with the matching constraints; a *goal state* is a state whose mapping is complete, i.e. when covers all the nodes in G_1. For each state s the algorithm keeps different sets: the *core sets* $\widetilde{M}_1(s)$ and $\widetilde{M}_2(s)$ containing the nodes of G_1 and G_2 that belongs to $\widetilde{M}(s)$, and two *feasibility sets* $\widetilde{T}_1(s)$ and $\widetilde{T}_2(s)$ containing the nodes of G_1 and G_2 connected to those in $\widetilde{M}_1(s)$ and $\widetilde{M}_2(s)$. Furthermore, we will denote as $\widetilde{V}_1(s)$ and $\widetilde{V}_2(s)$ the sets $V_1 - \widetilde{M}_1(s) - \widetilde{T}_1(s)$ and $V_2 - \widetilde{M}_2(s) - \widetilde{T}_2(s)$ respectively.

The algorithm starts from a state whose mapping is empty and explores the state space, using a depth-first strategy, till a goal state is reached. State transitions consist in adding a new pair of nodes (u_n, v_n) to the partial mapping of the current state s_c so as to generate a new state $s_n = s_c \cup (u_n, v_n)$, that becomes the new current state. The algorithm uses a set of rules, called *feasibility rules*, to check if the addition will generate a consistent state; if this is not the case, the new state is not explored. When no node pair remains that can be added to s_c for making a consistent state, the algorithm backtracks, i.e. it undoes the addition leading to s_c and restarts from its parent state s_p, looking for a different node pair to be added to it.

2.2   Making the SSR a Tree

The SSR is, by its nature, a graph. Indeed, if we consider a state s_c whose mapping $\widetilde{M}(s_c)$ contains k pairs, it can be generated using $k!$ different paths, involving the same pairs added in different orders. To avoid visiting the same state several times, either a memory-consuming data structure is needed to keep memory of the visited states, or the state space must be reduced to a tree, ensuring that each state is reachable from only one path. Since the order of the couples is not important to have a consistent mapping, the algorithm introduces a total order relationship over the nodes of G_1, and adds the couples so that their first component follows this order, making the state space a tree.

The algorithm defines the order relationship by computing the *node exploration sequence* N_{G_1}, that is a permutation of V_1. The order relationship is based on the idea to explore first the nodes that are more rare and constrained. As for the rareness of a node u, it is defined in terms of probability to find a node in G_2 that is suitable to generate a feasible couple. Such a probability $P_f(u)$ is obtained by combining $P_l(l)$ and $P_d(d)$ that are, respectively, the probability to find a node with label l and the probability to find a node with degree d in G_2. In the case of the subgraph isomorphism $P_f(u)$ is computed ad follows:

$$P_f(u) = P_l(\lambda_{V_1}(u)) \cdot \sum_{d' \geq d(u)} P_d(d') \tag{1}$$

where $\lambda_{V_1}(u)$ and $\lambda_{V_2}(v)$ are the labeling functions associating a label to each node. Note that the two probabilities are considered as independent instead of joint. This is a weak assumptions in some cases, but it reduces the worst-case complexity of the probability estimation from $O(N^3)$ to $O(N)$. As regards the constraints of a node u, they are computed by considering only its connections with the nodes already in the sequence N_{G_1}. To this aim, we defined the *node mapping degree* $d_M(u)$ as the number of edges connecting u to all the nodes that are already inside N_{G_1}.

Therefore, the procedure that computes N_{G_1} first uses, as the ordering criterion, d_M; if two or more nodes have the same d_M, they are sorted according to P_f; finally, if both d_M and P_f are equal, the nodes are sorted using their degree. If also the latter are equal, the choice is done randomly.

2.3   Checking for Feasibility

As introduced in Sect. 2.1 an important issue to reduce the search space is the exploration of only consistent states, i.e. states satisfying the constraints of the subgraph isomorphism problem. In addition, a further reduction is obtained by avoiding also consistent states that surely will not be part of a solution. Therefore, before generating a new state s_n from the current state s_c, the algorithm checks the candidate couple (u_n, v_n) using the feasibility rules F_s and F_t, to check the semantic and structural feasibility respectively. F_s analyzes only the labels of the two nodes and, if present, of the edges connecting them.

F_t analyzes the structural constraints given by the neighbors of u_n and v_n. To this aim, the nodes in the two graphs are partitioned into q *equivalence classes* using a classification function $\psi : V_1 \cup V_2 \to C = c_1, \ldots, c_q$; the classification function has the only constraint that it must ensure that if two nodes can be matched in a consistent mapping, they must be in the same class. The easiest way to define a classification function is to use the node labels, but other kind of information can be used if it makes sense for the problem at hand. For each class c_i we define as $\widetilde{T}_1^{c_i}(s)$ the restriction of $\widetilde{T}_1(s)$ to the nodes having this class; we define similarly $\widetilde{T}_2^{c_i}(s)$.

Now we can define the feasibility function F_t:

$$F_t(s_c, u_n, v_n) = F_c(s_c, u_n, v_n) \wedge F_{la1}(s_c, u_n, v_n) \wedge F_{la2}(s_c, u_n, v_n) \tag{2}$$

The rule $F_c(s_c, u_n, v_n)$ is called *core rule* and is responsible to verify the necessary and sufficient condition for the consistency. The latter verify that all the neighbors of u_n and v_n already in the mapping $\widetilde{M}(s_c)$ are mapped each other; more formally:

$$F_c(s_c, u_n, v_n) \Leftrightarrow \forall u' \in adj_1(u_n) \cap \widetilde{M}_1(s_c) \quad \exists v' = \tilde{\mu}(s_c, u') \in adj_2(v_n)$$
$$\wedge \; \forall v' \in adj_2(v_n) \cap \widetilde{M}_2(s_c) \quad \exists u' = \tilde{\mu}^{-1}(s_c, v') \in adj_1(u_n) \tag{3}$$

The other two rules $F_{la1}(s_c, u_n, v_n)$ and $F_{la2}(s_c, u_n, v_n)$, called *1-level and 2-level lookahead rules* respectively, check two additional necessary but not sufficient conditions for the (sub)graph isomorphism so as to guarantee that the new state will part of a solution. In particular, the rule $F_{la1}(s_c, u_n, v_n)$ counts the number of neighbors of u_n and v_n that are in the sets $\widetilde{T}_1^{c_i}(s_c)$ and $\widetilde{T}_2^{c_i}(s_c)$:

$$F_{la1}(s_c, u_n, v_n) \iff F_{la1}^1(s_c, u_n, v_n) \wedge \ldots \wedge F_{la1}^q(s_c, u_n, v_n) \tag{4}$$

where the functions F_{la1}^i, with $i = 1, \ldots, q$, are defined as follows:

$$F_{la1}^i(s_c, u_n, v_n) \iff |adj_1(u_n) \cap \widetilde{T}_1^{c_i}(s_c)| \leq |adj_2(v_n) \cap \widetilde{T}_2^{c_i}(s_c)| \tag{5}$$

The rule $F_{la2}(s_c, u_n, v_n)$ counts the remaining neighbors, i.e. those are neither in $\widetilde{M}(s_c)$ nor in the feasibility sets:

$$F_{la2}(s_c, u_n, v_n) \iff F_{la2}^1(s_c, u_n, v_n) \wedge \ldots \wedge F_{la2}^q(s_c, u_n, v_n) \tag{6}$$

where each F_{la2}^i, with $i = 1, \ldots, q$, is defined as:

$$F_{la2}^i(s_c, u_n, v_n) \iff |adj_1(u_n) \cap \widetilde{V}_1^{c_i}(s_c)| \leq |adj_2(v_n) \cap \widetilde{V}_2^{c_i}(s_c)| \tag{7}$$

It is worth noting that the rules have been shown only for undirected graphs, but they can be easily extended to the case of directed graphs by considering incoming and outgoing edges separately.

3   The VF3 Algorithm

When it was introduced, in [6], VF2Plus brought a great performance improvement to the VF2 algorithm, especially on large graphs. VF3 optimizes and refines some of the novelties introduced by VF2Plus, so as to further improve its performances when the density and the size of the graphs increase. VF3 inherits the structure of VF2Plus and introduces two main novelties: a new procedure to pre-process the pattern graph and a new criterion to select the next candidate couples.

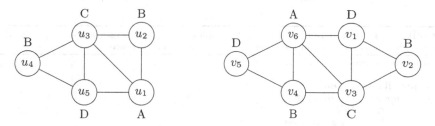

Fig. 1. Graphs, G_1 and G_2, used as an example.

3.1   State Space Precalculation

The exploration sequence N_{G_1}, provided by the sorting procedure, makes the algorithm able to pre-process the graph G_1 and compute, before starting the matching, the sets used to explore it: $\widetilde{M_1}(s)$, $\widetilde{T}_1^{c_i}(s)$, for $i = 1, \ldots, q$. Furthermore, during the pre-preprocessing VF3 computes, together with the feasibility sets, a spanning tree of G_1, hereinafter the *parent tree* (see Fig. 2), that associates a parent to each node of G_1. As explained in more details below, this tree will be used during the matching process to select the next candidate node from G_2.

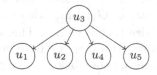

Fig. 2. Parent tree of G_1, for the example of Fig. 1.

The idea behind the pre-processing, is that the sequence N_{G_1} fixes, for each level of the depth-first search, the candidate node of G_1. So that, the algorithm is able to determine the exact composition of each feasibility set of G_1 for all the possible states. For instance, if we consider the exploration sequence $N_{G_1} = \{u_3, u_1, u_5, u_2, u_4\}$, when VF3 is exploring a state s_c that belongs to the *second level* of the SSR, the first node of the candidate couple (u_n, v_n) will always be the one is at the *third position* of N_{G_1}, i.e. u_5. Thus, the mapping $\widetilde{M}$ of all the states belonging to the third level of the SSR contains the couples (u_3, v_i), (u_1, v_j)

Table 1. Core and feasibility sets of G_1 (see Fig. 1), computed for each level of the search.

Level	$\widetilde{M}_1(s)$	$\widetilde{T}_1^{c_1}$	$\widetilde{T}_1^{c_2}$	$\widetilde{T}_1^{c_3}$	$\widetilde{T}_1^{c_4}$
0	$\{\}$	$\{\}$	$\{\}$	$\{\}$	$\{\}$
1	$\{u_3\}$	$\{u_5\}$	$\{u_2, u_4\}$	$\{\}$	$\{u_1\}$
2	$\{u_3, u_1\}$	$\{u_5\}$	$\{u_2, u_4\}$	$\{\}$	$\{\}$
3	$\{u_3, u_1, u_5\}$	$\{\}$	$\{u_2, u_4\}$	$\{\}$	$\{\}$
4	$\{u_3, u_1, u_5, u_2\}$	$\{\}$	$\{u_4\}$	$\{\}$	$\{\}$
5	$\{u_3, u_1, u_5, u_2, u_4\}$	$\{\}$	$\{\}$	$\{\}$	$\{\}$

Algorithm 1. Procedure to preprocess the graph G_1 given the sequence N_{G_1}. The procedure computes the feasibility sets and the parent tree ($Parent(u)$ in the procedure).

```
1:  function PREPROCESSGRAPH(G₁,N_G₁)
2:      i = 0
3:      for all u ∈ N_G₁ do
4:          for all u' ∈ adj₁(u) do
5:              cᵢ = ψ(u')
6:              Put u' in M̃₁ at level i
7:              if u' ∉ T̃₁^cⁱ then
8:                  Put u' in T̃₁^cⁱ at level i
9:                  Parent(u') = u
10:             i = i + 1
11:     return Parent
```

and (u_5, v_k). The order and the composition of these couples always follows the sequence N_{G_1}. The nodes v_i, v_j and v_k, belonging to G_2 are dynamically determined during the candidate selection step. Since $\widetilde{M}_1(s)$ is known, from G_1 and $\widetilde{M}_1(s)$ it is possible to precompute the sets $\widetilde{T}_1^{c_i}(s)$ for each SSR level. The time saved by this precomputation depends on how many states are at each level of the SSR; in general it increases with the density of the graphs. Notice that with a naive encoding of the $\widetilde{T}_1^{c_i}(s)$ sets, they would occupy a space that is $O(N_1^2)$ (where N_1 is the size of G_1), since there are N_1 levels, and at each level the $\widetilde{T}_1^{c_i}(s)$ sets have a size that is $O(N_1)$. This would be a problem when working with very large graphs. However, we have demonstrated that for each node of G_1, the levels at which the node belongs to a given $\widetilde{T}_1^{c_i}$ form a (possibly empty) interval; thus we are able to represent all the $\widetilde{T}_1^{c_i}(s)$ sets with a single table that for each node reports the first and the last level at which it is in $\widetilde{T}_1^{c_i}(s)$, with a space occupation that is just $O(N_1)$. Table 1 shows the result of the pre-preprocessing on the graph G_1 in Fig. 1.

3.2   Candidate Selection

Another relevant difference between VF2Plus and VF3 is the way they select the candidate node from the graph G_2. In the previous section we have clarified that VF3 defines, before the matching begins, the candidates of G_1 for each possible state in the SSR. However, this is not possible for the graph G_2, so VF3 has to select the candidate node for each new state. Thus, being u_n, the candidate node of G_1, the algorithm analyses neighborhood of the node $\widetilde{v}$ mapped to the parent of u_n (hereinafter $Parent(u_n)$) and select the first unmapped node belonging to the same class of u_n. If the node u_n has no parent (eg. it is the first node of the sequence N_{G_1}), VF3 will select u_v from the unmapped nodes of G_2 belonging to the same class of u_n. More formally, when the u_n has not a parent VF3 will consider the set $R_2 \subset V_2$; the latter is composed of the nodes in G_2 that are not in the mapping $\widetilde{M_2}(s_c)$ of the current state s_c and belong to the same class of the node u_n.

$$R_2(s_c, \psi(u_n)) = \{v_n \in V_2 : v_n \notin \widetilde{M_2}(s_c) \wedge \psi(v_n) = \psi(u_n)\}. \tag{8}$$

In the other case, when the node $Parent(u_u)$ exists, the algorithm will consider the subset of R_2^{adj} containing only the neighbors of $\widetilde{v}$.

$$R_2^{adj}(s_c, \psi(u_n), \widetilde{v}) = \{v_n \in V_2 : v_n \in adj_2(\widetilde{v}) \cap R_2(s_c, u_n)\} \tag{9}$$

The candidate selection procedure is shown in details in Algorithm 2. As it will be shown in Sect. 4, this difference has a great impact especially on dense graphs, because it greatly reduces the number of possible candidate nodes.

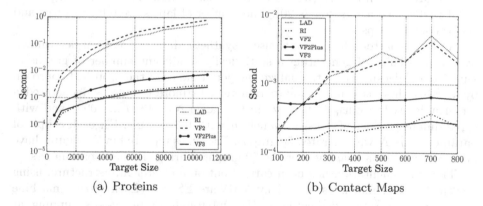

(a) Proteins   (b) Contact Maps

Fig. 3. Matching times for Proteins graphs (a) and Contact Map graphs (b) in the Biological dataset.

4   Experiments

The proposed approach has been tested over two different datasets: a biological dataset (hereinafter *Biological dataset*) and a synthetic dataset, composed by randomly generated graphs (hereinafter *Random dataset*).

As for the former, we have used a real dataset, recently introduced within the *International Contest on Pattern Search in Biological Databases* [4].

Algorithm 2. Procedure to generate the next candidate couple. The inputs are the current state s_c, the last inserted couple (u_c, v_c), the exploration sequence N_{G_1}, the set $Parent$, and the graphs G_1 and G_2. The procedure returns a candidate couple (u_n, v_n) to be checked for the feasibility or a null couple (ϵ, ϵ) if there are no more couples to explore.

```
 1:  function SELECTCANDIDATE(s_c, (u_c, v_c), N_{G_1}, Parent, G_1, G_2)
 2:      if u_c = ε then
 3:          u_n = GETNEXTINSEQUENCE(N_{G_1}, s_c)
 4:          if u_n = ε then                                    ▷ The sequence is finished
 5:              return (ε, ε)
 6:      else
 7:          u_n = u_c
 8:      if Parent(u_n) = ε then                                ▷ u_n has not a parent node
 9:          v_n = GETNEXTNODE(v_c, R_2(s_c, ψ(u_n)))
10:          return (u_n, v_n)
11:      else
12:          ṽ = μ̃(s_c, Parent(u_n))                           ▷ Get the node matched to Parent(u_n)
13:          if u_n in adj_1(Parent(u_n)) then                  ▷ u_n is predecessor of Parent(u_n)
14:              v_n = GETNEXTNODE(v_c, R_2^{adj}(s_c, ψ(u_n), ṽ))
15:              return (u', v')
16:      return (ε, ε)                                         ▷ There is not a pair for u_n
```

The dataset is composed of Contact Map and Protein graphs, extracted from the Protein Data Bank [1]. Protein graphs are very large and sparse: the number of nodes ranges from 500 to 10000 and the average degree is 4; contact maps graphs have a medium size (from 150 to 800 nodes) and are denser than protein graphs (their average degree is 20). The number of labels are 6 for proteins and 20 for contact maps.

As for the latter, the choice to use a synthetic dataset has been determined by the possibility of generating a statistically significant number of graphs of any size and density, so as to analyze the performance of the proposed approach by varying these two important parameters. In more details, we generated a set of unlabeled graphs, with a size in the range $N = \{300, ..., 1000\}$ and with three different values of density, namely $\eta = 0.2, 0.3$ and 0.4. For each couple of parameters $\eta - N$ we generated 50 graphs. The times reported in the figures have been obtained by averaging over the 50 graphs sharing the same parameters.

The experimentation has been carried out on a cluster infrastructure, using identical virtual machines hosted by VMWare ESXi 5. Each virtual machine is provided with two dedicated AMD Opteron 6376 processors running at 2300 MHz, with 2 Mb of cache and 4 Gb of RAM. In order to confirm the effectiveness of the proposed approach, we have compared it with its previous versions, VF2 and VF2Plus, and with two other state of the art algorithms, namely RI and LAD. The results are reported in Figs. 3 and 4 for Biological and Random datasets, respectively.

From Fig. 3, we can note that VF3 outperforms, independently on the size of the target graphs, LAD, VF2 and VF2Plus. The most interesting comparison, however, is between VF3 and RI, the last one being the winner of the Contest on Pattern Search in Biological Databases. We can note that VF3 is particularly suited for challenging graphs, since it overcomes RI around 2000 nodes for proteins and around 500 nodes in case of contact maps.

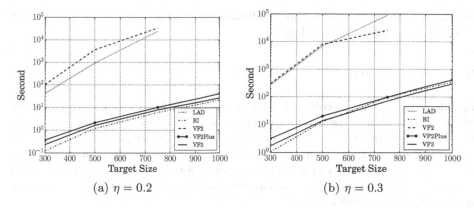

(a) $\eta = 0.2$ (b) $\eta = 0.3$

Fig. 4. Matching times for $\eta = 0.2$ (a), $\eta = 0.3$ (b) for the graphs in the Random dataset.

This consideration is confirmed by the results obtained over random graphs, reported in Figs. 4 and 5. Indeed, we can note that VF3 improves over VF2 and LAD of around two orders of magnitude. In practice, this is a very interesting result, since it means to be able to solve, for instance, a graph with 500 nodes and having $\eta = 0.2$ in around two seconds, with respect to more than 1000 s required by VF2 and LAD over the same graph. Furthermore, VF3 improves also VF2Plus with all the η values, confirming the effectiveness of the novelties introduced in the proposed approach.

The improvement with respect to RI becomes more and more evident by increasing the size of the graphs and the η value. Indeed, the overtaking of VF3 with respect to RI is around 550 nodes for graphs having $\eta = 0.3$ and

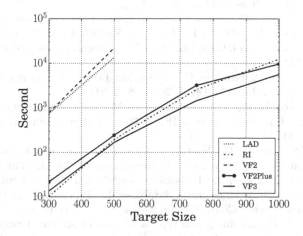

Fig. 5. Matching times for $\eta = 0.4$ (c) for the graphs in the Random dataset.

around $\eta = 450$ for $\eta = 0.4$, thus confirming that VF3 is particularly suited for challenging graphs, big and dense.

5 Conclusions

In this paper we have presented VF3, a new algorithm for (sub)graph isomorphism. VF3 extends the previously introduced VF2Plus algorithm, improving its ability to deal with larger and denser graphs. An experimental evaluation on different datasets show a consistent performance improvement, that increases as the graphs become larger or denser. The new algorithm has also been compared with two other state-of-the-art algorithm, and has shown to be the fastest one in almost all the conditions.

References

1. RCSB: Protein data bank web site (2017). http://www.rcsb.org/pdb
2. Aittokallio, T., Schwikowski, B.: Graph-based methods for analysing networks in cell biology. Brief. Bioinform. **7**(3), 243 (2006). http://dx.doi.org/10.1093/bib/bbl022
3. Bonnici, V., Giugno, R.: On the variable ordering in subgraph isomorphism algorithms. IEEE/ACM Trans. Comput. Biol. Bioinform. PP(99) (2016)
4. Carletti, V., Foggia, P., Vento, M., Jiang, X.: Report on the first contest on graph matching algorithms for pattern search in biological databases. In: Liu, C.-L., Luo, B., Kropatsch, W.G., Cheng, J. (eds.) GbRPR 2015. LNCS, vol. 9069, pp. 178–187. Springer, Cham (2015). doi:10.1007/978-3-319-18224-7_18
5. Carletti, V., Foggia, P., Vento, M.: Performance comparison of five exact graph matching algorithms on biological databases. In: Petrosino, A., Maddalena, L., Pala, P. (eds.) ICIAP 2013. LNCS, vol. 8158, pp. 409–417. Springer, Heidelberg (2013). doi:10.1007/978-3-642-41190-8_44
6. Carletti, V., Foggia, P., Vento, M.: VF2 plus: an improved version of VF2 for biological graphs. In: Liu, C.-L., Luo, B., Kropatsch, W.G., Cheng, J. (eds.) GbRPR 2015. LNCS, vol. 9069, pp. 168–177. Springer, Cham (2015). doi:10.1007/978-3-319-18224-7_17
7. Conte, D., Foggia, P., Sansone, C., Vento, M.: Thirty years of graph matching in pattern recognition. IJPRAI **18**(3), 265–298 (2004)
8. Cordella, L., Foggia, P., Sansone, C., Vento, M.: A (sub)graph isomorphism algorithm for matching large graphs. IEEE Trans. Pattern Anal. Mach. Intell. **26**, 1367–1372 (2004)
9. Foggia, P., Percannella, G., Vento, M.: Graph matching and learning in pattern recognition on the last ten years. J. Pattern Recognit. **28**(1), 1450001 (2014)
10. Han, W., Lee, J.h., Lee, J.: TurboISO: towards ultrafast and robust subgraph isomorphism search in large graph databases. In: Proceedings of the 2013 ACM SIGMOD International Conference on Management of Data, pp. 337–348 (2013)
11. Huan, J., et al.: Comparing graph representations of protein structure for mining family-specific residue-based packing motif. J. Comput. Biol. **12**(6), 657–671 (2005)
12. Lacroix, V., Fernandez, C., Sagot, M.: Motif search in graphs: application to metabolic networks. Trans. Computat. Biol. Bioinform. **4**, 360–368 (2006)

13. Lehmann, J., Isele, R., Jakob, M., Jentzsch, A., Kontokostas, D., Mendes, P.N., Hellmann, S., Morsey, M., van Kleef, P., Auer, S., Bizer, C.: DBpedia - a large-scale, multilingual knowledge base extracted from Wikipedia. Semant. Web J. $6(2)$, 167–195 (2015)
14. McGregor, J.: Relational consistency algorithms and their application in finding subgraph and graph isomorphisms. Inf. Sci. $19(3)$, 229–250 (1979)
15. Solnon, C.: Alldifferent-based filtering for subgraph isomorphism. Artif. Intell. $174(12–13)$, 850–864 (2010)
16. Ullmann, J.R.: An algorithm for subgraph isomorphism. J. Assoc. Comput. Mach. 23, 31–42 (1976)
17. Ullmann, J.: Bit-vector algorithms for binary constraint satisfaction and subgraph isomorphism. J. Exp. Algorithm. (JEA) $15(1)$ (2010)
18. Vento, M.: A long trip in the charming world of graphs for pattern recognition. Pattern Recognit. $48(1)$, 11 (2014)
19. Wasserman, S., Faust, K.: Social Network Analysis: Methods and Applications, vol. 8. Cambridge University Press, Cambridge (1994)
20. Zampelli, S., Deville, Y., Solnon, C.: Solving subgraph isomorphism problems with constraint programming. Constraints $15(3)$, 327–353 (2010)
21. Zhang, S., Li, S., Yang, J.: GADDI: Distance Index Based Subgraph Matching In Biological Networks. In: Proceedings of the 12th International Conference on Extending Database Technology: Advances in Database Technology (2009)

Large Graphs and Social Networks

Node Matching Computation Between Two Large Graphs in Linear Computational Cost

Pep Santacruz, Shaima Algabli, and Francesc Serratosa$^{(\boxtimes)}$

Universitat Rovira i Virgili, Catalonia, Spain
{joseluis.santacruz, francesc.serratosa}@urv.cat,
shaima.ahmed@estudiants.urv.cat

Abstract. Error-tolerant graph matching has been demonstrated to be an NP-problem, for this reason, several sub-optimal algorithms have been presented with the aim of making the runtime acceptable in some applications. Some well-known sub-optimal algorithms have 6th, cubic or quadratic cost with respect to the order of the graphs. When applications deal with large graphs (social nets), these costs are not acceptable. For this reason, we present an error-tolerant graph-matching algorithm that it is linear with respect to the order of the graphs. Our method needs an initial seed, which is composed of one or several node-to-node mappings. The algorithm has been applied to analyse the friendship variability of social nets.

Keywords: Large graph matching · Graph edit distance · Bipartite graph matching

1 Introduction

Recently, we have seen an increase in the number of people enrolled in the social nets and also the number of different social networks. In some applications, for instance, personalised publicity, it would be interesting to locate people from one net at the other net, with the aim of increasing the knowledge we have from these people. It is worth noting that in some cases, we know the node in each net that represents the same person, due to we have this knowledge from other sources of information. Nevertheless, this is not the most common case because of several people could have the same name in the net or people suggest different alias in each net.

Attributed graphs are good models to represent social nets, thus, if we want to correlate two nets, what we have to do is to find a matching between nodes of the graphs that represent these nets. The methods that return a distance between two graphs and a matching between their nodes are called error-tolerant graph matching [1].

Error-tolerant graph matching has been demonstrated to be an NP-problem [2], for this reason, several algorithms have been presented that apply some heuristics with the aim of reducing the computational cost [3–5]. Nevertheless, sub-optimal algorithms

This research is supported by the spanish projects TIN2016-77836-C2-1-R and ColRobTransp MINECO DPI2016-78957-R AEI/FEDER EU.

© Springer International Publishing AG 2017
P. Foggia et al. (Eds.): GbRPR 2017, LNCS 10310, pp. 143–153, 2017.
DOI: 10.1007/978-3-319-58961-9_13

have been presented that deduce a distance and a matching between nodes in poly-nomial time. For instance, the Graduated assignment [6], the Bipartite graph matching [7, 8] or the Greedy edit distance algorithm [9, 10]. All of these algorithms define a bi-dimensional matrix in which the number of rows or columns is the graph order.

The aim of this paper is to present an error-tolerant graph-matching algorithm designed to match huge graphs. Thus, we have imposed two main restrictions. Firstly, any bi-dimensional matrix cannot be defined in which the number of rows or columns is the graph order (or, a vector with a quadratic length with respect to the order of the graphs). Secondly, the computational cost has to be linear with respect to the order of the graphs. Moreover, we assume that some (few) initial mappings between nodes of both nets are given. We call these initial mappings as seeds since, as we will see in the following sections, they are the seeds from which the algorithm begins to spread its knowledge of the partial matching.

The rest of paper is going to be as follows. In the next section, we introduce the attributed graphs, the graph-edit distance and the graph-matching algorithms that computethe graph edit distance. In Sect. 3, we move on explaining our graph-matching algorithm that we have called Belief propagation graph matching. In Sect. 4, we show the experimental section. It is composed of two parts. In the first one, we have ran-domly generated some graphs and we compare our method to the state of the art methods. In the second part, we show how we have used our method to map people of two social nets. In Sect. 5, we conclude the paper.

2   Attributed Graphs and Graph Matching

Let $G = (\Sigma_v, \Sigma_e, \gamma_v, \gamma_e)$ and $G' = (\Sigma'_v, \Sigma'_e, \gamma'_v, \gamma'_e)$ be two attributed graphs. $\Sigma_v = \{v_i | i = 1, \ldots, n\}$ is the set of vertices and $\Sigma_e = \{e_{i,j} | i, j \in 1, \ldots, n\}$ is the set of edges. Functions $\gamma_v : \Sigma_v \to \Delta_v$ and $\gamma_e : \Sigma_e \to \Delta_e$ assign attribute values in any domain to vertices and edges. Coherent definitions hold for $G' = (\Sigma'_v, \Sigma'_e, \gamma'_v, \gamma'_e)$.

A local structure of a node is the set of edges and nodes of the graph adjacent to it. The influence on selecting different local structures was analysed in [10, 11]. The most common local structures are the *Node*, the *Degree* and the *Star* (also called *Clique* in some papers). In the *Node*, the local sub-structure is composed of only a node and any edges or other nodes are not considered. In the *Degree*, the local sub-structure is composed of a node and its connecting edges. Finally, in the *Star*, the local sub-structure is composed of a node, its connecting edges and the nodes that these edges connect. These structures are defined as attributed graphs with their specific node and edge structure. Larger structures are not used due to its matching computational cost.

There are some applications in which comparing graphs is needed. For instance, in the classification procedures such that elements are represented by attributed graphs. In these cases, a distance between attributed graphs has to be applied. One of the most widely used methods to deduce a distance between graph and to extract a "logical" matching between them is the Graph edit distance [1]. In the Graph edit distance; the distance is defined as the minimum amount of required distortion to transform one graph into the other. To this end, a number of distortion or edit operations, consisting of insertion, deletion or substitution of nodes and edges are defined. Edit cost functions are introduced to quantitatively evaluate the edit operations. The basic idea is to assign a penalty cost to each edit operation according to the amount of distortion that it introduces in the transformation. To allow the maximum flexibility in the matching process, both graphs are theoretically extended with null nodes and edges to have the same order n. The null nodes and edges are assigned at the set $\hat{\Sigma}_v$ and $\hat{\Sigma}_e$ for graph G and $\hat{\Sigma}'_v$ and $\hat{\Sigma}'_e$ for graph G'. Thus, deletion and insertion operations are transformed to assignations of a non-null node of the first or second graph to a null node of the second or first graph. Substitutions simply indicate node-to-node assignments. Using this transformation, given two graphs G and G' and a bijective matching between their nodes f, the graph edit cost, $EditCost(G, G', f)$, is computed. It is based on the following constants and functions: C_{vs} is a function that represents the cost of substituting node v_i of G by node $f(v_i)$ of G'. C_{es} is a function that represents the cost of substituting edge $e_{i,k}$ of G by edge $f(e_{i,k})$ of G'. C_{vd} and C_{vi} are the costs of deleting node v_i of G (mapping it to a null node) or inserting node v'_j of G' (or being mapped from a null node). Likewise, C_{ed} and C_{ei} are the costs of assigning edge $e_{i,k}$ of G to a null edge of G' or assigning edge $e'_{j,p}$ of G' to a null edge of G. Note that we have not considered the cases in which two null nodes or null edges are mapped; this is because this cost is zero by definition. The expression $EditCost$ is formally described as follows,

$$
\begin{aligned}
EditCost\left(G, G', f\right) = &\sum_{\substack{v_a \in \Sigma_v - \hat{\Sigma}_v \\ v'_i \in \Sigma'_v - \hat{\Sigma}'_v}} C_{vs}\left(v_a, v'_i\right) + \sum_{\substack{e_{ab} \in \Sigma_e - \hat{\Sigma}_e \\ e'_{ij} \in \hat{\Sigma}'_e - \hat{\Sigma}'_e}} C_{es}\left(e_{ab}, e'_{ij}\right) \\
+ &\sum_{\substack{v_a \in \hat{\Sigma}_v \\ v'_i \in \Sigma'_v - \hat{\Sigma}'_v}} C_{vd}\left(v_a, v'_i\right) + \sum_{\substack{e_{ab} \in \hat{\Sigma}_e \\ e'_{ij} \in \Sigma'_e - \hat{\Sigma}'_e}} C_{ed}\left(e_{ab}, e'_{ij}\right) \\
+ &\sum_{\substack{v_a \in \Sigma_v - \hat{\Sigma}_v \\ v'_i \in \hat{\Sigma}'_v}} C_{vi}\left(v_a, v'_i\right) + \sum_{\substack{e_{ab} \in \Sigma_e - \hat{\Sigma}_e \\ e'_{ij} \in \hat{\Sigma}'_e}} C_{ei}\left(e_{ab}, e'_{ij}\right)
\end{aligned}
$$

$$(1)$$

Where $f(v_a) = v'_i$ and $f(v_b) = v'_j$.

The Graph edit distance $EditDist$ is defined as the minimum cost under any bijection in T:

$$EditDist(G, G') = \min_{f \in T}\{EditCost(G, G', f)\} \qquad (2)$$

Computing the Graph edit distance and the optimal matching is an NP-problem [2]. For this reason, several optimal algorithms have been defined to compute it and deduce the matching that obtains the minimum cost applying different search strategies [14]. Nevertheless, due to runtime reasons, applications use to apply suboptimal algorithms that search for a suboptimal distance and a matching in polynomial time. One of the classical ones is the Graduated assignment [6] that has a $O(n^6)$ computational cost. Nowadays, one of the most used algorithms is the Bipartite graph matching [7, 8] that has a $O(n^3)$ computational cost. Finally, it is worth to mention the Greedy edit distance algorithm [9, 10] that returns a distance in $O(n^2)$ computational cost.

The whole mentioned algorithms have a first step in which a matrix is filled with the distances between all combinations of the local structures of both graphs (*Node*, *Degree* or *Star*). The computational cost of this step is approximated by $O((s \cdot n)^2)$ where s is the computational cost of computing the distance between local structures. Then, there is a second step in which the bijective matching is obtained. The computational cost of this second step is $O(n^6)$, $O(n^3)$ or $O(n^2)$ depending whether the matching is deduced by the Graduated assignment [6] Bipartite graph matching [7, 8] or Greedy edit distance [9, 10], respectively. In the case of the Bipartite graph matching, the problem at hand is seen as a minimisation of the sum of linear assignation given the cost matrix and it is usually solved through the Munkers algorithm [12] or the Jonker-Volgenant algorithm [13].

When applications handle huge graphs (more than 100.000 nodes) the matching process becomes an important handicap not only from the runtime point of view but also from the storage space. The graph-matching algorithm that we present in this paper is designed to match huge graphs. Thus, we have imposed two main restrictions. Firstly, any bi-dimensional matrix cannot be defined in which the number of rows or columns is the graph order. Secondly, the computational cost has to be linear with respect to the order of the graphs. These restrictions suppose that the algorithm does not perform the two previously mentioned steps and the distance between the whole combinations of the local structures of both graphs is never completely performed.

3 Belief Propagation Graph Matching

Algorithm 1 shows the pseudo-code of our error-tolerant graph-matching algorithm that we have called Belief Propagation Graph Matching. The input of the algorithm is composed of a pair of graphs, the edit cost functions and some initial mappings between nodes of both graphs, which we call seeds. The output is the deduced matching between both graphs.

The core of the algorithm is the distance between *Stars*. In any moment of the iterative process, the algorithm keeps a set of mappings between *Stars*. Thus, in each iteration, the algorithm considers the mapping in the set that its Star distance is the minimum one as a correctand definitive mapping between two nodes. Then, the algorithm computes the entire*Star* distances between the mapped neighbour nodes and introduces them into theset.

The algorithm uses the following four sets:

Seeds: Set of initial mappings between nodes of both graphs that are supposed to be ground-truth mappings. Each initial mappings is represented by $[Seed, Seed']$ being $Seed \in \Sigma_v$ and $Seed' \in \Sigma'_v$.

Matching: The output of the program. Each element is a mapping between a node of each graph $[v, v']$, $v \in \Sigma_v$ and $v' \in \Sigma$ and it represents a bijective function. During the execution of the algorithm, this set always increases since any pair of nodes is never deleted from itand it represents the current partial matching.

Pending: A Setofregisters composed of three elements: A pair of mapped nodes $[v, v']$, $v \in \Sigma_v$ and $v' \in \Sigma'_v$ and also the Graph edit distance D and the matching f between the Stars that they are the central nodes. This distance and matching is computed through function $(D, f) = Match\_Star([S, S'])$, where S and S' are the Stars of v and v', respectively. In each iteration of the algorithm, a mapping with the minimum distance is extracted and erased from *Pending*. The algorithm finishes when *Pending* is empty, which means that the algorithm has explored all the mappings that it believes they might be correct.

Computed: A Set of pairs of nodes such that $Match\_Star([S, S'])$ has been computed. It is necessary in order not to compute this function several times with the same pair of nodes. Note that this set always increases since any pair of nodes is never deleted from it. The algorithm is composed of two main parts. In the first one (lines 1–8), the imposed mappings are introduced into *Pending*. In the second one (the rest of lines), the algorithm iteratively extracts the mappings $[Map, Map']$ from *Pending* that have the minimum *Star* distance (line 10). These mappings are always considered part of the final matching (line 11). Any mapping in *Pending* that have one of the selected nodes in line 10 are deleted from Pending to force the matching to be bijective (line 12). Symbol $\sim$ means any value. Line 13 selects each mapping $[N\_Map, N\_Map']$ of the matching f from the current mapping obtained in line 10. The aim of the loop in lines 13–23 is to compute the distance between *Stars* of the mapped neighbourhood nodes $[N\_Map, N\_Map']$ and insert them into *Computed* and *Pending*. Note that this action is performed only if they have not been previously computed (line 14) and if the involved nodes of both graphs do not form part of the partial current matching (line 15). This is to assure we obtain a bijective mapping. For this reason, for sure, the maximum of node-to-node comparison is n and this makes the algorithm to be linear with respect to the number of nodes, n.

Algorithm 1. Belief Graph Matching (G, G', Seeds)
1. Initialize Pending, Matching and Computed to the empty set
2. For each register in Seeds: $[Seed, Seed']$
3.   S=Star(G, Seed)
4.   S'=Star(G', Seed')
5.   (D, f) = Match_Star (S, S')
6.   Insert $[Seed, Seed']$ into Computed
7.   Insert $\{[Seed, Seed'], D, f\}$ into Pending
8. End for
9. While Pending not empty
10. $\{[Map, Map'], D, f\}$ = Min_Distance {Pending}
11.   Insert $[Map, Map']$ into Matching
12.   Delete $\{[Map, \sim], \sim, \sim\}$ and $\{[\sim , Map'], \sim, \sim\}$ from Pending
13.   For each mapping $[N\_Map, N\_Map']$ such that $N\_Map = f(N\_Map')$
14.   If $[N\_Map, N\_Map']$ not in Computed
15.   If not ([N_Map, ~] or [~ , N_Map']) in Matching
16.   S=Star(G, N_Map)
17.   S'=Star(G', N_Map')
18.   (D, f) = Match_Star (S, S')
19.   Insert $[N\_Map, N\_Map']$ into Computed
20.   Insert $\{[N\_Map, N\_Map'], D, f\}$ into Pending
21.   End if
22.   End if
23.   End for
24. End while
25. Return Matching

4   Experimental Validation

In the first part of this section we validate and analyse our algorithm using synthetic graphs whereas in the second part we show a real application of it. Graphs in the first part are small because we want to compare our algorithm to other non-linear algorithms. Nevertheless, in the second part, we only use our new algorithm and so we have used large graphs.

4.1   Validation Using Synthetic Graphs

The aim of this section is to validate our new proposal from the quality of the matching and also from the runtime point of view. As it can be deduced from Eq. 2, the lower the edit cost, the better we consider the matching is. Thus, we have compared our method with the Bipartite graph matching [7, 8] that has a cubic cost and the Greedy graph matching [9, 10] that has a quadratic cost. Remember that our method has a linear computational cost but it needs an initial mapping that we have called Seed. We have not computed the optimal matching through an A* algorithm [14] due to runtime reasons. For this reason, we do not know which is the optimal distance given a pair of graphs. Algorithms are implemented in Matlab and they have been executed in a Windows i7. Experiments are publically available at [16].

The experiments have been set up as follows. First, we have randomly generated an attributed graph with only one attribute on the nodes (a value between 0 and 99), and a degree of approximately $0.2 \cdot n$, being n the order of graphs. From this graph, we have generated another one by first copying it and then deleting and inserting the 10% of nodes and edges and modifying the attribute value of other 10% of nodes. We assume that the optimal matching is the identity and for this reason, when we compute the Belief propagation algorithm, we impose the Seed = [1, 1]. Clearly, it may happen that there is another matching with lower cost due to the noise added to the second graph and so the mapping [1, 1] could not the best option. We consider this fact as part of the noise our algorithm has to deal with. The cost of deleting and inserting nodes and edges has been 25 (it is a ¼ of the maximum value [15]).

Figure 1 shows the average distance of 100 runs. In general, the Belief algorithm (with only one Seed) performs worse than the Greedy algorithm and the Bipartitegraph matching algorithm. Moreover, larger the graphsare, large the gap between these algorithms is.

Figure 2 shows the runtime of the three algorithms. As it is supposed to be, the fastest algorithm is the Belief propagation graph matching followed by the Greedy and

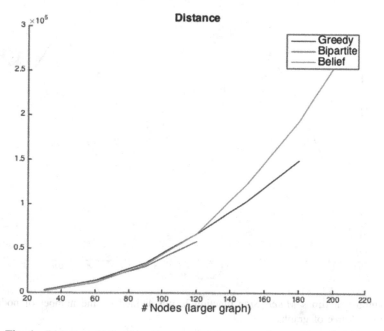

Fig. 1. Distance obtained by three error-tolerant graph-matching algorithms.

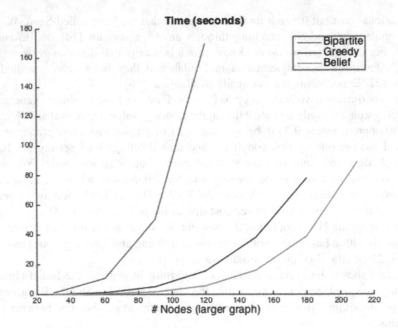

Fig. 2. Runtime in seconds spent by three error-tolerant graph-matching algorithms.

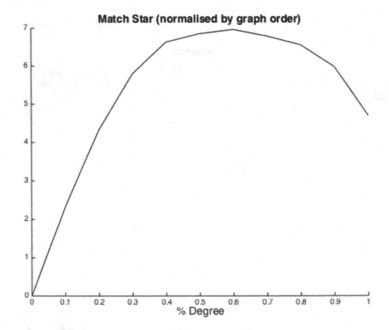

Fig. 3. Number of times *Match_Star* function is run divided by the number of nodes with respect to the degree of graphs.

the slowest one is the Bipartite algorithm. From the plots, we realise the computational cost of the three algorithms.

Having compared our algorithm with two other ones, we proceed now to analyse the behaviour of our proposal. The knowledge of the current and partial matching is spread through mapping the Stars. For this reason, the degree of the graphs is an important parameter to be considered in the computational cost. Note that given a pair of mapped nodes, the number of times the *Match_Star* function is executed is lower or equal than the amount of neighbours it has (line 13 in the algorithm). Figure 3 shows the number of times that function *Match_Star* has been executed (normalised by the order of the graphs) with respect to the degree of the graphs. We realise that the maximum value is achieved when the degree is 0.5. Moreover, although not shown in the paper, we performed the same tests considering different levels of noise and orders of graphs. In the whole tests, the function showed a similar behaviour although different maximum values. The shown test is the average of 100 runs. The order of the initial graph was 120 nodes. From this graph, we generated another one by copying it and then deleting 30 nodes and modifying the attribute of 30 other ones. Moreover the 20% of the edges were deleted and inserted.

Figure 4 shows the number of times *Match_Star* was executed (normalised by the graph order) and considering three levels of noise on the graphs. We realise that with low noise on the graphs, the plot is almost constant, and so, linear with respect to the number of nodes. With high ratios of noise, the plot increases when the order of graphs is low but it rapidly tends to stabilise.

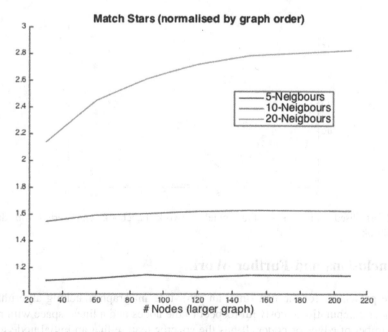

Fig. 4. Number of times *Match_Star* function is run divided by the number of nodes with respect to the degree of graphs.

4.2 Real Application

The final experiments have been conducted with a real graph database called *ego-Facebook* [17]. In this database, nodes represent people and edges arefriendships. The order of the graph is 4039 and the number of edges is 88234. Facebook data has been anonymised by replacing the Facebook-internal ids for each user with a new value. Also, while feature vectors from this dataset have been provided, the interpretation of those features has been obscured. For instance, where the original dataset may have contained a feature "political = Democratic Party", the new data would simply contain "political = anonymised feature 1". Thus, using the anonymised data it is possible to determine whether two users have the same political affiliations, but not what their individual political affiliations represent.

The aim of our application is to know the variability of the friendships. That is, we want to know which percentage of friendships change each time the social net is sampled. Due to the anonimisation, we do not know the nodemapping in each net sample. Thus, the matching between social nets (represented as attributed graphs) is deduced through our algorithm before computing the friendship variation. Thanks to the process of creation of the graphs that represent the social net samples, we know the mapping of the first node of each graph. Figure 5 shows the normalised histogram of the friendship variation, which has been computed as the difference between the degree of each node. We conclude that almost half of the population have a variation lower than 2 friends, which is considered very low.

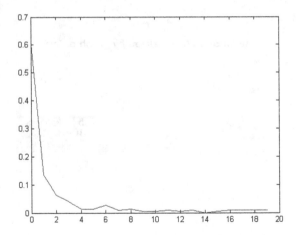

Fig. 5. Normalised variation of the social network Facebook considering the database ego-Facebook.

5 Conclusions and Further Work

We have presented, for the first time, an error-tolerant graph-matching algorithm that has a linear computational costwith respect to the nodes and a linear space with respect to the number of edges or nodes. It has the specific feature that an initial node-to-node mapping is needed to begin to spread the knowledge of the node-to-node matching.

Experimental validation shows that the algorithm is clearly faster than two of the most used algorithms although the average distance seems to be larger than these algorithms. Nevertheless, it is the first time that the matching between two large social netshas been deduced due to two main reasons. The linear runtime with respect to the number of nodes and also the fact thata bi-dimensional matrix, in which the number of rows or columns is the graph order, is not needed to be defined.

References

1. Riesen, K.: Structural Pattern Recognition with Graph Edit Distance. Advances in Computer Vision and Pattern Recognition. Springer, Cham (2015)
2. Garey, M., Johnson, D.: Computers and Intractability: A Guide to the Theory of NP-Completeness. W. H. Freeman, San Francisco (1979)
3. Conte, D., Foggia, P., Sansone, C., Vento, M.: Thirty years of graph matching in pattern recognition. IJPRAI **18**(3), 265–298 (2004)
4. Foggia, P., Percannella, G., Vento, M.: Graph matching and learning in pattern recognition in the last 10 years. Int. J. Pattern Recogn. Artif. Intell. (2013)
5. Solé, A., Serratosa, F., Sanfeliu, A.: On the graph edit distance cost: properties and applications. Int. J. Pattern Recogn. Artif. Intell. **26**(5), 1260004 (2012)
6. Gold, S., Rangarajan, A.: A Graduated assignment algorithm for graph matching. IEEE Trans. Pattern Anal. Mach. Intell. **18**(4), 377–388 (1996)
7. Riesen, K. Bunke, H.: Approximate graph edit distance computation by means of bipartite graph matching. Image Vis. Comput. **27**(7), 950–959 (2009)
8. Serratosa, F.: Fast computation of bipartite graph matching. Pattern Recogn. Lett. PRL **45**, 244–250 (2014)
9. Riesen, K., Ferrer, M., Dornberger, R., Bunke, H.: Greedy graph edit distance. In: Perner, P. (ed.) MLDM 2015. LNCS (LNAI), vol. 9166, pp. 3–16. Springer, Cham (2015). doi:10. 1007/978-3-319-21024-7_1
10. Cortés, X., Serratosa, F., Riesen, K.: On the relevance of local neighbourhoods for greedy graph edit distance. In: Robles-Kelly, A., Loog, M., Biggio, B., Escolano, F., Wilson, R. (eds.) S+SSPR 2016. LNCS, vol. 10029, pp. 121–131. Springer, Cham (2016). doi:10.1007/978-3-319-49055-7_11
11. Serratosa, F., Cortés, X.: Graph edit distance: moving from global to local structure to solve the graph-matching problem. Pattern Recogn. Lett. **65**, 204–210 (2015)
12. Kuhn, H.W.: The Hungarian method for the assignment problem. Naval Res. Logistics Q. **2**, 83–97 (1955)
13. Jonker, R., Volgenant, T.: Improving the Hungarian assignment algorithm. Oper. Res. Lett. **5**(4), 171–175 (1986)
14. Ferrer, M., Serratosa, F., Riesen, K.: Improving bipartite graph matching by assessing the assignment confidence. Pattern Recogn. Lett. **65**, 29–36 (2015)
15. Serratosa, F., Cortés, X., Moreno, C.-F.: Graph edit distance or graph edit pseudo-distance? In: Robles-Kelly, A., Loog, M., Biggio, B., Escolano, F., Wilson, R. (eds.) S+SSPR 2016. LNCS, vol. 10029, pp. 530–540. Springer, Cham (2016). doi:10.1007/978-3-319-49055-7_47
16. http://deim.urv.cat/∼francesc.serratosa/SW/
17. McAuley, J., Leskovec, J.: Learning to discover social circles in ego networks. In: NIPS 2012
18. Leskovec, J., Kleinberg, J., Faloutsos, C.: Graph evolution: densification and shrinking diameters. ACM Trans. Knowl. Discov. Data (ACM TKDD), **1**(1), 5–34 (2007)

Measuring Vertex Centrality Using the Holevo Quantity

Luca Rossi[1](✉) and Andrea Torsello[2]

[1] School of Engineering and Applied Science, Aston University, Birmingham, UK
l.rossi@aston.ac.uk
[2] DAIS, Università Ca' Foscari Venezia, Venice, Italy

Abstract. In recent years, the increasing availability of data describing the dynamics of real-world systems led to a surge of interest in the complex networks of interactions that emerge from such systems. Several measures have been introduced to analyse these networks, and among them one of the most fundamental ones is vertex centrality, which quantifies the importance of a vertex within a graph. In this paper, we propose a novel vertex centrality measure based on the quantum information theoretical concept of Holevo quantity. More specifically, we measure the importance of a vertex in terms of the variation in graph entropy before and after its removal from the graph. More specifically, we find that the centrality of a vertex v can be broken down in two parts: (1) one which is negatively correlated with the degree centrality of v, and (2) one which depends on the emergence of non-trivial structures in the graph when v is disconnected from the rest of the graph. Finally, we evaluate our centrality measure on a number of real-world as well as synthetic networks, and we compare it against a set of commonly used alternative measures.

Keywords: Complex networks · Vertex centrality · Quantum information

1 Introduction

A large number of real-world systems can be modelled and analysed by looking at the structure that emerges from the interaction between their components [5]. The resulting graph is called a complex network, and provides a powerful way to study the static and dynamic aspects of the underlying system. Typical examples of systems that are studied in network science include metabolic pathways [9], protein interactions [8], brain regions interactions [20] and scientific collaborations [13]. Complex networks often display non-trivial structural properties that distinguish them from Erdös-Rényi random graphs [4], such as small-worldness and a power-law distribution of vertex degrees [5].

In these large networks one of the key problems is that of identifying the set of most relevant nodes, also called *central* nodes. A number of centrality measure have been introduced in the literature [3,5–7,12,17], each of them capturing different but equally significant aspects of vertex importance. Commonly

© Springer International Publishing AG 2017
P. Foggia et al. (Eds.): GbRPR 2017, LNCS 10310, pp. 154–164, 2017.
DOI: 10.1007/978-3-319-58961-9_14

encountered examples include the degree centrality [7], the closeness centrality [21], and the betweenness centrality [7]. Let G be a graph with n nodes. In
the *degree centrality* [7] the normalised (degree) is taken as the centrality value
of a vertex, i.e.,

$$DC(v) = \frac{d_v}{\sum_{u=1}^{n} d_u},$$

where d_v denotes the degree of the vertex v. In other words, the number of
edges incident on a vertex is interpreted as a measure of its "popularity", or,
alternatively, as the risk of a node being infected in an epidemiological scenario.
The *closeness centrality* [21] links the importance of a vertex to its proximity to
the remaining vertices of the graph. More specifically, the closeness centrality is
defined as the inverse of the sum of the distance of a vertex v to the remaining
nodes of the graph, i.e.,

$$CC(v) = \frac{n-1}{\sum_{u=1}^{n} sp(u, v)},$$

where $sp(u, v)$ denotes the shortest path distance between nodes u and v. Finally,
the *betweenness centrality* [7] measures the extent to which a given vertex lies
on the (shortest) paths between the remaining vertices, i.e.,

$$BC(v) = \sum_{s,t \in V} \frac{\sigma(s, t|v)}{\sigma(s, t)},$$

where V is the set of nodes, $\sigma(s,t)$ and $\sigma(s,t|v)$ denote the number of shortest
paths between s and t and the number of shortest paths between s and t that
pass through v.

Recently, there has been an increasing interest in using concepts from quantum mechanics and quantum information theory to probe the structure of
graphs [11,18,19]. In [11], Lockhart et al. introduced an edge centrality index
based on quantum information theory, where the importance of an edge is measured in terms of its contribution to the Von Neumann entropy of the network [16]. This in turn relies on decomposing the edge set of a graph as follows.
Given an edge u, the original graph is decomposed into two graphs over the same
vertex set, but with different number of edges: (1) a graph where only the edge
e is present, and (2) a graph where all the original edges except e are present.
With this decomposition to hand, the centrality of e is measured as the Holevo
quantity of the associated decomposition [11].

In this paper, we show that a similar approach can be taken to measure the
centrality of a vertex. Given a vertex v, we propose to decompose the graph
into two graphs over the same vertex set but with edge sets as follows: (1) one
graph where only the edges incident to v are present, and (2) one graph where
all the original edges except those incident to v are present. Then, the centrality
of v is the Holevo quantity associated to the resulting graph ensemble. We show
that the centrality of a vertex v can be broken down in two parts: (1) one part
that is negatively correlated with the degree centrality of v, and (2) one part

that depends on the emergence of non-trivial structures in the graph when v is disconnected from the rest of the graph by removing all edges incident to it. Finally, we perform a series of experiments to evaluate the proposed edge centrality measure on real-world as well as synthetic graphs, and we compare it against a number of commonly used alternative measures.

The remainder of this paper is organised as follows: Sect. 2 reviews the necessary quantum mechanical background and the quantum information theoretical concepts that underpin our approach. Section 3 introduces the proposed vertex centrality measure, which is then analysed and compared to alternative measures in Sect. 4. Finally, Sect. 5 concludes the paper.

2 Quantum Information Theoretical Background

2.1 Quantum States and von Neumann Entropy

In quantum mechanics, a system can be either in a pure state or a mixed state. Using the Dirac notation, a *pure state* is represented as a column vector $|\psi_i\rangle$. A *mixed state*, on the other hand, is an ensemble of pure quantum states $|\psi_i\rangle$, each with probability p_i. The density operator of such a system is a positive unit-trace matrix defined as

$$\rho = \sum_i p_i |\psi_i\rangle \langle\psi_i|. \tag{1}$$

The *von Neumann entropy* [14] S of a mixed state is defined in terms of the trace and logarithm of the density operator ρ

$$S(\rho) = -\operatorname{Tr}(\rho \ln \rho) = -\sum_i \lambda_i \ln(\lambda_i) \tag{2}$$

where $\lambda_1, \ldots, \lambda_n$ are the eigenvalues of ρ. If $\langle\psi_i| \rho |\psi_i\rangle = 1$, i.e., the quantum system is a pure state $|\psi_i\rangle$ with probability $p_i = 1$, then the Von Neumann entropy $S(\rho) = -\operatorname{Tr}(\rho \ln \rho)$ is zero. On other hand, a mixed state always has a non-zero Von Neumann entropy associated with it.

2.2 A Mixed State from the Graph Laplacian

Let $G = (V, E)$ be a simple graph with n vertices and m edges. We assign the vertices of G to the elements of the standard basis of an Hilbert space $\mathcal{H}_G$, $\{|1\rangle, |2\rangle, \ldots, |n\rangle\}$. Here $|i\rangle$ denotes a column vector where 1 is at the i-th position. The *graph Laplacian* of G is the matrix $L = D - A$, where A is the adjacency matrix of G and D is the diagonal matrix with elements $d(u) = \sum_{v=1}^n A(u, v)$. For each edge $e_{i,j}$, we define a pure state

$$|e_{i,j}\rangle := \frac{1}{\sqrt{2}}(|i\rangle - |j\rangle). \tag{3}$$

Then we can define the mixed state $\{\frac{1}{m}, |e_{i,j}\rangle\}$ with density matrix

$$\rho(G) := \frac{1}{m} \sum_{\{i,j\} \in E} |e_{i,j}\rangle \langle e_{i,j}| = \frac{1}{2m} L(G). \tag{4}$$

Let us define the Hilbert spaces $\mathcal{H}_V \cong \mathbb{C}^V$, with orthonormal basis $\mathbf{a}_v$, where $v \in V$, and $\mathcal{H}_E \cong \mathbb{C}^E$, with orthonormal basis $\mathbf{b}_{u,v}$, where $\{u, v\} \in E$. It can be shown that the graph Laplacian corresponds to the partial trace of a rank-1 operator on $\mathcal{H}_V \otimes \mathcal{H}_E$ which is determined by the graph structure [2]. As a consequence, the Von Neumann entropy of $\rho(G)$ can be interpreted as a measure of the amount of entanglement between a system corresponding to the vertices and a system corresponding to the edges of the graph [2].

2.3 Holevo Quantity of a Graph Decomposition

Given a graph G, we can define an ensemble in terms of its subgraphs. Recall that a *decomposition* of a graph G is a set of subgraphs $H_1, H_2, ..., H_k$ that partition the edges of G, i.e., for all i, j, $\bigcup_{i=1}^{k} H_i = G$ and $E(H_i) \cap E(H_j) = \emptyset$, where $E(G)$ denotes the edge set of G. Notice that isolated vertices do not contribute to a decomposition, so each H_i can always be seen a subgraph that contains all the vertices. If we let $\rho(H_1), \rho(H_2), ..., \rho(H_k)$ be the mixed states of the subgraphs, the probability of H_i in the mixture $\rho(G)$ is given by $|E(H_i)|/|E(G)|$. Thus, we can generalise Eq. 4 and write

$$\rho(G) = \sum_{i=1}^{k} \frac{|E(H_i)|}{|E(G)|} \rho(H_i). \tag{5}$$

Consider a graph G and its decomposition $H_1, H_2, ..., H_k$ with corresponding states $\rho(H_1), \rho(H_2), ..., \rho(H_k)$. Let us assign $\rho(H_1), \rho(H_2), ..., \rho(H_k)$ to the elements of an alphabet $\{a_1, a_2, ..., a_k\}$. In quantum information theory, the classical concepts of uncertainty and entropy are extended to deal with quantum states, where uncertainty about the state of a quantum system can be expressed using the density matrix formalism. Assume a source emits letters from the alphabet and that the letter a_i is emitted with probability $p_i = |E(H_i)|/|E(G)|$. An upper bound to the accessible information is given by the *Holevo quantity* of the ensemble $\{p_i, \rho(H_i)\}$:

$$\chi(\{p_i, \rho(H_i)\}) = S\left(\sum_{i=1}^{k} p_i \rho(H_i)\right) - \sum_{i=1}^{k} p_i S(\rho(H_i)) \tag{6}$$

3   The Holevo Vertex Centrality

Given a graph $G = (V, E)$ and a vertex $v \in V$, we propose to measure the centrality of v as follows. Let $G_v = (V, E_v)$ denote the subgraph with vertex set V and edge set $E_v = \{(u, v) \in E | u \in V\}$, and $G_{\bar{v}} = (V, E_{\bar{v}})$ be the subgraph with vertex set V and edge set $E_v = \{(u,v) \in E | (u,v) \notin E_v\}$. In other words, $E = E_v \cup E_{\bar{v}}$ and $E_v \cap E_{\bar{v}} = \emptyset$. Hence, from Eq. 5, we can show that

$$\frac{|E_{\bar{v}}|}{|E|}\rho(G_{\bar{v}}) + \frac{|E_v|}{|E|}\rho(G_v) = \rho(G). \qquad (7)$$

With this decomposition to hand, we define the Holevo vertex centrality of v as

$$HC(v) = \chi\left(\left\{\left(\frac{|E_{\bar{v}}|}{|E|}, G_{\bar{v}}\right), \left(\frac{|E_v|}{|E|}, G_v\right)\right\}\right)$$

$$= S\left(\rho(G)\right) - \left(\frac{|E_{\bar{v}}|}{|E|}S\left(\rho(G_{\bar{v}})\right) + \frac{|E_v|}{|E|}S\left(\rho(G_v)\right)\right). \qquad (8)$$

Given a graph $G = (V, E)$ and a vertex $v \in V$, the first term in Eq. 8 (i.e., $S\left(\rho(G)\right)$) does not depend on the choice of v, and thus can be ignored when ranking the nodes of G according to their Holevo centrality. Moreover, note that we only need to compute the spectrum of $\rho(G_{\bar{v}})$, as the spectrum of $\rho(G_v)$ can be easily determined analytically. Recall that the star graph on n vertices $K_{1,n-1}$ has Laplacian spectrum

$$\{n^{[1]}, 1^{[n-2]}, 0^{[1]}\},$$

i.e., it has three eigenvalues n, 1, and 0 with multiplicity 1, $n - 2$, and 1, respectively. This in turn implies that the spectrum of the density matrix $\rho(K_{1,n-1})$ is

$$\{\frac{n}{2n-2}^{[1]}, \frac{1}{2n-2}^{[n-2]}, 0^{[1]}\},$$

as shown in [10]. Since adding disconnected vertices to a graph does not change its Von Neumann entropy [16], we have that the entropy of $\rho(G_v)$ is

$$S\left(\rho(G_v)\right) = -\frac{d_v + 1}{2d_v}\log\left(\frac{d_v + 1}{2d_v}\right) - \frac{d_v - 1}{2d_v}\log\left(\frac{1}{2d_v}\right)$$

$$= \frac{d_v + 1}{2d_v}\log\left(\frac{2d_v}{d_v + 1}\right) + \frac{d_v - 1}{2d_v}\log\left(2d_v\right)$$

$$= \frac{1}{2d_v}\left(2d_v\log\left(2d_v\right) - (d_v + 1)\log(d_v + 1)\right) \qquad (9)$$

where d_v denotes the degree of v. In other words, the entropy of $\rho(G_v)$ is completely determined by the degree of v. As a result, the computational complexity of computing the Holevo centrality of v is dominated by the cost of computing the eigendecomposition of $\rho(G_{\bar{v}})$.

Finally note that the Von Neumann entropy of a star graph is 0 when $d_v = 1$, and it grows logarithmically as a function of d_v. This in turn suggests that the Holevo centrality given by Eq. 8 could be negatively correlated with the degree centrality, however proving this would require finding general analytical form of the spectrum of $\rho(G_{\bar{v}})$. Moreover, it should be noted that the Von Neumann entropy of $\rho(G_{\bar{v}})$ depends on the presence of several non-trivial structural patterns, including paths, cliques, and connected components. Therefore the Holevo vertex centrality measures the importance of a vertex as a combination of its degree as well as the structural patterns that emerge after its removal.

4   Experimental Evaluation

We perform our experiments on two well known real-world networks, the Florentine families graph [15] and the Karate club network [22], as well as a number of synthetic graphs. We compare the proposed similarity measure the three commonly used alternative measures: (1) the degree centrality [7], (2) the closeness centrality [21], and (3) the betweenness centrality [7].

4.1   Synthetic Networks

Wheel Graph. The Wheel graph W_n on n nodes is the graph obtained by taking a cycle C_{n-1} on $n-1$ nodes and connecting each of the nodes of C_{n-1} to another node, i.e., the hub. Figure 1 shows 3 wheel graphs of increasing number of nodes n and the corresponding value of the Holevo centrality. Note that for small values of n the hub is the least central node. However as n grows the centrality of the hub remains constant while the centrality of the other nodes decreases, until the hub becomes the most central node.

Indeed, our centrality measure seems to capture the increasing redundancy of the nodes along the cycle C_{n-1} as n grows. Note that this implies that our measure is negatively correlated with the degree centrality for small values of n, but positively correlated for large values of n. While this may seem surprising given the negative correlation highlighted in Eq. 9, the observed behaviour is likely due to the other component in Eq. 8, i.e., $S(\rho(G_{\bar{v}}))$, as well as their respective weights.

Lollipop Graph. The (m, n)-lollipop graph on $m + n$ nodes is the graph obtained by joining the clique K_m on m nodes with the path graph P_n on n nodes. Figure 2 shows the value of the Holevo centrality for increasing size of the clique, while keeping the size of the path fixed. For small clique sizes, the most central node is the central node on the path. However, as the size of the clique increases, the centrality of the path node decreases, while the clique nodes become increasingly important. We observe a similar behaviour when the length

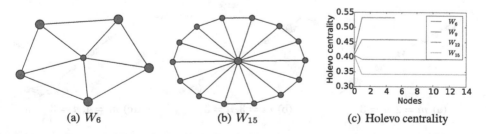

(a) W_6 (b) W_{15} (c) Holevo centrality

Fig. 1. Holevo centrality for the nodes of the Wheel graph on 6 (a) and 15 (b) nodes. The radius of each node is proportional to their Holevo centrality. In (c) we show a line plot of the Holevo centrality for increasing graph size. Here 0 denotes the hub node.

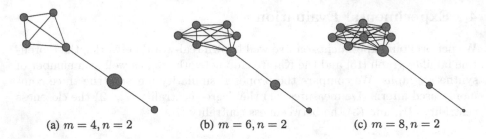

Fig. 2. Holevo centrality for the nodes of the (m, n)-lollipop graph for (a) $m = 4, n = 2$, (b) $m = 6, n = 2$, and (c) $m = 8, n = 2$. The radius of each node is proportional to their Holevo centrality.

of the path is increased, with the nodes belonging to it being the most central ones for small values of n. However, as we increase n, the most central node in the graph becomes the node the shared node between the clique and the path.

Similarly to the Wheel graph, the Holevo centrality measure seems to capture the importance of the nodes belonging to the path when the total number of nodes in the graph is small. However, as $m+n$ grows, our measure places most of the importance on the tightly connected nodes of the clique, while the centrality of the path is "diluted" as its length increases.

Barbell Graph. The Barbell graph is the graph obtained by joining two cliques K_m through a path P_n (i.e., a bridge between the two cliques). Note that when $m = 2$ the corresponding Barbell graph is the path graph P_{n+2m}. Figure 3 shows three Barbell graphs with n constant (i.e., the length of the bridge is 3 in all the graphs) and m equal to 2, 3, and 4, respectively. When $m = 2$ the graph is a path over 7 nodes. In this case, our centrality measures assigns the largest weight to the two nodes that connect the two ends of the path to the rest of the nodes. As the m increases, the weight of the cliques shifts the importance from the bridge to the cliques, with the node connecting the cliques to the path becoming then most central ones. If we increase the path length, we observe that the junctions

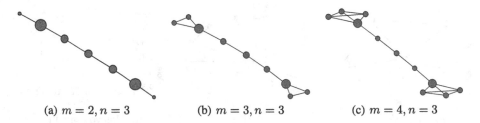

Fig. 3. Holevo centrality for the nodes of the Barbell graph joining two cliques K_m through a path P_n, for (a) $m = 2, n = 3$, (b) $m = 3, n = 3$, and (c) $m = 4, n = 3$. The radius of each node is proportional to their Holevo centrality.

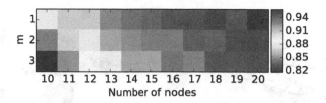

Fig. 4. Average correlation with the degree centrality for different realisation of the Barabási-Albert preferential attachment model [1]. Here we varied the number of nodes of the generated graph, as well as the number of edges k that are created from a new node to existing nodes.

between the cliques and the path remain the most central nodes. However we start to discriminate between the nodes along the bridge, with the nodes closer to the center of the bridge being assigned a higher centrality. Note that this is contrast with the degree centrality, which would assign the same weight to all the nodes on the bridge (since they all have the same degree).

Scale-Free Graph. Finally, we consider a set of scale-free graphs generated by the Barabási-Albert preferential attachment model [1]. Starting from an empty graph, we iteratively add nodes to it until a user-defined size n is reached. At each iteration, the new node is connected to at most m nodes chosen according to their degree, i.e., nodes with a higher degree are more likely to be selected. We let m and n vary between 1 and 3, and 10 and 20, respectively, and for each pair of parameters we generated 100 graphs.

We are interested in measuring the correlation between the Holevo vertex centrality and the degree centrality. In the previous subsections we have observed that the correlation with the degree centrality seems to increase as the size of the graph increases. Figure 4 confirms that this is the case. The figure also shows that the correlation increases as we increase the number of connections added per iteration. Note that when $m = 1$ the resulting graph is guaranteed to be a tree. Indeed, Fig. 4 seems to suggest that the correlation is particularly high on trees, however further investigation is needed to understand if the observed effect is instead due to the particular degree distribution of scale-free graphs or to the presence of high degree nodes.

4.2   Real-World Networks

We conclude our experimental evaluation by measuring the centrality of two well-known networks, the Florentine families graph [15] and the Karate club network [22]. Figures 5(a) and (b) show the two networks, where the radius of the nodes is proportional to their Holevo centrality. We also compute the degree (DC), closeness (CC), and betweenness (BC) centralities over these networks and we show the corresponding correlation matrices in Figs. 5(c) and (d).

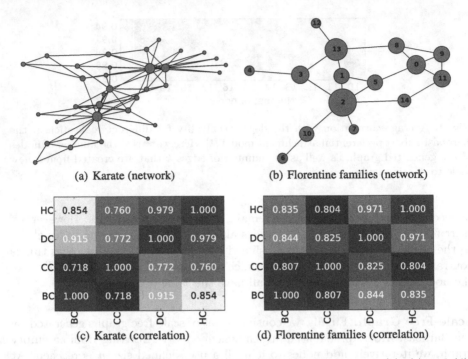

(a) Karate (network)   (b) Florentine families (network)

HC	0.854	0.760	0.979	1.000
DC	0.915	0.772	1.000	0.979
CC	0.718	1.000	0.772	0.760
BC	1.000	0.718	0.915	0.854
	BC	CC	DC	HC

HC	0.835	0.804	0.971	1.000
DC	0.844	0.825	1.000	0.971
CC	0.807	1.000	0.825	0.804
BC	1.000	0.807	0.844	0.835
	BC	CC	DC	HC

(c) Karate (correlation)   (d) Florentine families (correlation)

Fig. 5. The Karate network (a) and the Florentine families network (b). (c) and (d) show the correlation matrices between the Holevo (HC), degree (DC), closeness (CC), and betweenness (BC) centrality measures on the Karate and Florentine families networks, respectively. Each node of (b) is labeled with the index associated to the corresponding family in Table 1.

In these networks, the Holevo centrality measure shows a large positive correlation with all the other measure, in particular the degree centrality. However, if we look at the ranking induced by our measure there are some important differences. Table 1 shows the ranking given by the Holevo centrality, as well as the degree of each family in the network. Clearly, the degree centrality cannot distinguish between those families that have the same degree, while the Holevo centrality allows to define a more fine-grained ranking. For example, the Salviati

Table 1. The Holevo and the degree (in bold) centrality of the families of Florentine families network. The number next to the name of each family is the index of the corresponding node in Fig. 5(b).

Family	Centrality	Family	Centrality	Family	Centrality
Medici (2)	0.6582 (**0.150**)	Peruzzi (9)	0.4409 (**0.075**)	Barbadori (14)	0.3586 (**0.050**)
Guadagni (13)	0.5728 (**0.100**)	Bischeri (8)	0.4403 (**0.075**)	Pazzi (6)	0.2615 (**0.025**)
Strozzi (0)	0.5057 (**0.100**)	Ridolfi (5)	0.4259 (**0.075**)	Ginori (4)	0.2501 (**0.025**)
Albizzi (3)	0.4834 (**0.075**)	Tornabuoni (1)	0.4242 (**0.075**)	Lamberteschi (12)	0.2424 (**0.025**)
Castellani (11)	0.4615 (**0.075**)	Salviati (10)	0.4075 (**0.050**)	Acciaiuoli (7)	0.2305 (**0.025**)

family (node 10) is ranked higher than the Barbadori family (node 14), although both families have degree two. However the Salviati family is the only node connecting the Pazzi family (node 6) to the Medici family (node 2) and the rest of the graph, and therefore its importance is higher.

5 Conclusion

In this paper we have proposed a novel vertex centrality measure based on the quantum information theoretical notion of Holevo quantity. The idea underpinning our approach is that the importance of a vertex is proportional to the variation in the information content of the network before and after its removal. We have shown that the centrality of a vertex v can be broken down in two parts, one which is negatively correlated with the degree centrality of v, and one which depends on the emergence of non-trivial structures in the graph when v is disconnected from the rest of the graph. Finally, we have evaluated the Holevo centrality measure on a number of synthetic as well as real-world networks, and we have compared it against commonly used alternative measures. Future work will be aimed at investigating further the structural pattern that influence this centrality measure.

References

1. Barabási, A.L., Albert, R.: Emergence of scaling in random networks. Science **286**(5439), 509–512 (1999)
2. de Beaudrap, N., Giovannetti, V., Severini, S., Wilson, R.: Interpreting the von Neumann entropy of graph Laplacians, and coentropic graphs. Panorama Math. Pure Appl. **658**, 227 (2016)
3. Bonacich, P.: Power and centrality: a family of measures. Am. J. Sociol. **92**, 1170–1182 (1987)
4. Erdös, P., Rényi, A.: On random graphs. Publ. Math. Debrecen **6**, 290–297 (1959)
5. Estrada, E.: The Structure of Complex Networks. Oxford University Press, New York (2011)
6. Freeman, L.C.: A set of measures of centrality based on betweenness. Sociometry **40**(1), 35–41 (1977)
7. Freeman, L.C.: Centrality in social networks conceptual clarification. Soc. Netw. **1**(3), 215–239 (1979)
8. Ito, T., Chiba, T., Ozawa, R., Yoshida, M., Hattori, M., Sakaki, Y.: A comprehensive two-hybrid analysis to explore the yeast protein interactome. Proc. Nat. Acad. Sci. **98**(8), 4569 (2001)
9. Jeong, H., Tombor, B., Albert, R., Oltvai, Z., Barabási, A.: The large-scale organization of metabolic networks. Nature **407**(6804), 651–654 (2000)
10. Li, J.Q., Chen, X.B., Yang, Y.X.: Quantum state representation based on combinatorial Laplacian matrix of star-relevant graph. Quantum Inf. Process. **14**(12), 4691–4713 (2015)
11. Lockhart, J., Minello, G., Rossi, L., Severini, S., Torsello, A.: Edge centrality via the Holevo quantity. In: Robles-Kelly, A., Loog, M., Biggio, B., Escolano, F., Wilson, R. (eds.) S+SSPR 2016. LNCS, vol. 10029, pp. 143–152. Springer, Cham (2016). doi:10.1007/978-3-319-49055-7_13

12. Newman, M.E.: A measure of betweenness centrality based on random walks. Social Netw. **27**(1), 39–54 (2005)
13. Newman, M.: Scientific collaboration networks. i. network construction and fundamental results. Phys. Rev. E **64**(1), 016131 (2001)
14. Nielsen, M.A., Chuang, I.L.: Quantum Computation and Quantum Information. Cambridge University Press, New York (2010)
15. Padgett, J.F., Ansell, C.K.: Robust action and the rise of the medici, 1400–1434. Am. J. Sociol. **98**(6), 1259–1319 (1993)
16. Passerini, F., Severini, S.: Quantifying complexity in networks: the von Neumann entropy. Int. J. Agent Technol. Syst. (IJATS) **1**(4), 58–67 (2009)
17. Rossi, L., Torsello, A., Hancock, E.R.: Node centrality for continuous-time quantum walks. In: Fränti, P., Brown, G., Loog, M., Escolano, F., Pelillo, M. (eds.) S+SSPR 2014. LNCS, vol. 8621, pp. 103–112. Springer, Heidelberg (2014). doi:10.1007/978-3-662-44415-3_11
18. Rossi, L., Torsello, A., Hancock, E.R.: Measuring graph similarity through continuous-time quantum walks and the quantum Jensen-Shannon divergence. Phys. Rev. E **91**(2), 022815 (2015)
19. Rossi, L., Torsello, A., Hancock, E.R., Wilson, R.C.: Characterizing graph symmetries through quantum Jensen-Shannon divergence. Phys. Rev. E **88**(3), 032806 (2013)
20. Sporns, O.: Network analysis, complexity, and brain function. Complexity **8**(1), 56–60 (2002)
21. Stanley, W., Faust, K.: Social Network Analysis: Methods and Applications. Cambridge University, Cambridge (1994)
22. Zachary, W.W.: An information flow model for conflict and fission in small groups. J. Anthropol. Res. **33**(4), 452–473 (1977)

On the Interplay Between Strong Regularity and Graph Densification

Marco Fiorucci[1], Alessandro Torcinovich[1], Manuel Curado[2],
Francisco Escolano[2(✉)], and Marcello Pelillo[1,3]

[1] DAIS, Ca' Foscari University, Via Torino 155, 30172 Venezia Mestre, Italy
[2] DCCIA, University of Alicante, 03690 San Vicente del Raspeig, Alicante, Spain
sco@dccia.ua.es
[3] ECLT, Ca' Foscari University, S. Marco 2940, 30124 Venezia, Italy

Abstract. In this paper we analyze the practical implications of Szemerédi's regularity lemma in the preservation of metric information contained in large graphs. To this end, we present a heuristic algorithm to
find regular partitions. Our experiments show that this method is quite
robust to the natural sparsification of proximity graphs. In addition, this
robustness can be enforced by graph densification.

Keywords: Graph algorithms · Regular partition · Commute time ·
Graph densification

1 Introduction

A crucial role in the development of machine learning and pattern recognition is
played by the tractability of large graphs, which is intrinsically limited by their
size. In order to overcome this limit, the input graph can be compressed into a
reduced version by means of Szemerédi's regularity lemma [16], which is "one
of the most powerful results of extremal graph theory" [10]. Basically, it states
that any sufficiently large (dense) graph can almost entirely be partitioned into
a bounded number of random-like bipartite graphs, called regular pairs. Komlós
et al. [9,10] introduced an important result, the so-called key lemma. It states
that, under certain conditions, the partition resulting from the regularity lemma
gives rise to a *reduced graph* which inherits many of the essential structural properties of the original graph. This result provides a solid theoretical framework
for the exploitation of the regularity lemma to summarize large graphs, and can
be regarded as a manifestation of the all-pervading dichotomy between structure
and randomness. The regularity lemma is an existential, non-constructive pred-
icate, but during the last decades different constructive algorithms have been
proposed.

In this paper we use an approximate approach of the exact algorithm introduced by Alon et al. [1], who proposed a constructive version of the original
(strong) regularity lemma useful only for large *dense* graphs. This is a crucial
limit in practical applications considering that real large graphs not only are

P. Foggia et al. (Eds.): GbRPR 2017, LNCS 10310, pp. 165–174, 2017.
DOI: 10.1007/978-3-319-58961-9_15

often very sparse, but also become sparser and sparser as the dimensionality d of the data increases.

The aim of this work is to analyze the *ideal density regime* where the regularity lemma can find useful applications. In particular, we use the regularity lemma to reduce an input graph and we then exploit the key lemma to obtain an expanded version which preserves some topological properties of the original graph. If we are out of the ideal density regime, we have to densify the graph before applying the regularity lemma. Among the many topological measures we test the effective resistance (or equivalently the scaled commute time), one of the most important metrics between the vertices in the graph, which has been very recently questioned. In [12] it is argued that this measure is meaningless for large graphs. However, recent experimental results show that the graph can be pre-processed (densified) to provide some informative estimation of this metric [4,5]. Therefore, in this paper, we analyze the practical implications of the key lemma in the estimation of commute time in large graphs.

2   Regular Partitions and the Key Lemma

In essence, Szemerédi's regularity lemma states that for every $\varepsilon > 0$, every sufficiently dense graph G can almost entirely be partitioned into $k(\varepsilon)$ random-like bipartite graphs, where the deviation from randomness is controlled by ε. In particular, the lemma deals with vertex subsets that shows a sort of regular behaviour which is expressed in terms of edge density. To state Szemerédi's regularity lemma, some terminology is required.

Let $G = (V, E)$ be an undirected graph without self-loops. The *edge density* d of a pair (X, Y) of two disjoint subsets of V is defined as $d(X,Y) = e(X,Y)/(|X||Y|)$, where $e(X,Y)$ is the number of edges with an endpoint in X and the other in Y.

A pair is said to be ε-*regular* with $\varepsilon > 0$ if, given $A, B \subseteq V$ such that A and B are disjoint, then for each pair of subsets X, Y such that $X \in A$ and $Y \in B$ the following inequality is satisfied:

$$|d(X,Y) - d(A,B)| < \varepsilon \tag{1}$$

This means that the edges in an ε-regular pair are distributed fairly uniformly, where the deviation from uniform distribution is controlled by the parameter ε.

Further, a partition of V into pairwise disjoint classes $C_0, C_1, ..., C_k$ is called *equitable* if all the classes C_i $(1 \le i \le k)$ have the same cardinality. Thus we can define an ε-regular partition as follows

Definition 1 (ε-regular partition). *An equitable partition $C_0, C_1, ..., C_k$, with C_0 being the exceptional set is called ε-regular if:*

1. *$|C_0| < \varepsilon|V|$*
2. *all but at most εk^2 of the pairs (C_i, C_j) are ε-regular $(1 \le i < j \le k)$*

The regularity lemma states that every sufficiently large dense graph admits an ε-regular partition.

Lemma 1 (Szemerédi's regularity lemma [2]). *For every positive real ε and for every positive integer m, there are positive integers $N = N(\varepsilon, m)$ and $M = M(\varepsilon, m)$ with the following property: for every graph $G = (V, E)$, with $|V| \geq N$, there is an ε-regular partition of G into $k + 1$ classes such that $m \leq k \leq M$.*

The lemma allows us to specify a lower bound m on the number of classes. A large value of m ensures that the partition classes C_i are sufficiently small, thereby increasing the proportion of (inter-class) edges subject to the regularity condition and reducing the intra-class ones. The upper bound M on the number of partitions guarantees that for large graphs the partition sets are large too. Finally, it should be noted that a singleton partition is ε-regular for every value of ε and m.

An ε-regular partition resulting from the regularity lemma gives rise to a *reduced graph* which is basically a graph $R = (V(R), E(R))$ whose vertices represents the classes of the regular partition, and an edge joins two vertices if the corresponding pair of classes is ε-*regular*, with density greater than a given threshold d. The reduced graph R plays an important role in most applications of the regularity lemma, relying on the Komlós and Simonovits's "key lemma" [10]. It states that many structural properties of the original graph G are inherited by R.

Before presenting the key lemma, another kind of graph needs to be defined, namely the *t-fold reduced graph*, which is a graph $R(t)$ obtained from R by replacing each vertex $x \in V(R)$ by a set V_x of t independent vertices, and joining $u \in V_x$ to $v \in V_y$ if and only if (x, y) is an edge in R. $R(t)$ is a graph in which every edge of R is replaced by a copy of the complete bipartite graph K_{tt}.

The key lemma asserts that, under certain conditions, the existence of a subgraph in $R(t)$ implies its existence in G.

Lemma 2 (Key lemma). *Given the reduced graph R, $d > \varepsilon > 0$, a positive integer m, let construct a graph G by replacing every vertex of R by m vertices, and replacing the edges of R with ε-regular pairs of density at least d. Let H be a subgraph of $R(t)$ with h vertices and maximum degree $\Delta > 0$ and let $\delta = d - \varepsilon$ and $\varepsilon_0 = \delta^\Delta/(2 + \Delta)$. If $\varepsilon \leq \varepsilon_0$ and $t - 1 \leq \varepsilon_0 m$, then H is embeddable into G (i.e., G contains a subgraph isomorphic to H). In fact, we have:*

$$||H \to G|| > (\varepsilon_0 m)^h \tag{2}$$

where $||H \to G||$ denotes the number of labeled copies of H in G.

Thus, the reduced graph R can be considered as a summary of the graph G, which inherits many structural properties of the original graph G.

The constructive version of the regularity lemma introduced by Alon et al. [1] has been formalized in the following theorem:

Theorem 1 (Alon et al. [1]). *For every $\varepsilon > 0$ and every positive integer t there is an integer $Q = Q(\varepsilon, t)$ such that every graph with $n > Q$ vertices has an ε-regular partition into $k + 1$ classes, where $t \le k \le Q$. For every fixed $\varepsilon > 0$ and $t \ge 1$ such a partition can be found in $O(M(n))$ sequential time, where $M(n) = O(n^{2.376})$ is the time for multiplying two $n \times n$ matrices with $0,1$ entries over the integers.*

The proof of Theorem 1 provides a deterministic polynomial time algorithm for finding a regular partition of an input dense graph. In the following, a sketch of the proof and the resulting algorithm are presented.

Let H be a bipartite graph with classes A, B such that $|A| = |B| = n$, then the *neighbourhood deviation* of a pair of different vertices $y_1, y_2 \in B$ is defined as:

$$\sigma(y_1, y_2) = |N(y_1) \cap N(y_2)| - \frac{d^2}{n} \tag{3}$$

where $N(x)$ is the neighbourhood of x. The *deviation* of a subset $Y \subseteq B$ is defined as follows:

$$\sigma(Y) = \frac{\sum_{y_1, y_2 \in Y} \sigma(y_1, y_2)}{|Y|^2} \tag{4}$$

The following lemma states the conditions to check the regularity of a pair:

Lemma 3 (Alon et al. [1]). *Let H be a bipartite graph with equal classes $|A| = |B| = n$ and let d denote the average degree of H. Let $0 < \varepsilon < 1/16$. If there exists $Y \subseteq B, |Y| > \varepsilon n$ such that $\sigma(Y) \ge \varepsilon^3 n/2$, then at least one of the following cases occurs:*

1. *$d < \varepsilon^3 n$ (which implies that H is ε-regular);*
2. *there exists in B a set of more than $\frac{1}{8}\varepsilon^4 n$ vertices whose degrees deviate from d by at least $\varepsilon^4 n$;*
3. *there are subsets $A' \subset A, B' \subset B, |A'| \ge \frac{\varepsilon^4}{4} n, |B'| \ge \frac{\varepsilon^4}{4} n$ and $|d(A', B') - d(A, B)| \ge \varepsilon^4$.*

Conditions 1 and 2 can be easily checked in $O(n^2)$ time. The third condition involves a matrix squaring of H to compute the quantities $\sigma(y, y'), \forall y, y' \in B$, thus requiring $O(M(n)) = O(n^{2.376})$ time.

Finally, the algorithm to find a regular partition for a graph $G = (V, E)$ with n vertices is described as follows:

1. **Create the initial partition:** Arbitrarily divide the vertices of G into an equitable partition P_1 with classes $C_0, C_1, ..., C_b$ where $|C_1| = \lfloor \frac{n}{b} \rfloor$ and hence $|C_0| < b$. Denote $k_1 = b$
2. **Check regularity:** For every pair (C_r, C_s) of P_i, verify if it is ε-regular or find $X \subseteq C_r, Y \subseteq C_s, |X| \ge \frac{\varepsilon^4}{16}|C_1|, |Y| \ge \frac{\varepsilon^4}{16}|C_1|$, such that

$$|d(X, Y) - d(C_s, C_t)| \ge \varepsilon^4$$

3. **Count regular pairs:** If there are at most $\varepsilon \binom{k_i}{2}$ pairs that are not verified as ε-regular, then halt. P_i is an ε-regular partition

4. **Refine:** Apply the refinement algorithm (Lemma 2) where $P = P_i$, $k = k_i$, $\gamma = \frac{\varepsilon^4}{16}$ and obtain a partition P' with $1 + k_i 4^{k_i}$ classes
5. **Iterate:** Let $k_{i+1} = k_i 4^{k_i}$, $P_{i+1} = P'$, $i = i + 1$, and go to step (2)

The above mentioned algorithm has a polynomial worst-case complexity in the size of the underlying graph, but it also has a hidden tower-type dependence on an accuracy parameter, which is necessary in order to ensure a regular partition for *all* graphs [7]. The latter is a crucial limit in the application of regular partitions to practical problems. The main obstacle concerns Step 2 and Step 4: in Step 2, in fact, the algorithm finds all possible irregular pairs in the graph, which leads to an exponential growth, while in Step 4 the cardinality of the refined partition increases according to the tower-type dependence. To avoid such problems, Sperotto and Pelillo [15] proposed for each class to limit the number of irregular pairs containing it to at most one, possibly chosen randomly among all irregular pairs. The introduction of such heuristics allowed to divide the classes into a constant, instead of an exponential number of subclasses. These approximations made this algorithm truly applicable in practice. In this heuristic framework, an additional implementation is used in this paper. More details, as well as the code, are available in the following repository [6]. Finally, it is worth noting that in the past few years, different algorithms explicitly inspired by the regularity lemma have been applied in pattern recognition, bioinformatics and social network analysis. The reader can refer to [13] for a survey of these emerging algorithms.

3   Motivation of the Experimental Setup

In this section, we analyze the *ideal density regime*, defined as the range of densities of the input graph G such that our heuristic algorithm outputs a reduced graph G' preserving some topological properties of G. We use the effective resistance to assess to what extent G' retains the metric information that can be inferred from G.

As we noted in the introduction, the effective resistance is a metric between the vertices in G, whose stability is theoretically constrained by the size of G. In particular, von Luxburg et al. [12] derived the following bound for any connected, undirected graph that is not bipartite:

$$\left| \frac{1}{vol(G)} C_{ij} - \left(\frac{1}{d_i} + \frac{1}{d_j} \right) \right| \leq \frac{1}{\lambda_2} \frac{2}{d_{min}} \tag{5}$$

where C_{ij} is the commute time between vertices i and j, $vol(G)$ is the volume of the graph, λ_2 is the so called *spectral gap* and d_{min} is the minimum degree in G. Since $C_{ij} = vol(G)R_{ij}$, where R_{ij} is the effective resistance between i and j, this bound leads to $R_{ij} \approx \frac{1}{d_i} + \frac{1}{d_j}$. This means that, in large graphs, effective resistances do only depend on local properties, i.e. degrees. However, some of the authors of this work have recently argued that looking at the density of the graph can be a way of mitigating the devastating effects of the bound in Eq. 5.

In particular, Escolano et al. [5] showed that densifying G significantly decreases the spectral gap which in turn enlarges the von Luxburg bound. As a result, effective resistances do not depend only on local properties and become meaningful for large graphs provided that these graphs have been properly densified. As defined in [8] and revisited in [4], *graph densification* aims to significantly increase the number of edges in G while preserving its properties as much as possible. One of the most interesting properties of large graphs is their fraction of *sparse cuts*, that are cuts where the number of pairs of vertices involved in edges is a small fraction of the overall number of pairs associated with any subset $S \subset V$, i.e. sparse cuts stretch the graphs, thus leading to small conductance values, which in turn reduce the spectral gap. This is exactly what is accomplished by the state-of-the-art strategies for graph densification, including anchor graphs [11].

In light of these observations, our experiments aim to answer two questions:

- *Phase transition:* What is the expected behaviour of our heuristic algorithm when the input graph is locally sparse?
- *Commute times preservation:* Given a densified graph G, to what extent does our algorithm preserve its metrics in the expanded graph G'?

To address them we perform experiments both with synthetic and real datasets. Experiments on synthetic datasets allow us to control the degree of both intra-cluster and inter-cluster sparsity. On the other hand, the use of real datasets, such as NIST, leads to understand the so called *global density scenario*. Reaching this scenario in realistic data sets may require a proper densification, but once it is provided, the regularity lemma becomes a powerful structural compression method.

4   Results

Since we are exploring the practical effect of combining regularity and key lemmas to preserve metrics in large graphs, our performance measure relies on the so called relative deviation between the measured effective resistance and the von Luxburg et al. local prediction [12]: $RelDev(i, j) = \left| R_{ij} - \left(\frac{1}{d_i} + \frac{1}{d_j} \right) \right| / R_{ij}$. The larger $RelDev(i, j)$ the better the performance. For a graph, we retain the average $RelDev(i, j)$, although the maximum and minimum deviations can be used as well.

4.1   Synthetic Experiments

For these experiments we designed a Ground Truth (GT) consisting of k cliques linked by $O(n)$ edges. Inter-cluster links in the GT were only allowed between class k and $k + 1$, for $k = 1, \ldots, k - 1$. Then, each experiment consisted of modifying the GT by either removing intra-cluster edges (sparsification) and/or adding inter-cluster edges and then looking at the reconstructed GT after the application of our

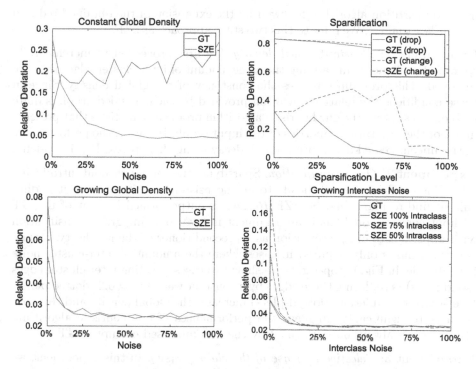

Fig. 1. Top: experiments 1, 2. Bottom: experiment 3 ($n = 200$, $k = 10$ classes).

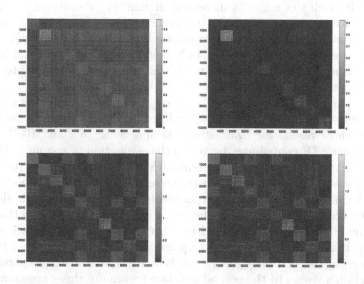

Fig. 2. Reconstruction from R. From left to right: Original similarity matrix W with $\sigma = 0.0248$, its reconstruction after compressing-decompressing, sparse matrix obtained by densifying W and its reconstruction.

heuristic partition algorithm followed by the expansion of the obtained reduced graph (key lemma). We refer to this two stage approach as SZE.

Experiment 1: *Constant global density.* We first proceeded to incrementally sparsify the cliques while adding the same amount of inter-cluster edges that are removed. This procedure assures the constancy of the global density. Since in these conditions the relative deviation provided by the expanded graph is quite stable, we can state the our heuristic algorithm produces partitions that preserve many of the structural properties of the input graph. However, the performances of the uncompressed-decompressed GT decay along this process Fig. 1(top-left).

Experiment 2: *Only sparsification.* Sparsifying the cliques without introducing inter-cluster edges typically leads to an inconsistent partition, since it is difficult to find regular pairs. So SZE *RelDev* is outperformed by that of the GT without compression. This is an argument in favor of using graph densification with approximate cut-preservation as a preconditioner of the regularity lemma. However, this is only required in cases where the amount of inter-cluster noise is negligible. In Fig. 1 (top-right) we show two cases: deleting inter-cluster edges (solid plots) *vs* replacing these edges by a constant weight $w = 0.2$ (dotted plots). Inter-cluster completion (dotted-plots) increases the global density and this contributes to significantly increase the performances of our heuristic algorithm, although it is always outperformed by the uncompressed corrupted GT.

Experiment 3: *Selective increase of the global density.* In this experiment, we increase the global density of the GT as follows. For Fig. 1 (bottom-left), each noise level x means the fraction of intra-cluster edges removed, while the same fraction of inter-cluster edges is increased. Herein, the density of x is $D(x) = (1-x)\#_{In} + x\#_{Out}$, where $\#_{In}$ is the maximum number of intra-cluster links and $\#_{Out}$ is the maximum number of inter-cluster links. Since $\#_{Out} \gg \#_{In}$ we have that $D(x)$ increases with x. However, only moderate increases of $D(x)$ lead to a better estimation of commute times with SZE, since adding many inter-cluster links destroys the cluster structure.

However, in Fig. 1 (bottom-right) we show the impact of increasing the fraction x' of inter-cluster noise (add edges) while the intra-cluster fraction is fixed. We overlay three results for SZE: after retaining 50%, 75% and 100% of $\#_{In}$. We obtain that SZE contributes better to the estimation of commute times for small fractions on $\#_{In}$ which is consistent with Experiment 2. Then, the optimal configuration for SZE is: low inter-cluster noise and moderate sparsified clusters.

As a conclusion of the synthetic experiments, we can state that our algorithm is robust to a high amount of intra-clustering sparsification provided that a certain number of inter-cluster edges exists. This answers the first question (phase transition). It also partially ensures the preservation of commute times provided that the density is high enough or it is kept constant during a sparsification process, which answers to the second question (commute times preservation).

4.2 Experiments with the NIST Dataset

When analyzing real datasets, NIST (herein we use $10,000$ samples with $d = 86$) provides a nice amount of intra-cluster sparsity and inter-cluster noise (both due to ambiguities). We compare our two stage approach (SZE) either applied to the original graph (for a given σ) or to an anchor graph obtained with a nested MDL strategy relying on our EBEM clustering method [3]. In Fig. 2, we show a NIST similarity matrix W (with $O(10^7)$ edges) obtained using the negative exponentiation method. Even with $\sigma = 0.0248$ we obtain a dense matrix due to inter-cluster noise. Let $R(W)$ be the reduced graph of W. After expanding this graph we obtain a locally dense matrix, which suggests that our algorithm plays the role of a cut densifier. We also show the behaviour of compression-decompression for densified matrices in Fig. 2. The third graph in this figure corresponds to $D(W)$, namely the selective densification of W (with $O(2 \times 10^6)$ edges). From $R(D(W))$ the key lemma leads to a reconstruction with a similar density but with more structured inter-cluster noise. Finally, it is worth noting that the compression rate in both cases is close to 75%.

5 Conclusions

In this paper, we have explored the interplay between regular partitions and graph densification. Our synthetic experiments show that the proposed heuristic version of Alon et al.'s algorithm is quite robust to intra-cluster sparsification provided that the graph is globally dense. This behavior has a good impact in similarity matrices obtained from negative exponentiation, since this implementation of the regularity lemma plays the role of a selective densifier. Regarding the effect of compression-decompression in non-densified matrices, the reconstruction preserves the structure of the input matrix. This result suggests that graph densification acts as a preconditioner to obtain reliable regular partitions. Future work may include the study of the reduced graph as a source of selective densification.

Acknowledgments. We are grateful to I. Elezi for his advice on our code implementation, and to the anonymous reviewers for their constructive feedback. Francisco Escolano and Manuel Curado are funded by the Project TIN2015-69077-P of the Spanish Government.

References

1. Alon, N., Duke, R.A., Lefmann, H., Rödl, V., Yuster, R.: The algorithmic aspects of the regularity lemma. J. Algorithms **16**(1), 80–109 (1994)
2. Alon, N., Fischer, E., Krivelevich, M., Szegedy, M.: Efficient testing of large graphs. Combinatorica **20**(4), 451–476 (2000)
3. Benavent, A.P., Escolano, F.: Entropy-based incremental variational Bayes learning of Gaussian mixtures. IEEE Trans. Neural Netw. Learn. Syst. **23**(3), 534–540 (2012). http://dx.doi.org/10.1109/TNNLS.2011.2177670

4. Escolano, F., Curado, M., Hancock, E.R.: Commute times in dense graphs. In: Robles-Kelly et al. [14], pp. 241–251
5. Escolano, F., Curado, M., Lozano, M.A., Hancock, E.R.: Dirichlet graph densifiers. In: Robles-Kelly et al. [14], pp. 185–195
6. Fiorucci, M., Torcinovich, A.: Alonszemerediregularitylemma github repository (2013). https://github.com/MarcoFiorucci/AlonSzemerediRegularityLemma
7. Gowers, T.: Lower bounds of tower type for Szemerédi's uniformity lemma. Geom. Func. Anal. **7**(2), 322–337 (1997)
8. Hardt, M., Srivastava, N., Tulsiani, M.: Graph densification. In: Innovations in Theoretical Computer Science 2012, Cambridge, MA, USA, 8–10 January 2012, pp. 380–392 (2012). http://doi.acm.org/10.1145/2090236.2090266
9. Komlós, J., Shokoufandeh, A., Simonovits, M., Szemerédi, E.: The regularity lemma and its applications in graph theory. In: Khosrovshahi, G.B., Shokoufandeh, A., Shokrollahi, A. (eds.) Theoretical Aspects of Computer Science: Advanced Lectures, pp. 84–112. Springer, Berlin (2002)
10. Komlós, J., Simonovits, M.: Szemerédi's regularity lemma and its applications in graph theory. In: Miklós, D., Szonyi, T., Sós, V.T. (eds.) Combinatorics, Paul Erdós is Eighty, pp. 295–352. János Bolyai Mathematical Society, Budapest (1996)
11. Liu, W., He, J., Chang, S.: Large graph construction for scalable semi-supervised learning. In: Proceedings of the 27th International Conference on Machine Learning (ICML-2010), Haifa, Israel, 21–24 June 2010, pp. 679–686 (2010)
12. von Luxburg, U., Radl, A., Hein, M.: Hitting and commute times in large random neighborhood graphs. J. Mach. Learn. Res. **15**(1), 1751–1798 (2014)
13. Pelillo, M., Elezi, I., Fiorucci, M.: Revealing structure in large graphs: Szemerédi's regularity lemma and its use in pattern recognition. Pattern Recogn. Lett. **87**, 4–11 (2017)
14. Robles-Kelly, A., Loog, M., Biggio, B., Escolano, F., Wilson, R. (eds.): S+SSPR 2016. LNCS, vol. 10029. Springer, Cham (2016)
15. Sperotto, A., Pelillo, M.: Szemerédi's regularity lemma and its applications to pairwise clustering and segmentation. In: Proceedings of the 6th International Conference on Energy Minimization Methods in Computer Vision and Pattern Recognition, EMMCVPR 2007, Ezhou, China, 27–29 August 2007, pp. 13–27 (2007)
16. Szemerédi, E.: Regular partitions of graphs. In: Colloques Internationaux CNRS 260–Problèmes Combinatoires et Théorie des Graphes, pp. 399–401. Orsay (1976)

Mining and Clustering

Mining Frequent Patterns in 2D+t Grid Graphs for Cellular Automata Analysis

Romain Deville[1,2]([⊠]), Elisa Fromont[1], Baptiste Jeudy[1], and Christine Solnon[2]

[1] UJM, CNRS, LaHC UMR 5516, 42000 Saint-Etienne, France
romain.deville@insa.lyon.fr
[2] Université de Lyon, INSA-Lyon, LIRIS UMR 5205, 69621 Villeurbanne, France

Abstract. A 2D grid is a particular geometric graph that may be used to represent any 2D regular structure such as, for example, pixel grids, game boards, or cellular automata. Pattern mining techniques may be used to automatically extract interesting substructures from these grids. 2D+t grids are temporal sequences of grids which model the evolution of grids through time. In this paper, we show how to extend a 2D grid mining algorithm to 2D+t grids, thus allowing us to efficiently find frequent patterns in 2D+t grids. We evaluate scale-up properties of this algorithm on 2D+t grids generated by a classical cellular automaton, *i.e.*, the game of life, and we show that the extracted spatio-temporal patterns may be used to analyze this kind of cellular automata.

1 Introduction

A 2D grid is a particular geometric graph that may be used to model any 2D regular structure such as, for example, grids of pixels (*i.e.*, images), game boards, or cellular automata. To characterize these grids, we may mine them to extract recurrent patterns [6]. In some applications, we use temporal sequences of grids (*i.e.*, 2D+t grids) to model the evolution of grids through time. This is the case, for example, of videos, or sequences of actions in board games. In this paper, we motivate and illustrate our work on Cellular Automata (CA) used to model the temporal evolution of ecosystems [3,12,13]. Indeed, biodiversity of ecosystems is increasingly recognized as an important element of global change. CA-based models are used to understand, predict and control spatio-temporal spread of species which is a key issue to preserve biodiversity [9]. A CA is a regular grid of cells. Each cell has a state which evolves through time, depending on the state of its neighbours in the grid. One of the most famous CA is the Game of Life [5]. In this CA, the grid is in 2 dimensions (on toric grids), and each cell has 8 neighbours (horizontally, vertically, and diagonally). Initially (at time $t = 0$), each cell is either alive or dead. The state at time $t + 1$ of a cell depends on its state and on the state of its 8 neighbours at time t. It is computed by applying the following rules: (1) if the cell is alive at time t and has 2 or 3 living neighbours, then it is alive at time $t + 1$, otherwise it becomes dead; (2) if the cell is dead at time t and has exactly 3 living neighbours, then it becomes alive at time $t + 1$, otherwise it stays dead. When executing a CA from a given initial state, one

© Springer International Publishing AG 2017
P. Foggia et al. (Eds.): GbRPR 2017, LNCS 10310, pp. 177–186, 2017.
DOI: 10.1007/978-3-319-58961-9_16

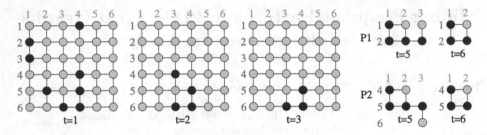

Fig. 1. Left: First three states of a 6×6 game of life modelled with a $2D+t$ grid. Living (resp. dead) cells are displayed in black (resp. gray). (x, y) coordinates are displayed in green and blue, respectively. Temporal edges are not displayed, but there is a temporal edge between each pair of nodes (i, j) such that $x_i = x_j$, $y_i = y_j$, and $|t_i - t_j| = 1$. Right: Two examples of spatio-temporal patterns (temporal edges are not displayed). P1 is isomorphic to a subgrid of the grid on the left (with translation $T = (2, 4, -4)$ and rotation $\theta = -\pi/2$). P2 is also isomorphic to a subgrid of the grid on the left. P2 is not isomorphic to P1 because the angle between edges (a, b) and (b, c) with $x_a = y_a = y_b = 2, x_b = x_c = 3$, and $y_c = 1$ in P1 is not preserved in P2. (Color figure online)

may observe the emergence of spatio-temporal patterns, and these patterns are characteristic of different ecosystem outcomes. [12] distinguishes four possible outcomes: (1) development of a homogeneous fixed pattern, (2) development of a periodic pattern, (3) development of a chaotic pattern, and (4) development of patterns composed of homogeneous regions and regions containing complex localized structures.

In this paper, we present an efficient algorithm for extracting spatio-temporal patterns in 2D+t grids. This algorithm may be used, for example, to extract meaningful spatio-temporal patterns in CA. When CA are used to model ecosystems, these patterns could be used by ecologists to better understand and control the dynamics of the ecosystems. For example, [3] explains that we can foresee the future of an ecosystem by identifying recurring patterns. Ecologists are also interested in understanding how dependent the patterns and the initial state are.

Our algorithm is an extension of GRiMA [6], an algorithm for mining 2D grids which has been designed to tackle real-life applications such as image classification. This algorithm is recalled in Sect. 2. In Sect. 3, we show how to extend it to mine 2D+t grids. In Sect. 4, we evaluate scale-up properties of our new algorithm for mining game-of-life CA, and we show that the extracted spatio-temporal patterns are relevant for classification purposes.

2   Background on 2D Grids and GriMA

Definition of 2D Grids, and 2D Subgrid Isomorphism. A 2D grid is a special case of graph such that each node has a 2D coordinate which is a couple of integer values, and each edge connects nodes which are neighbours on a grid. More formally, a **grid** is defined by $G = (N, E, L, x, y)$ such that N is a set of nodes, $E \subseteq N \times N$ is a set of edges, $L : N \cup E \to \mathbb{N}$ is a labeling function which associates a label $L(c)$ with every component (node or edge) $c \in N \cup E$, and $x : N \to \mathbb{Z}$ and $y : N \to \mathbb{Z}$ map each node $u \in N$ to its 2D coordinates (x_u, y_u), and $\forall (u, v) \in E$, $|x_u - x_v| + |y_u - y_v| = 1$. A **subgrid** of a grid $G = (N, E, L, x, y)$ is a grid $G' = (N', E', L', x', y')$ such that $N' \subseteq N$, $E' \subseteq E \cap N' \times N'$ and L', x', and y' are the restrictions of L, x, and y to $N' \cup E'$, N', and N', respectively.

Looking for patterns in a grid amounts to searching for subgrid isomorphisms. Patterns should be invariant to translations and rotations. More formally, the **translation** of $G = (N, E, L, x, y)$ by a vector $T \in \mathbb{Z}^2$, denoted $G + T$, is the grid obtained by moving all its nodes with respect to T, i.e., $\forall u \in N$, (x_u, y_u) becomes $(x_u, y_u) + T$. Let $\Theta = \{ \left(\begin{smallmatrix} 1 & 0 \\ 0 & 1 \end{smallmatrix} \right), \left(\begin{smallmatrix} 0 & -1 \\ 1 & 0 \end{smallmatrix} \right), \left(\begin{smallmatrix} -1 & 0 \\ 0 & -1 \end{smallmatrix} \right), \left(\begin{smallmatrix} 0 & 1 \\ -1 & 0 \end{smallmatrix} \right) \}$ be the set of rotation matrices of respective angles 0, $\pi/2$, π and $3\pi/2$. The **rotation** of G with respect to $\theta \in \Theta$, denoted θG, is the grid obtained by rotating all its nodes with respect to θ, i.e., $\forall u \in N, (x_u, y_u)$ becomes $(x_u, y_u)\theta$. Two grids G_1 and G_2 are **grid isomorphic** if there exist a translation $T \in \mathbb{Z}^2$ and a rotation $\theta \in \Theta$ such that $G_1 = T + \theta G_2$. Finally, G_1 is **sub-grid-isomorphic** to G_2 if there exists a subgrid of G_2 which is isomorphic to G_2 (see Fig. 1 for an example).

Graph Mining. Given a database D of graphs and a frequency threshold σ, the goal of the graph mining problem is to output all frequent subgraphs in D, i.e., all graphs G such that there exist at least σ graphs in D to which G is sub-isomorphic. This problem may be solved by GSPAN [14], and all similar general exhaustive graph mining algorithms [8]. However, as the subgraph isomorphic problem is $\mathcal{NP}$-complete, these algorithms do not scale well. On the other hand, PLAGRAM [11] and FREQGEO [1] are graph mining algorithms dedicated to special cases of graphs for which subgraph isomorphism becomes polynomial, i.e., plane graphs for PLAGRAM and geometric graphs for FREQGEO. These algorithms have better scale-up properties. However, PLAGRAM only mines patterns composed of faces and the smallest possible subgraph pattern is a single face, i.e., a cycle with 3 nodes. Using PLAGRAM to mine grids is possible but the problem needs to be transformed such that each grid node becomes a face in the graph tackled by PLAGRAM. This transformation artificially increases the number of nodes and edges which causes a scalability problem for PLAGRAM. Also, grids are special cases of geometric graphs. Therefore, FREQGEO may be used to mine grids. However, it has a higher time-complexity than GRIMA, the 2D grid mining algorithm introduced in [6].

Description of GRIMA. GRIMA follows the same basic principle as GSPAN, PLAGRAM, and FREQGEO to avoid generating the same pattern multiple times: It uses codes to represent grids. This code is a list of edges encountered when performing a traversal of the grid. A grid may have several codes but one of them is chosen as the signature: The canonical code, which is the largest code wrt lexicographic order. GRIMA explores the search space of all canonical codes in a depth-first recursive way. It first computes all frequent edges and then calls an Extend function for each of these frequent extensions. Extend has one input parameter: A pattern code P which is frequent and canonical. It outputs all frequent canonical codes P' such that P is a prefix of P'. To this aim, it first computes the set E of all possible valid extensions of all occurrences of P in the database D of grids: A valid extension is the code e of an edge such that $P.e$ occurs in D. Finally, Extend is recursively called for each extension e such that $P.e$ is frequent and canonical. Hence, at each recursive call, the pattern grows.

3   2D+t Grid Mining Algorithm

A 2D+t grid is defined by a tuple (N, E, L, x, y, t) such that (N, E, L, x, y) is a 2D grid graph and $t : N \to \mathbb{Z}$ is a function that maps nodes to temporal coordinates, i.e., $\forall u \in N$, t_u is the temporal coordinate of node u. Also, edges are enforced to connect neighbour nodes in the grid, i.e., $\forall (u, v) \in E$, $|x_u - x_v| + |y_u - y_v| + |t_u - t_v| = 1$. We distinguish two different kinds of edges: spatial edges (such that $|x_u - x_v| + |y_u - y_v| = 1$) and temporal edges (such that $|t_u - t_v| = 1$).

2D grid isomorphism is defined so that isomorphism is invariant to translations and rotations. When extending this definition to 2D+t grids, we still ensure that isomorphism is invariant to translations wrt all axis. However, as time is an oriented dimension, we allow rotations only along the temporal axis. Hence, we consider the set $\Theta = \{\left(\begin{smallmatrix}1&0&0\\0&1&0\\0&0&1\end{smallmatrix}\right), \left(\begin{smallmatrix}0&-1&0\\1&0&0\\0&0&1\end{smallmatrix}\right), \left(\begin{smallmatrix}-1&0&0\\0&-1&0\\0&0&1\end{smallmatrix}\right), \left(\begin{smallmatrix}0&1&0\\-1&0&0\\0&0&1\end{smallmatrix}\right)\}$ of rotation matrices of respective angles 0, $\pi/2$, π and $3\pi/2$ along the temporal axis.

To extend GRIMA to 2D+t grids, we have to define the canonical code of a 2D+t grid. A **code** $C(G)$ of a 2D+t grid G is a sequence of n edge codes ($C(G) = \langle ec_0, ..., ec_{n-1} \rangle$) which is associated with a depth-first traversal of G starting from a given initial node. During this traversal, each edge is traversed once, and nodes are numbered: The initial node has number 0; each time a new node is discovered, it is numbered with the smallest integer not already used in the traversal. Each edge code corresponds to a different edge of G and the order of edge codes in $C(G)$ corresponds to the order edges are traversed. Hence, ec_k is the code associated with the kth traversed edge. This edge code ec_k is the tuple $(\delta, i, j, a, L_i, L_j, L_{(i,j)})$ where:

- i and j are the numbers associated with the nodes of the kth traversed edge.
- $\delta \in \{0, 1\}$ is the direction of the kth traversed edge:
  - $\delta = 0$ if it is forward, i.e., j is a new node reached for the first time;
  - $\delta = 1$ if it is backward, i.e., j already appears in $\langle ec_0, ..., ec_{k-1} \rangle$.
- $a \in \{-2, -1, 0, 1, 2, 3\}$ is the angle value of the kth traversed edge (i, j):
  - if (i, j) is a temporal edge, then $a = -2$ if $t_i = t_j + 1$, and $a = -1$ if $t_i = t_j - 1$;
  - else, (i, j) is a spatial edge.
    * If (i, j) is the first spatial edge encountered since the beginning of the traversal, then $a = 0$.
    * Else, let (l, m) be the first spatial edge in $\langle ec_0, ..., ec_{k-1} \rangle$ such that $x_i = x_m$ and $y_i = y_m$. We have $a = 2A/\pi$ where $A \in \{0, \pi/2, \pi, 3\pi/2\}$ is the angle between (l, m) and (i, j) in the x, y plane.
- L_i, L_j, $L_{(i,j)}$ are labels of i, j, and (i, j), respectively.

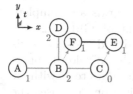

	Code 1		Code 2		Code 3	
edge	$\delta\ i\ j\ a\ L_i L_j L_{ij}$	edge	$\delta\ i\ j\ a\ L_i L_j L_{ij}$	edge	$\delta\ i\ j\ a\ L_i L_j L_{ij}$
(D,B)	0 0 1 0 2 2 0	(D,B)	0 0 1 0 2 2 0	(C,E)	0 0 1 -1 2 1 0
(B,C)	0 1 2 3 2 0 0	(B,A)	0 1 2 1 2 1 0	(E,F)	0 1 2 0 1 1 0
(C,E)	0 2 3 -1 0 1 0	(B,C)	0 1 3 3 2 0 0	(F,B)	0 2 3 -2 1 2 0
(E,F)	0 3 4 0 1 1 0	(C,E)	0 3 4 -1 0 1 0	(B,D)	0 3 4 1 2 2 0
(F,B)	1 4 1 -2 1 2 0	(E,F)	0 4 5 0 1 1 0	(B,A)	0 3 5 2 2 1 0
(B,A)	0 1 5 1 2 1 0	(F,B)	1 5 1 -2 1 2 0	(B,C)	1 3 0 3 2 0 0

Fig. 2. Left: A 2D+t grid (temporal edges are displayed in red, node labels are displayed next to nodes, and all edges have the same label 0). Right: 3 codes for this grid (other codes may be built by changing the traversal). (Color figure online)

For example, let us consider code 1 in Fig. 2. Let us explain how the code of the fourth traversed edge (E,F) is built. $\delta = 0$ because (E,F) is a forward edge (F has not been reached before). (E,F) is a spatial edge, and the first spatial edge (l,m) such that m has the same spatial coordinates as E is (B,C). The angle between (E,F) and (B,C) is 0. So, $a = 0$. For the fifth edge of code 1, (F,B), $\delta = 1$ because B has already been reached before (backward edge). As (F,B) is a temporal edge, $a = -2$.

Given a code, we can reconstruct the corresponding grid since edges are listed in the code together with angles and labels. However, there exist different possible codes for a given grid, as illustrated in Fig. 2: Each code corresponds to a different traversal (starting from a different initial node and choosing edges in a different order). As we did for GRiMA, we define a total order on the set of all possible codes that may be associated with a given grid by considering a lexicographic order (all code components have integer values). Among all the possible codes for a grid, the largest one according to this order is the canonical code of this grid and it is unique. For example, in Fig. 2, code 1 is canonical: It is greater than codes 2 and 3, and it is also greater than all other possible codes for this grid (not shown here).

Note that it is not necessary to exhaustively build all codes when computing a canonical code. We use heuristics to first build large codes (by first choosing spatial edges with $3\pi/2$ angles, such as for (D,B) and (B,C), for example). Also, when building a code, we stop the traversal as soon as the corresponding code becomes smaller than the largest current code.

This canonical code for 2D+t allows us to extend GRiMA to mine 2D+t grids in a straightforward way, and we can show that the resulting mining algorithm, called GRiMA2D+t, is both *correct* (it only outputs frequent subgrids) and *complete* (it cannot miss any frequent subgrid). The proof (not detailed due to lack of space) basically shows that every prefix of a canonical code is canonical.

GRiMA2D+t enumerates all frequent patterns in $\mathcal{O}(kn^2.|P|^2) = \mathcal{O}(kn^4)$ time per pattern P, where k is the number of grids in the set D of input grids, n the size of the largest grid $G_i \in D$ (in number of edges) and $|P|$ the number of edges in a pattern P.

Node-induced GRiMA*2D+t*. In our application, the mined grids are complete and have no label on edges. Thus, we designed a variant of GRiMA2D+t, called node-induced-GRiMA2D+t, which computes node-induced grids, i.e. grids induced by their node sets. This corresponds to a "node-induced" closure operator on graphs where, given a pattern P, we add all possible edges to P without adding new nodes. We have shown in [6] that this optimization decreases the number of extracted patterns and the extraction time.

Limitation on Edge Extension. Moreover, to avoid mining patterns that only contain dead cells, we also limit the extension procedure of our mining process. In the Extend function, we forbid extension with edges linking two dead cells. As a consequence, every edge (i, j) in a mined pattern is such that either i, or j, or both i and j correspond to living cells.

4   Experiments

We study the scale-up properties of GRiMA*2D+t* and assess the relevance of the mined patterns on a classification task related to the behavior of a CA, *i.e.*, the *Game of Life* described in Sect. 1. More precisely, given the k first cell states, with $k \in \{1, 2, 5, 10, 20\}$, the goal is to forecast the outcome at time $t = 1000$, where we only consider two possible outcomes: *dead* (if all cells are dead at time $t = 1000$), or *alive* (if at least one cell is alive at time $t = 1000$).

Dataset. We consider four sizes of grids $n \times n$, with $n \in \{20, 30, 40, 50\}$. For each size n, we randomly choose the initial state (dead or alive) of each $n \times n$ cell wrt to a cell probability p. We have chosen p in such a way that the outcome at time $t = 1000$ is *dead* or *alive* with equal probabilities. This way, we ensure during our dataset generation process that there is no bias towards one of the two classes. This imposes a cell probability p of 74%, 78%, 80%, and 81% for $n = 20, 30, 40$, and 50, respectively. Besides, to avoid trivial predictions of the class *dead*, due to the fact that all cells may be dead before the kth iteration, we only select initial states such that there is at least one cell alive at the 50th iteration. For each size $n \in \{20, 30, 40, 50\}$, we generate a set S_n of 2000 initial states such that the outcome at time $t = 1000$ is *dead* for half of them (S_n^d), and *alive* for the other half (S_n^a). We split each set S_n^d and S_n^a into two equal parts for learning (L_n^d and L_n^a) and training (T_n^d and T_n^a).

2D+t Grids. For each state $s_i \in S_n$ (with $n \in \{20, 30, 40, 50\}$), and for each temporal horizon $k \in \{1, 2, 5, 10, 20\}$, we build a 2D+t grid $G(s_i, k)$ which is a temporal sequence of k 2D grids: The first one corresponds to the state s_i, and the next $k - 1$ ones correspond to states obtained by iteratively applying the game-of-life rules starting from s_i. Each node is labeled with either 0 (*dead* cell) or 1 (cell *alive*), and all edges have the same label.

Mining Process. For each size $n \in \{20, 30, 40, 50\}$ and each temporal horizon $k \in \{1, 2, 5, 10, 20\}$, we mine frequent patterns in the learning sets. This is done for each class separately: We compute the set $F_{n,k}^d$ (resp. $F_{n,k}^a$) of frequent patterns in all $G(s_i, k)$ with $s_i \in S_n^d$ (resp. $s_i \in S_n^a$). We consider two different frequency threshold $\sigma \in \{50\%, 100\%\}$: When $\sigma = 50\%$ (resp. $\sigma = 100\%$), a pattern is frequent if it is present in half of the grids (resp. all the grids). Note that, the higher the frequency, the lower the number of mined patterns and the more efficient the mining process. Each mining process has been limited to 12 h of CPU time: If the mining process is not completed after 12 h, we stop it and consider the subset of patterns that have been extracted within this time limit.

Classification Process. For each size $n \in \{20, 30, 40, 50\}$ and each temporal horizon $k \in \{1, 2, 5, 10, 20\}$, we build the set $F_{n,k} = F_{n,k}^a \cup F_{n,k}^d$ that contains all frequent patterns (in the two classes). Then, for each state $s_i \in L_n^d \cup L_n^a$, we count the number of occurrences of each pattern of $F_{n,k}$ in $G(s_i, k)$, and build a frequency vector that gives the frequency of each pattern. Hence, each state is represented by a histogram of frequent substructures.

We report two sets of experiments: One with histograms created using all the patterns mined on both classes (which can be very sparse) and one with a selected subset of 100 patterns. This post-processing selection is performed using the relevance score and the greedy selection algorithm presented in [7]. To fasten the preprocessing step, we delete at each of the 100 iterations of the greedy algorithm, the patterns with the 10% lowest scores.

Frequency vectors (of length $|F_{n,k}|$ or 100) are used to train a binary Support Vector Machine (SVM) to discriminate between the two classes. We use the Libsvm [4] library with the intersection kernel presented in [10] (known to be good on histograms).

Finally, we use the trained model to forecast the class of each state in our training set: For each state $s_i \in T_n^d \cup T_n^a$, we count the number of occurrences of each pattern of $F_{n,k}$ (or the 100 selected patterns of $F_{n,k}$) in $G(s_i, k)$, and build a frequency vector which is used by the SVM model to forecast an outcome (dead or alive) which is compared to the true outcome (dead for states coming from T_n^d and alive for states coming from T_n^a). We report accuracy results, *i.e.*, the percentage of states for which the forecasted outcome is equal to the true outcome.

Accuracy Results. We report accuracy results in Table 1. When increasing the temporal horizon k (*i.e.*, the temporal size of the mined grids), accuracy results are improved. This shows the relevance of the GRIMA2D+t algorithm compared to GRIMA. However, when increasing k, the mining process needs more time and we often had to stop the mining process after 12 h (red cells) for the largest values of k. In this case, we only explored part of the substructure search space.

Also, the larger the grid size n, the better the results. It is well known that, for the game of life, large grids have higher probabilities of containing stable patterns that may characterize *alive* outcomes. However, as with k, we often

Table 1. Accuracy results for the classification of states in $T_n^d \cup T_n^a$. For each size $n \in \{20, 30, 40, 50\}$, the first line reports the number of frequent patterns $|F_{n,k}|$, and the cell is colored in red if the 12 h time-out has been reached and green otherwise; the second and third line report accuracy results with vectors of size $|F_{n,k}|$ and 100, respectively (if $|F_{n,k}| < 100$, results are not given for vectors of size 100). Each line gives results for $k = 1, 2, 5, 10$, and 20, and with $\sigma = 50\%$ and 100%.

k		1		2		5		10		20	
n	σ	50%	100%	50%	100%	50%	100%	50%	100%	50%	100%
	$\lvert F_{n,k}\rvert$	673	22	23589	64	824616	2478	707743	96762	417724	213861
20	All Pat	72.40	70.70	77.30	72.50	83.40	85.20	85.00	88.70	88.30	91.30
	100 Pat	72.70		75.80		83.80	83.50	84.10	87.80	85.60	89.50
	$\lvert F_{n,k}\rvert$	662	18	28795	68	783701	2472	688546	99827	355381	252891
30	All Pat	77.00	68.40	81.60	76.60	84.80	87.80	88.00	89.10	92.70	92.80
	100 Pat	74.10		79.90		84.40	86.90	88.40	89.40	91.80	92.20
	$\lvert F_{n,k}\rvert$	667	27	38103	77	786619	4620	634501	178710	403411	246885
40	All Pat	79.10	70.90	86.50	82.10	89.90	92.50	91.60	93.10	95.50	96.00
	100 Pat	77.10		84.40		86.80	90.10	89.40	93.60	94.30	96.80
	$\lvert F_{n,k}\rvert$	740	26	46235	79	906209	4171	720779	206508	373600	282003
50	All Pat	78.60	72.70	82.80	79.70	89.30	90.90	91.80	93.20	96.40	95.80
	100 Pat	77.90		84.20		88.50	89.40	89.90	92.90	91.90	96.40

had to stop the mining process after 12 h for the largest grids. This shows the necessity of efficient algorithms to tackle real-life problems.

The number of mined patterns $|F_{n,k}|$, is smaller when the frequency threshold $\sigma = 100\%$ than when it is 50%. For small temporal horizons $k \in \{1, 2\}$, the number of mined patterns is not large enough (smaller than 27 for $k = 1$ and than 79 for $k = 2$). In this case, the results obtained with $\sigma = 100\%$ are worse than those obtained with $\sigma = 50\%$. However, for larger time horizons $k \in \{5, 10, 20\}$, the number of mined patterns becomes large enough for $\sigma = 100\%$ while it becomes so large for $\sigma = 50\%$ that the mining process is never completed. As we only have a subset of the frequent patterns in this case, it may be possible that some relevant patterns have not be found. We observe that in this case the results are worse with $\sigma = 50\%$ than with $\sigma = 100\%$.

Finally, let us compare the results obtained when all patterns of $F_{n,k}$ are used for the classification (All Pat) with the results obtained when we only use the 100 first patterns selected by the post-processing process (100 Pat): The difference is usually rather small, and in some cases it improves results (e.g., $k = 20, n = 40$) whereas in some other cases it degrades them (e.g., $k = 20, n = 20$). However, the post-processing improves the efficiency of the counting step: The process of counting all occurrences of all patterns of $F_{n,k}$ (to create histograms used as inputs for the SVMs) takes on average 0.002 s when $k = 1$ and up to 45 s when $k = 20$, whereas it takes 0.0005 s when $k = 1$ and up to 0.008 s when $k = 20$ if we only count occurrences of the 100 patterns selected by post-processing.

Table 2. Average and Maximum (in parenthesis) depth and number of cells for all patterns of $F_{n,k}$ or only those selected with post processing.

k		1		2		5		10		20	
n		All Pat	100 Pat	All Pat	100 Pat	All Pat	100 Pat	All Pat	100 Pat	All Pat	100 Pat
20	Depth	0 (0)	0 (0)	0.9 (1)	0.9 (1)	2.1 (4)	2.0 (4)	5.3 (9)	4.5 (9)	11.7 (19)	9.0 (19)
	NbCell	6.4 (11)	6.6 (10)	7.9 (15)	7.2 (11)	11.3 (25)	11.8 (20)	14.8 (28)	15.3 (24)	18.3 (32)	14.7 (25)
30	Depth	0 (0)	0 (0)	0.9 (1)	0.9 (1)	2.3 (4)	2.4 (4)	6.4 (9)	6.0 (9)	14.5 (19)	11.1 (19)
	NbCell	6.4 (11)	6.4 (9)	8.4 (17)	7.7 (12)	11.7 (23)	11.0 (18)	16.8 (35)	15.3 (29)	21.5 (32)	16.6 (25)
40	Depth	0 (0)	0 (0)	0.9 (1)	0.9 (1)	2.3 (4)	2.2 (4)	6.8 (9)	6.4 (9)	16.2 (19)	14.1 (19)
	NbCell	6.4 (12)	6.3 (9)	9.0 (18)	7.6 (13)	12.1 (24)	11.1 (17)	20.2 (39)	21.3 (35)	23.4 (34)	19.6 (27)
50	Depth	0 (0)	0 (0)	1.0 (1)	0.9 (1)	2.3 (4)	2.3 (4)	6.7 (9)	6.5 (9)	16.5 (19)	15.6 (19)
	NbCell	6.5 (12)	6.7 (11)	9.2 (19)	7.5 (13)	13.2 (27)	12.3 (21)	19.4 (39)	20.5 (33)	24.1 (34)	21.8 (30)

Overall, those results show that taking into account the structural information along the spatio-temporal grids can be used for the prediction of the outcome of cellular automata and the extension to temporal dimension of our grid mining algorithm can be used to tackle spatio-temporal problems.

Patterns Statistics. Table 2 reports some statistics about the mined patterns (all patterns in $F_{n,k}$, or the 100 ones selected by post-processing). We report the average and maximum (in parenthesis) number of nodes of each pattern as well as their depth, i.e., the number of temporal steps on which the patterns are present (spatial patterns have a depth of 0). The average number of nodes and the depth of the patterns selected by post-processing are usually less important than the same statistics for all the mined patterns. This may come from the fact that deep patterns are not diverse enough to be selected by the post-processing step which in turn suggests that, when the timeout is reached, the diversity of the mined pattern is not high enough. To further increase this diversity, stochastic search methods such as Monte-Carlo Tree Search [2] could be integrated in our algorithm.

5   Conclusion and Future Work

We have presented GRIMA2D+t, an algorithm to mine temporal sequences of 2D regular structures called grids. We have shown on experiments on a classical cellular automaton, the game of life, that GRIMA2D+t can effectively extract spatio-temporal patterns in temporal grids. We have also shown that those patterns can be used as new features for classification algorithms and, in particular, to successfully predict the outcomes of cellular automata. This opens interesting new paths in the automatic analysis of the evolution of ecosystems and, in particular, to predict and control spatio-temporal spread of species in order to preserve biodiversity.

To further increase the efficiency of the classification process and scale to larger problems, we proposed to use a post-processing step that allows us to select a good subset of the mined patterns. In future work, we planned to directly mined a relevant subset of the possible patterns by using Monte-Carlo tree search methods. We also plan to apply this algorithm to analyze other temporal structures such as videos.

Acknowledgements. This work has been supported by the ANR project SoLStiCe (ANR-13-BS02-0002-01).

References

1. Arimura, H., Uno, T., Shimozono, S.: Time and space efficient discovery of maximal geometric graphs. In: Corruble, V., Takeda, M., Suzuki, E. (eds.) DS 2007. LNCS, vol. 4755, pp. 42–55. Springer, Heidelberg (2007). doi:10.1007/978-3-540-75488-6_6
2. Bosc, G., Raïssi, C., Boulicaut, J.-F., Kaytoue, M.: Any-time diverse subgroup discovery with Monte Carlo tree search. CoRR (2016)
3. Breckling, B., Pe'er, G., Matsinos, Y.G.: Cellular automata in ecological modelling. In: Jopp, F., Reuter, H., Breckling, B. (eds.) Modelling Complex Ecological Dynamics: An Introduction into Ecological Modelling for Students, Teachers & Scientists, pp. 105–117. Springer, Heidelberg (2011)
4. Chang, C.-C., Lin, C.-J.: LIBSVM: a library for support vector machines. ACM-TIST **2**, 27:1–27:27 (2011). http://www.csie.ntu.edu.tw/~cjlin/libsvm
5. Conway, J.: The game of life. Sci. Am. **223**(4), 4 (1970)
6. Deville, R., Fromont, E., Jeudy, B., Solnon, C.: GriMa: a grid mining algorithm for bag-of-grid-based classification. In: Robles-Kelly, A., Loog, M., Biggio, B., Escolano, F., Wilson, R. (eds.) S+SSPR 2016. LNCS, vol. 10029, pp. 132–142. Springer, Cham (2016). doi:10.1007/978-3-319-49055-7_12
7. Fernando, B., Fromont, É., Tuytelaars, T.: Mining mid-level features for image classification. IJCV **108**(3), 186–203 (2014)
8. Jiang, C., Coenen, F., Zito, M.: A survey of frequent subgraph mining algorithms. KER **28**, 75–105 (2013)
9. Marco, D.E., Páez, S.A., Cannas, S.A.: Species invasiveness in biological invasions: a modelling approach. Biol. Invasions **4**(1), 193–205 (2002)
10. Odone, F., Barla, A., Verri, A.: Building kernels from binary strings for image matching. IEEE-TIP **14**(2), 169–180 (2005)
11. Prado, A., Jeudy, B., Fromont, E., Diot, F.: Mining spatiotemporal patterns in dynamic plane graphs. IDA **17**, 71–92 (2013)
12. Wolfram, S.: Cellular automata as models of complexity. Nature **311**(5985), 419–424 (1984)
13. Wootton, J.T.: Local interactions predict large-scale pattern in empirically derived cellular automata. Nature **413**(6858), 841–844 (2001)
14. Yan, X., Han, J.: gSpan: graph-based substructure pattern mining. In: ICDM, pp. 721–724 (2002)

Density Normalization in Density Peak Based Clustering

Jian Hou[1,2(✉)] and Hongxia Cui[3]

[1] College of Engineering, Bohai University, Jinzhou 121013, China
dr.houjian@gmail.com
[2] ECLT, Università Ca' Foscari Venezia, 30124 Venezia, Italy
[3] College of Information Science, Bohai University, Jinzhou 121013, China

Abstract. As a promising clustering approach, the density peak (DP) based algorithm utilizes the data density and carefully designed distance to identify cluster centers and cluster members. The key to this approach is the density calculation, which has a significant impact on the clustering results. However, the original DP algorithm applies the local density to identify cluster centers directly, and fails to take into account the density difference among clusters. As a result, large-density clusters may be partitioned into multiple parts and small-density clusters are likely to be merged with other clusters. In this paper we introduce a density normalization step to deal with this problem, and show that the normalized density can be used to characterize cluster centers more accurately than the original one. In experiments on various datasets, our method is shown to improve the performance of different density kernels evidently.

Keywords: Density peak · Clustering · Density normalization · Density kernel

1 Introduction

As an important unsupervised learning approach, data clustering has been studied extensively for decades, and a lot of algorithms have been proposed [1,2,5,18,21]. Among various branches of clustering algorithms, graph based clustering has been attracting increasing attention due to the impressive performance. Graph based algorithms use as input the pairwise data similarity (distance) matrix, which captures rich information of the data distribution. Different algorithms have been proposed to make use of the data distribution information. Spectral clustering [12,13] performs dimensionality reduction based on the eigenstructure of the similarity matrix and then accomplish the clustering in the data space of fewer dimensions. As an instance of spectral clustering, the normalized cuts algorithm (NCuts) [18] has become a standard baseline of image segmentation techniques, and important advances in this area include the algorithm based on robust graph [21]. The affinity propagation algorithm (AP) [2] identifies cluster centers and cluster members iteratively by passing among data the affinity

© Springer International Publishing AG 2017
P. Foggia et al. (Eds.): GbRPR 2017, LNCS 10310, pp. 187–196, 2017.
DOI: 10.1007/978-3-319-58961-9_17

message encoded in the similarity matrix. Different from both spectral clustering and AP, the dominant sets algorithm (DSets) [14,16] extends the clique concept in graph theory to edge-weighted graph, and defines dominant set as a graph based concept of a cluster. By treating a dominant set as a cluster, the DSets algorithm extracts clusters sequentially with game dynamics and obtains the number of clusters automatically. Further works on DSets include [8,9,15,19].

Another graph based clustering approach worth mentioning is the density peak based algorithm (DP) proposed in [17]. With the pairwise data distance matrix as input, the DP algorithm firstly calculates the local density ρ of each data, which is then used to calculate the distance δ denoting the distance between one data and its nearest neighbor with larger density. It is found that cluster centers are often with both large ρ's and large δ's and correspond to the density peaks of the dataset, whereas the non-center data are usually with either small ρ's or small δ's. Consequently, the cluster centers are presented as the outliers of the dataset in the so-called ρ-δ decision graph, and can be identified relatively easily. After the cluster centers are identified, each non-center data is assigned the same label as its nearest neighbor with larger density. Considering that identifying cluster centers with the ρ-δ decision graph involves two thresholds, [17] further proposes to use $\gamma = \rho\delta$ as the single measure to describe the data, and select the data with largest γ's as the cluster centers. The DP algorithm is reported to generate superior clustering results in [17].

Unfortunately, [17] fails to provide a reliable method to identify cluster centers. While cluster centers are assumed to have both large σ's and large δ's, or equivalently, large γ's, there is no clear distinction between the large and small values of these features. As a result, it is not easy to determine how many data should be identified as cluster centers. Even if the number of cluster centers are given, it is possible that not all of the cluster centers with largest γ's are really cluster centers. In the case that the densities of different clusters are similar, the centers of all the clusters can be identified and the method works well. However, in the case that different clusters have a large density difference, even the density peak of a small-density cluster may have a small ρ. Consequently, a large-density cluster may have multiple density peaks being identified as cluster centers, whereas no data in a small-density cluster is selected as the cluster center. As a result, a large-density cluster is likely to be partitioned into several clusters, and a small-density cluster may be merged with other clusters. In summary, using the original density ρ to identify cluster centers may not be appropriate as the density difference among clusters is not taken into account. On the basis of the work in [10], we propose to use density normalization to account for the density difference among clusters. We show that the normalized density can be used to describe cluster centers more accurately, and therefore improves clustering results in comparison to the original one.

The remainder of this paper is organized as follows. Section 2 provides a brief introduction of the DP algorithm. Then in Sect. 3 we discuss the problem of the DP algorithm and present normalized density as a better alternative to the original one. The detailed experimental validation of the proposed method is reported in Sect. 4. Finally, Sect. 5 concludes this paper.

2 Density Peak Clustering

The DP algorithm is proposed on the basis of the following assumptions. First, cluster centers are local density peaks in their neighborhoods. This means that cluster centers have larger ρ's than the neighboring non-center data. Second, with δ denoting the distance between one data and its nearest neighbor with larger density, cluster centers are with large δ's. This assumption is supported by the observation that cluster centers are surrounding non-center data with smaller density. In contrast, as in practice few data have identical density, it is easy to find a neighbor with larger density for a non-center data. Consequently, the δ's of non-center data are usually small. Third, the label of one non-center data is the same as that of its nearest neighbor with larger density. The first two assumptions point out the difference between cluster centers and non-center data, and are used to identify cluster centers and determine the labels of center data. After the center of each cluster is determined, the third assumption is used to group non-center data into respective clusters. While the three assumptions have no theoretical foundation, they are consistent with human intuition and are shown to be effective in experiments.

The original DP algorithm is described in the following. Given the pairwise distance matrix, the first step is to calculate the local density ρ of each data. Two density kernels are used in [17] for this purpose, namely cutoff and Gaussian kernel. With the cutoff kernel, the density of the data i is calculated by

$$\rho_i = \sum_{j \in S, j \neq i} \chi(d_c - d_{ij}), \tag{1}$$

where S denotes the dataset for clustering, $d_c \in R$ is the cutoff distance, $d_{ij} \in R$ measures the distance between data i and j, and

$$\chi(x) = \begin{cases} 1, & x > 0, \\ 0, & x < 0. \end{cases} \tag{2}$$

Intuitively, the cutoff kernel uses the number of data in the neighborhood of radius d_c to measure the local density. With the Gaussian kernel, the density is computed by

$$\rho_i = \sum_{j \in S, j \neq i} exp(-\frac{d_{ij}^2}{d_c^2}). \tag{3}$$

As to the parameter d_c involved in both kernels, [17] recommends to determine d_c so that on average 1% to 2% of all data are included in the neighborhood. After the local density is obtained, the distance δ_i of the data i is calculated by definition as

$$\delta_i = \min_{j \in S, \rho_j > \rho_i} d_{ij}. \tag{4}$$

With the ρ and δ of each data available, we are able to represent the data as points in the $\rho - \delta$ space and obtain the $\rho - \delta$ decision graph. Taking the Aggregation dataset [7] for example, we show its $\rho - \delta$ decision graph with the cutoff kernel in Fig. 1(b), where d_c is determined by including 1.7% of the data in the neighborhood on average. It is evident that a few data are with both large ρ and γ and isolated from the majority of the dataset. Considering that with the $\rho - \delta$ decision graph we need two thresholds to identify the cluster centers, we further sort the data in the decreasing order according to their γ's, with $\gamma_i = \rho_i \delta_i$, and show the data in the γ decision graph in Fig. 1(c). With the γ decision graph the data with the largest γ's will be selected as cluster centers.

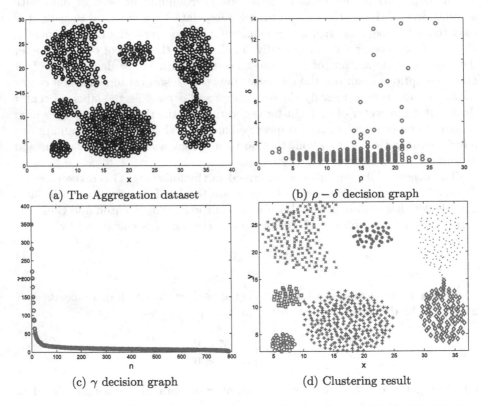

(a) The Aggregation dataset   (b) $\rho - \delta$ decision graph

(c) γ decision graph   (d) Clustering result

Fig. 1. The Aggregation dataset, two decision graphs with the cutoff kernel and cluster result.

While it is true that cluster centers are separated from non-center data with the $\rho - \delta$ decision graph or γ decision graph, we also notice that it is still difficult to select cluster centers, if the number of clusters are not given. In Fig. 1(b) more than 10 data can be regarded as cluster centers, and in Fig. 1 there are also 10 data being isolated from the majority of the dataset. In both cases the obtained numbers of clusters are different from the real value (7). The reason is that there is no clear distinction between the large and small values of features, and therefore the differentiation between center and non-center data are ambiguous. Unfortunately, [17] fails to provide a reliable solution to this problem.

Considering that γ has been proposed to as the sole feature to measure the qualification of data as cluster centers, in this paper we assume that the number of clusters is given by user, as in the case of the k-means algorithm. With the specified number N of clusters, we simply use the N data with largest γ's as the cluster centers. The following clustering of non-center data can then be accomplished easily. The clustering result of the Aggregation dataset is shown in Fig. 1(d). Evidently, the clustering result is very close to the ground truth.

3   Our Algorithm

In the last section we show that with the cutoff kernel the DP algorithm is able to generate very good clustering result on the Aggregation dataset. However, this does not mean that the DP clustering results will be satisfactory with other density kernels and datasets, even if the number of clusters is given. In Fig. 2 we use two examples to illustrate this problem. Figure 2(a) shows the distribution of the 7 selected cluster centers with Gaussian kernel, where we observe that there is one cluster with two centers and one cluster without centers. Correspondingly, the clustering result is not satisfactory, as shown in Fig. 2(c). The two selected cluster centers with the cutoff kernel on the Jain dataset [11] are shown in Fig. 2(b), where both centers appear in the large-density cluster. As a result, the small-density cluster is merged with part of the large-density one in Fig. 2(d).

In our opinion, the incorrectly selected cluster centers shown in Fig. 2 can be attributed to the following reasons. First, while it is generally reasonable to regard density peaks as cluster centers, the adopted density kernel and involve parameters have significant influence on the estimated densities, and then on the selected cluster centers and clustering results. In this sense, it is not strange to see that the cutoff kernel performs well on the Aggregation dataset, and the Gaussian kernel generates unsatisfactory result on the same dataset. Second, in the case that the density difference among clusters is large, even the density peaks of small-density clusters have relatively small ρ's. In this case, applying $\gamma = \rho\delta$ to select cluster centers directly may reduce the chance of data in small-density clusters being selected. This problem is illustrated quite evidently on the Jain dataset in Fig. 2. In addition, these two reasons often interact with some other factors, including the data distribution and cluster size and shapes, and they together result in the incorrectly selected cluster centers.

The basic idea of the DP algorithm is to identify cluster centers and then group non-center data into respective clusters. For this purpose, the features ρ and δ are proposed to select cluster centers and the density relationship among data are used to group non-center data. Since the grouping of non-center data is a simple procedure based on cluster centers and a reasonable assumption, we focus our discussion on the cluster center identification. The key to identifying cluster centers is to highlight the difference between cluster centers and non-center data in some feature space. In the DP algorithm we use $\gamma = \rho\delta$ to select

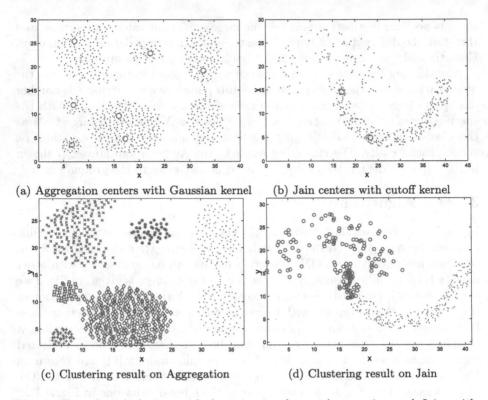

(a) Aggregation centers with Gaussian kernel   (b) Jain centers with cutoff kernel

(c) Clustering result on Aggregation   (d) Clustering result on Jain

Fig. 2. The selected clusters and clustering results on Aggregation and Jain, with different density kernels.

cluster centers. Noticing that δ is calculated based on ρ, we further limit our analysis to the density ρ. It is generally reasonable to use density peaks as the candidates of cluster centers. However, as we use γ as the criterion of cluster center selection, there exists the possibility that the centers of small-density clusters are not selected because of the small density, as shown in the Jain dataset in Fig. 2. In order to make each cluster have one and only one data being selected as cluster center, we need a better feature than the original ρ to characterize cluster centers.

The DP algorithm uses ρ as a feature of the data and regards density peaks as candidates of cluster centers. Here we make use of only the density relationship between cluster centers and non-center data, i.e., cluster centers have larger density than neighboring non-center data. However, in applying $\gamma = \rho\delta$ to select cluster centers, the absolute values of density, but not the density relationship, are what really works. This inconsistency between purpose and implementation is at the root of the problems observed in Fig. 2. In order to relieve this problem, we propose to replace the original ρ by normalized density ρ' in cluster center identification, and use $\gamma' = \rho'\delta$ to select cluster centers. The normalized density can be simply obtained as the ration of the original density and the average density of neighboring data, i.e.,

$$\rho'_i = \frac{\rho_i}{\frac{1}{|D_{inn}|} \sum_{j \in D_{inn}} \rho_j},\tag{5}$$

where D_{inn} is a subset consisting of the 30 nearest neighbors of i, and $|D_{inn}|$ is the subset size. The ρ' defined this way eliminates the influence of density difference among clusters to some extent. It should be noted that the normalized density is only used in selecting the cluster centers. The grouping of non-center data is still based on the relationship of the original density, which reflects the relationship between individual data more accurately. With this simple density normalization step, the selected cluster centers and clustering results on Aggregation and Jain datasets are shown in Fig. 3. We observe from Fig. 3 that both cluster center selection and clustering results are improved significantly. On the other hand, while the two cluster centers are really in two clusters of the Jain dataset, a considerable amount of data are still grouped into the wrong cluster.

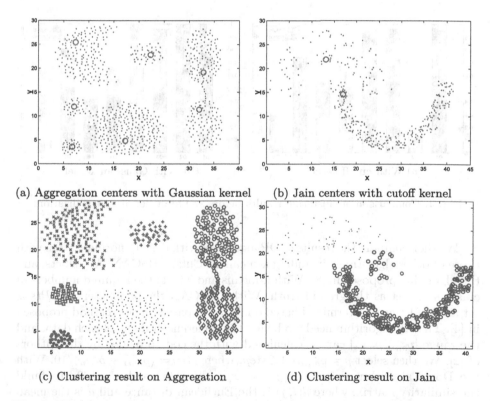

(a) Aggregation centers with Gaussian kernel   (b) Jain centers with cutoff kernel

(c) Clustering result on Aggregation   (d) Clustering result on Jain

Fig. 3. The selected clusters and clustering results on Aggregation and Jain, based on normalized density.

This can be attributed to the effect of several factors, including density kernel and parameters, cluster shape and complex data distribution in clusters.

4   Experiments

We firstly validate the effectiveness of density normalization in improving clustering results quantitatively. The experiment is conducted on eight datasets, including Aggregation, Spiral [3], D31 [20], R15 [20], Jain, Flame [6] and two UCI datasets Iris and Breast. The cutoff kernel and Gaussian kernel are used to calculate the density ρ, and with both kernels the parameter d_c is determined by including 1.7% of the data in the neighborhood of radius d_c. With Normalized Mutual Information (NMI) as the clustering result evaluation criterion, we report the comparison of the results with and without density normalization in Fig. 4. Evidently on all the datasets, the density normalization either improves the clustering results, or keep the results unchanged. This observation shows that the density normalization is effective in solving the problems caused by density difference among clusters.

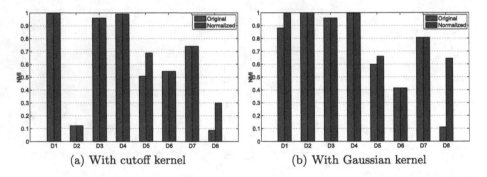

(a) With cutoff kernel   (b) With Gaussian kernel

Fig. 4. Comparison of clustering results before and after density normalization.

We then compare the results of DP algorithm with density normalization with those of other algorithms, including k-means, NCuts, DBSCAN, AP, DSets, and the algorithm proposed in [8]. With k-means and NCuts the required numbers of clusters are set as the ground truth. With DBSCAN the parameter $MinPts$ is set as 3 and Eps is determined based on $MinPts$ based on the method proposed in [4]. The AP algorithm needs to be fed the preference value p of each data, and the range $[p_{min}, p_{max}]$ can be calculated with the code provided by the authors of [2]. We then select $p = p_{min} + 9.2step$, where $step = (p_{max} - p_{min})/10$. With the DSets algorithm we use $s(i, j) = exp(-d(i, j)/\sigma)$ and $\sigma = 20\overline{d}$ to build the similarity matrix, where $d(i, j)$ is the Euclidean distance and $\overline{d}$ is the mean of all pairwise distances. Finally, we also use average distances to evaluate the data density in the form of $\rho_i = d_{max}/d_{imean}$, where d_{max} is the maximum of all pairwise distances, and d_{imean} is the average distance between i and its 30

Table 1. Comparison of cluster results (NMI) among different algorithms on eight datasets.

	DSets	k-means	NCuts	DBSCAN	AP	[8]	cutoff	Gaussian	dmean
Aggregation	0.86	0.85	0.77	0.92	0.82	0.89	0.99	0.99	0.99
Spiral	0.14	0.00	0.00	0.71	0.00	0.66	0.44	0.73	1.00
D31	0.85	0.92	0.96	0.84	0.59	0.67	0.96	0.96	0.96
R15	0.83	0.91	0.99	0.87	0.74	0.91	0.99	0.99	0.99
Jain	0.43	0.36	0.33	0.73	0.46	0.87	0.28	0.68	0.51
Flame	0.60	0.45	0.42	0.83	0.57	0.90	0.78	0.41	1.00
Iris	0.65	0.74	0.74	0.75	0.79	0.60	0.71	0.81	0.72
Breast	0.54	0.74	0.80	0.62	0.57	0.54	0.30	0.64	0.40
Average	0.61	0.62	0.63	0.78	0.57	0.75	0.68	0.78	0.82

nearest neighbors. The comparison of these algorithms are shown in Table 1, where cutoff, Gaussian and dmean are used to denote DP algorithms with the three density kernels. With all the three DP algorithms, normalized density is used to replace the original one.

Table 1 shows that in terms of average clustering results, the cutoff kernel performs the worst in the three kernels of the DP algorithm. This is not strange as this kernel is based on only the number of data in a neighborhood and much distance information is discarded. Even in this case, we find it performs better than DSets, k-means, NCuts and AP. The Gaussian kernel performs much better than the cutoff kernel, and the average distance based kernel further generates the best results in all the algorithms. We believe these observations highlight the potential of the DP algorithm, and also indicates the necessity to explore better density kernels for further performance improvement.

5 Conclusions

Density peak based clustering is a promising clustering approach. In this paper we study the influence of density kernels on the clustering results, and find that existing density kernels are not able to deal with the density difference among clusters. We analyze the reason behind this observation and propose to use density normalization to relieve this problem. In experiments we show that density normalization is able to improve the clustering results evidently. In comparison with other clustering algorithms, the density peak clustering with normalized density generates the best results on average.

Acknowledgement. This work is supported in part by National Natural Science Foundation of China under Grant No. 61473045 and No. 41371425, and in part by China Scholarship Council.

References

1. Ankerst, M., Breunig, M.M., Kriegel, H.P., Sander, J.: Optics: ordering points to identify the clustering structure. In: ACM SIGMOD International Conference on Management of Data, pp. 49–60 (1999)
2. Brendan, J.F., Delbert, D.: Clustering by passing messages between data points. Science **315**, 972–976 (2007)
3. Chang, H., Yeung, D.Y.: Robust path-based spectral clustering. Pattern Recogn. **41**(1), 191–203 (2008)
4. Daszykowski, M., Walczak, B., Massart, D.L.: Looking for natural patterns in data: Part 1. density-based approach. Chemometr. Intell. Lab. Syst. **56**(2), 83–92 (2001)
5. Ester, M., Kriegel, H.P., Sander, J., Xu, X.W.: A density-based algorithm for discovering clusters in large spatial databases with noise. In: International Conference on Knowledge Discovery and Data Mining, pp. 226–231 (1996)
6. Fu, L., Medico, E.: Flame, a novel fuzzy clustering method for the analysis of dna microarray data. BMC Bioinform. **8**(1), 1–17 (2007)
7. Gionis, A., Mannila, H., Tsaparas, P.: Clustering aggregation. ACM Trans. Knowl. Discov. Data **1**(1), 1–30 (2007)
8. Hou, J., Gao, H., Li, X.: Dsets-dbscan: a parameter-free clustering algorithm. IEEE Trans. Image Process. **25**(7), 3182–3193 (2016)
9. Hou, J., Liu, W., Xu, E., Cui, H.: Towards parameter-independent data clustering and image segmentation. Pattern Recogn. **60**, 25–36 (2016) .
10. Hou, J., Pelillo, M.: A new density kernel in density peak based clustering. In: International Conference on Pattern Recognition, pp. 463–468 (2016)
11. Jain, A.K., Law, M.H.C.: Data clustering: a user's dilemma. In: International Conference on Pattern Recognition and Machine Intelligence, pp. 1–10 (2005)
12. von Luxburg, U.: A tutorial on spectral clustering. Stat. Comput. **17**(4), 395–416 (2007)
13. Ng, A., Jordan, M., Weiss, Y.: On spectral clustering: analysis and an algorithm. In: Advances in Neural Information Processing Systems, pp. 849–856 (2002)
14. Pavan, M., Pelillo, M.: A graph-theoretic approach to clustering and segmentation. In: IEEE International Conference on Computer Vision and Pattern Recognition, pp. 145–152 (2003)
15. Pavan, M., Pelillo, M.: Efficient out-of-sample extension of dominant-set clusters. In: Advances in Neural Information Processing Systems, pp. 1057–1064 (2005)
16. Pavan, M., Pelillo, M.: Dominant sets and pairwise clustering. IEEE Trans. Pattern Anal. Mach. Intell. **29**(1), 167–172 (2007)
17. Rodriguez, A., Laio, A.: Clustering by fast search and find of density peaks. Science **344**, 1492–1496 (2014)
18. Shi, J., Malik, J.: Normalized cuts and image segmentation. IEEE Trans. Pattern Anal. Mach. Intell. **22**(8), 167–172 (2000)
19. Torsello, A., Bulo, S.R., Pelillo, M.: Grouping with asymmetric affinities: a game-theoretic perspective. In: IEEE International Conference on Computer Vision and Pattern Recognition, vol. 1, pp. 292–299 (2006)
20. Veenman, C.J., Reinders, M., Backer, E.: A maximum variance cluster algorithm. IEEE Trans. Pattern Anal. Mach. Intell. **24**(9), 1273–1280 (2002)
21. Zhu, X., Loy, C.C., Gong, S.: Constructing robust affinity graphs for spectral clustering. In: IEEE International Conference on Computer Vision and Pattern Recognition, pp. 1450–1457 (2014)

Fast Nearest Neighbors Search in Graph Space Based on a Branch-and-Bound Strategy

Zeina Abu-Aisheh[✉], Romain Raveaux, and Jean-Yves Ramel

Laboratoire d'Informatique (LI), Université François Rabelais, 37200 Tours, France
{zeina.abu-aisheh,romain.raveaux,jean-yves.ramel}@univ-tours.fr
http://www.li.univ-tours.fr/

Abstract. When using k-nearest neighbors, an unknown object is classified by comparing it to all the prototypes stored in the training database. When the size of the database is large, and especially if prototypes are represented by graphs, the search of k-nearest neighbors can be very time consuming. On this basis, some researchers have tried to propose optimization techniques to speed up or to approximate the search of the nearest neighbors of a query. However, these studies pay attention only to the case of vector space. In this paper, we propose an optimization technique dedicated to structural pattern recognition. We take advantage of a recent branch-and-bound graph edit distance approach in order to speed up the classification stage. Instead of considering each graph edit distance problem as an independent search tree, the search trees whose purpose is to classify an unknown graph are considered as a one search tree. Results showed that this approach drastically outperformed the classical one under limited time constraints. Moreover, this approach beat fast graph matching algorithms in terms of average execution time.

Keywords: Graph classification · Graph edit distance · K-nearest neighbors · Branch-and-bound · Optimization

1 Introduction

Due to the long time needed by the k-nearest neighbors (KNN) classifier when the size of the database (i.e., prototypes and unknown graphs) is large, some optimization techniques have been proposed to speed up or to approximate the search of KNN [14]. These studies pay attention only to vector space. In this paper, we propose an optimization technique for structural pattern recognition (graph space). Similar to the works proposed for the vector space, when the objective is to classify unknown graphs, questions like "why do we need to consider each graph matching problem as an independent one?" and "cannot we consider all the graph matching problems used to classify the unknown graph as a single problem?" become of crucial interest.

To classify unknown objects using the KNN paradigm, one needs to define a metric that measures the distance between the unknown object and the elements in the learning set. In the context of attributed graphs, the distortion and noise

© Springer International Publishing AG 2017
P. Foggia et al. (Eds.): GbRPR 2017, LNCS 10310, pp. 197–207, 2017.
DOI: 10.1007/978-3-319-58961-9_18

are taken into account during the matching process. One of the most well-known and used approaches to compute a distance (dissimilarity) between two graphs taking distortion into account is Graph Edit Distance (GED). GED is achieved by finding a set of graph edit operations: insertions, deletions and substitutions of vertices as well as edges in order to transform a graph into another with the minimal cost. GED between two graphs can be computed by different manners but it is usually time consuming especially if we want an exact solution.

On the basis of all these hypotheses and remarks, we propose to take advantage of Branch-and-Bound (BnB) based algorithms to elaborate the idea of dealing with the classification of each unknown graph as a global problem instead of independently solving successive GED computations. In this paper, we couple a recent anytime GED algorithm with the KNN classifier. Instead of considering each search tree of the unknown graph compared to a training graph as an independent one, we group all of them in a single search tree. The comparisons are achieved in a sequential manner and the best upper bound found so far is used as an initial upper bound of the next GED problem. For instance, the solution of the first GED problem is considered as an upper bound of the second one. Results, on 3 different datasets, showed that the proposed approach drastically outperformed the classical one and beat fast graph matching algorithms in terms of average execution time under limited time constraints.

The rest of the paper is organized as follows: In Sect. 2, the problem statement is presented in details after defining the KNN classifier and the GED problem. At the end of Sect. 2, the selection of a GED method is made. In Sect. 3, the BnB strategy proposed to speed up the KNN search in graph space based on a recent GED algorithm is described. Section 4 is dedicated to the experiments, protocol and results that show the efficiency of the extended approach. Section 5 is devoted to conclusions and perspectives.

2   Problem Statement

2.1   K-Nearest Neighbors Problem

KNN is a simple and precise classifier. It is non-parametric and thus it does not need knowledge about the distribution of classes. Moreover, the associated algorithm is quite simple to implement. When there are enough training patterns, classification error will be smaller than twice the Bayes error [7].

Unlike other classifiers which are considered as black-box models [8], when the metric is defined, KNN can provide an explanation of the classification results. However, KNN requires enormous computation time that is proportional to the number of training samples and the number of dimensions of feature vector. Thus, because of its simplicity and precision, many researches have tried to speed-up the algorithm in vector space. Even if optimization methods in graph space are different from the ones in vector space, we can shed light on the two categories of methods proposed in vector space. The first category provides exact neighbors and thus tries to reduce, in an off-line way, the number of samples for distance calculation by finding an effective subset from training data set [11] or by constructing a new set used for classification [6]. The second category provides approximate neighbors and tries to limit the search space according to the query. Consequently, the number of problems for distance calculation becomes smaller, and the computation time is decreased. In some cases of this category, there is no guarantee that the results using subsets or using selected new sets are the same as the results using the original KNN rule. One can notice that some methods reduce computation time and space complexity, and some others reduce only time complexity.

Fukunaga et al. in [10] and Omachi et al. in [20] have proposed fast search methods based on a BnB strategy. They achieve a fast search by skipping the search of subtrees that are unnecessary to explore. Then, the search efficiency strongly depends on the structure of search tree (i.e., height of the tree, number of children of one node, etc.) that is to say on the propriety of the construction method (i.e., clustering algorithm).

To solve KNN in graph space, we formulate the problem as follows: Given a set S of n samples and a query element q, find a subset $S_0 \subset S$ of $k \leq n$ elements such that for any elements $p_1 \in S_0$ and $p_2 \in S - S_0$, $dist(q; p_1) \leq dist(q; p_2)$.

2.2   Graph Edit Distance as a Metric

To classify graphs using KNN, one needs to define a metric that measures the distance between the graphs. In the context of attributed graphs, distortion and noise are taken into account during the graph matching process. One of the most well-known approaches to compute distance (dissimilarity) between graphs taking distortion into account is Graph Edit Distance (GED) [17,21].

Let $G_1 = (V_1, E_1, \mu_1, \xi_1)$ and $G_2 = (V_2, E_2, \mu_2, \xi_2)$ be two graphs with $V_1 = (u_1, ..., u_n)$ and $V_2 = (v_1, ..., v_m)$ the sets of vertices of G_1 and G_2, respectively. E_1 and E_2 represent the edges of G_1 and G_2, successively, whereas the terms μ and ζ refer to the attributes on vertices and edges, respectively. In error-tolerant GM, a measurement of the cost of matching vertices and/or edges of two graphs G_1 and G_2, referred to as penalty cost, is applicable on both graph structures and attributes. The basic idea is to assign a penalty cost to each matching operation. When (sub)graphs differ in their attributes or structures, a high penalty cost is added during the matching process. Such a cost prevents dissimilar (sub)graphs from being matched since they are different. Likewise, when (sub)graphs are similar, a small penalty cost is added to the overall cost. This cost includes matching, inserting and/or deleting vertices/edges.

Formally saying, GED is based on a set of edit operations o_i where $i = 1 \ldots k$ and k is the number of edit operations. This set is referred to as *Edit Path* in the literature [17].

Definition 1. *Edit Path*
A set $\{o_1, \cdots, o_k\}$ of k edit operations that completely transform G_1 into G_2 is called a (complete) edit path between G_1 and G_2. A partial edit path refers to a subset of $\{o_1, \cdots, o_q\}$ that partially transforms G_1 into G_2.

Formally saying, GED between two graphs is defined as follows:

Definition 2. *Graph Edit Distance*
Let $G_1 = (V_1, E_1, \mu_1, \zeta_1)$ and $G_2 = (V_2, E_2, \mu_2, \zeta_2)$ be two graphs, the graph edit distance between G_1 and G_2 is defined as:

$$d_{\lambda_{min}}(G_1, G_2) = \min_{\lambda \in \Gamma(G_1, G_2)} \sum_{o \in \lambda} c(o) \qquad (1)$$

where $c(o)$ denotes the cost function measuring the cost of an edit operation o and $\Gamma(G_1, G_2)$ denotes the set of all edit paths transforming G_1 into G_2. The exact correspondence is one of the correspondences that obtains the minimum cost (i.e., $d_{\lambda_{min}}(G_1, G_2)$).

2.3 Formulation of K-Nearest Neighbors Coupled with Graph Edit Distance Computation

For simplicity, we will formalize the KNN problem such that $k = 1$. However, this problem can be easily generalized for $K > 1$.

Let L be the set of training graphs and G_q is a query graph, the KNN problem is defined as follows:

$$G^* = arg \min_{G \in L} d(G_q, G) \qquad (2)$$

where d is a dissimilarity measure between graphs and G^* is the graph that minimizes the distance to G_q. The complexity of the problem grows linearly with the size of L. However, the problem of computing d in graph space is NP-Hard, see Eq. 1. Solving Eq. 2 implies independently solving the problem of Eq. 1 $|L|$ times. Instead of separating the $|L|$ sub-problems, we propose to unify them and redefine the problem of Eq. 2 as follows:

$$G^* = arg \min_{\lambda_G \in \{\Gamma(G_q, G)\}} \sum_{o \in \lambda_G} c(o) \; \forall G \in L \qquad (3)$$

where $\{\Gamma(G_q, G)\} \; \forall G \in L$ is the set of all the possible matchings between G_q and each graph G in L.

2.4 Possible Techniques for GED Computation

We aim at choosing a GED algorithm to solve the problem described in Eq. 3. Thus, in this section, we explore the state-of-the-art methods dedicated to solving GED.

Techniques for the GED computation vary in their way of solving GED and their complexities. In this section, we globally divide them into three main categories. First, the exact GED category such as the A^* algorithm [13]. A^* is a foundation work that is based on a best-first search and thus it is memory consuming. Recently, a depth-first GED algorithm (DF) has been proposed in [3] to tackle the memory consumption of A^* using the depth-first paradigm with a preprocessing and a BnB steps. However, in a classification context, A^* and DF are relatively slow, especially when matching large graphs.

The second category represents approximate GED methods. Beam-Search (BS) [12] has been put forward to reduce the complexity of A^*. The purpose of BS is to prune the search tree via a parameter that keeps the x most promising partial edit paths. However, such an algorithm cannot always find the exact matching. Riesen et al. in [16] reformulated the assignment problem as finding an exact matching in a complete bipartite graph in order to reduce the quadratic assignment problem (of GED computation) to an instance of a linear sum assignment problem. This method was then sped up in [19]. These approaches take local rather than global relationships into consideration. To go beyond the local structure problem, few works have been proposed [5,9,18], to name a few of them. Recently, two approaches based on Integer Projected Fixed Point and Graduated Non-Convexity and Concavity methods have been proposed in [4]. All these approaches cannot speed up the search of KNN since a precedent comparison cannot be used to prune the search space of the current one.

Recently, a new category has been added to GED, this category is referred to as *anytime* [2]. From an initial solution, anytime algorithms provide successive solutions during the enumeration of the search tree, they can be used to get good approximate distances (under limited time constraints) as well as exact ones (with plenty of available time). A first algorithm proposed in this category to solve GED is referred to as *anytime* depth-first (ADF). In this paper, we selected ADF since it can take advantage of the previous graph comparisons, in an iterative manner, to prune the search space.

3   Fast Nearest Neighbors in Graph Space

Traditionally, when classifying G_q using GED approaches, the class of the nearest neighbor $G \in L$ (when $K = 1$) is assigned to G_q after comparing it with each $G \in L$. Furthermore, the distances between G_q and all the graphs in L are computed independently by running a GED algorithm $|L|$ times. This solution is naive as each comparison is launched and not stopped until its end of execution and the initial upper bound (UB) of each comparison is set to ∞. If we have no extra information, this step is essential to find the correct solution. But in the KNN problem, the result of the previous problem may give us some clues for a better pruning in the next search whose aim is to classify G_q. It is an evident fact that with a smaller UB, BnB algorithms become more efficient during the successive GED computations.

In this paper, we propose to consider the classification of G_q as a single problem by coupling the KNN classifier with ADF. Algorithm 1 depicts the main steps of the proposed approach, called *One-Tree-ADF*. First, the initialization step (lines 1 to 3) starts. Second, the refined ADF is applied on G_q and each G_i in the training set L (line 5). UB and so the class assigned to G_q (*i.e.*, C_q) are modified if d_{min} is better than the current UB

Algorithm 1. One-Tree-ADF Algorithm

Input: The set L of labeled graphs (i.e., train set): $\{(G_1, C_1), \cdots, (G_l, C_l)\}$ and the unknown graph G_q

Output: the class label assigned to G_x

1: $d_{min} = \infty$ ▷ An initial distance between two graphs
2: $UB = \infty$ ▷ The initial upper bound
3: $C_q = \phi$ ▷ The class assigned to G_x
4: **for** $i = 1$ to $|L|$ **do**
5:      d_{min}=ADF(G_q,G_i,UB)
6:      **if** $UB > d_{min}$ **then**
7:          UB= d_{min}
8:          $C_q = C_i$
9:      **end if**
10: **end for**
11: Return C_q

(lines 7 and 8). The outputted distance d_{min} is used as UB of the next comparison (i.e., $ADF(G_q, G_{i+1})$) (line 7). This algorithm finally terminates by outputting C_q.

Note that for the sake of simplicity, we considered $K = 1$. The extension of the algorithm for $K > 1$ is straightforward, UB is the distance obtained by the current k^{th} nearest neighbor. When a distance is calculated (line 5), it is compared to the distance of the k nearest neighbors and the current table of the

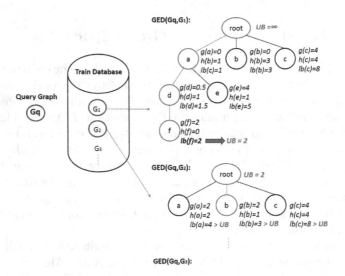

Fig. 1. *One-Tree-ADF.* Given a query graph G_q and graphs in the training set, the problems GED(G_q,G_1), GED(G_q,G_2) and GED(G_q,G_3) are considered as sub-trees of the global tree (T_{G_q}). The sub-tree of GED(G_q,G_2) is pruned thanks to UB that is found via GED(G_q,G_1).

k nearest neighbors is updated when it is necessary, discarding a sample from the table when finding a better distance (i.e., a better nearest neighbor).

Figure 1 highlights the idea of *One-Tree-ADF*. Given a query graph G_q and a learning database L, the idea is to consider each search tree S of the GED(G_q,G_i) as a sub-tree of the global tree dedicated to G_q and referred to as T_{G_q}. For instance, in Fig. 1, one can see that the first *UB* found while exploring the sub-tree S of GED(G_q,G_1) is 2. *UB* is then used as an initial *UB* of the sub-tree S' of GED(G_q,G_2) and so on. Such an operation helps in pruning the sub-trees as fast as possible while searching for the nearest neighbor of G_q.

4   Protocol and Experiments

4.1   Selected Datasets

In the experiments, GREC, Protein and Mutagenicity of the IAM database [15] are selected. **GREC** data set consists of 1100 graphs where graphs are uniformly distributed between 22 symbols. 286 graphs are included in the training set while 528 graphs are included in the test set. In **Mutagenicity**, 4337 elements are represented in this data set (2401 mutagen elements and 1936 non-mutagen elements) which are divided into: a training set of size 1500, a test set of size 2337 and the rest of elements are in the validation set. For simplicity, we refer to this database as Muta. 600 **Proteins** are uniformly distributed over 100 classes. The size of each of the training and the test sets is 200. The cost functions of the selected datasets and their parameters can be found in [1].

4.2   Chosen Methods

On the exact method side, we chose the *DF* algorithm since it outperforms the A^* algorithm in terms of running time. On the approximate side, we included *BS-1* (i.e., the greedy algorithm) and *BS-100*. We also chose the bipartite matching algorithm *BP* [16] since it has been shown to be one of the most efficient approximate algorithms so far. In addition, we selected a fast version of *BP* [19], referred to as *FBP* in the literature.

4.3   Environment and Constraints

The experiments were conducted on a computer with a 24-core Intel i5 processor at 2.10 GHz and 16 GB of memory. The time constraint used for all the datasets is fixed to 500 milliseconds (ms) which is the maximum time needed by *BP* and *FBP* to output a solution. That is, any GED algorithm that needs more than 500 ms is stopped and the best answer found so far is outputted. Note that *One-Tree-ADF* and *DF* are exact algorithms without time constraints.

The order of the training graphs of each of GREC, Protein and Muta is randomized. Four different orders are generated. The reason is that we did not want *One-Tree-ADF* to be influenced by the ordered lists that are given in IAM. Note than one can extensively study the influence of the order of the training graphs on the accuracy and the total execution time of *One-Tree-ADF*.

4.4 Results

In Table 1, the results achieved on all datasets are presented. Note that the computation time corresponds to the average time needed to classify graphs G_q.

We ran the experiments on the 4 different orders of train graphs (see Sect. 4.3). The results of these orders are quite similar. Thus, the measured time (in milliseconds) is the average time of the 4 different orders. The results show that on the 3 datasets, *One-Tree-ADF* was always faster than the classical *DF* approach. It also improved the classification rate of *DF* on both Protein and Muta. This is due to the fact that *One-Tree-ADF* could improve *UB* while moving from one comparison to another. As a consequence, it pruned unfruitful parts of some sub-trees and found a better *UB*. On the other hand, when comparing *One-Tree-ADF* to *BP*, one can see that *One-Tree-ADF* was 4 times faster (on GREC) and 2 times faster (on Muta), it also improved the classification rate on Muta. However, on Protein, it was less accurate than *BP*. *FBP* was faster than *One-Tree-ADF* on Protein, however, the accuracy of *FBP* was lower.

Table 1. Classification results on GREC, Protein and MUTA. The best results are marked in bold style.

	GREC		Protein		Muta	
	Acc	*t*	*Acc*	*t*	*Acc*	*t*
One-Tree-ADF	**98.5**	**15483.16**	47	58321.20	**71.28**	**104183.21**
DF	98.5	140675.0	42	124361.61	70	1139134.29
BS-1	98.5	69236.34	24	129571.76	55.5	1015688
BS-100	58.7	83928.20	26	141265.41	55.5	1383838.66
BP	98.5	62294.60	**52**	59041.84	70	528546.64
FBP	98.5	27922.65	38.5	**39425.69**	70	376135.51

Figure 3 depicts two examples, for classifying two graphs G_q taken from GREC and Muta. Note that the value of *UB* is outputted after each $ADF(G_q, G)$, see line 5 in Algorithm 1. One can see that improving *UB* is easier in the first few milliseconds, however, after a certain time, *UB* could keep stable for a longer time, depending on whether or not the next comparisons are fruitful.

For the same examples illustrated in Fig. 2, we divided the comparisons into intervals, and measured the average time needed in each interval, see Fig. 3. Results showed that, on both datasets, the first interval needed around 400 ms while the other intervals needed less time. This shows the ability of *UB* in pruning the search tree. One could also notice that the average time of the intervals on Muta is higher than the one on GREC; that is due to the difficulty of search trees of Muta.

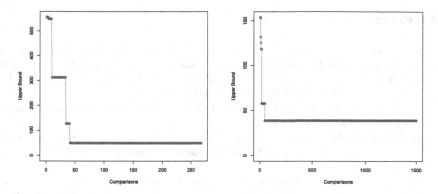

Fig. 2. The upper bound found at the end of each comparison $ADF(G_q,G)$ and needed to find the 1NN of G_q. (Left: GREC, Right: Muta)

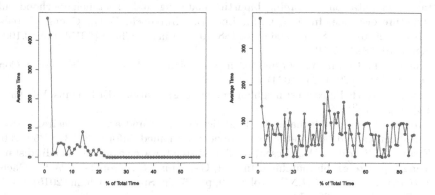

Fig. 3. The evolution of the time needed to compare two graphs while exploring the search tree of G_q. (Left: GREC, Right: Muta)

5   Conclusions and Perspectives

In this paper, a fast nearest neighbor approach dedicated to graph classification was proposed. This approach, referred to as *One-Tree-ADF* takes advantage of a recent BnB-based algorithm dedicated to solving GED. Instead of considering the comparison of graph G_q and a graph G in the training set as a single problem, *One-Tree-ADF* groups trees whose objective is to classify a graph G_q in one search tree. Such an approach aims at improving the upper bound as fast as possible and thus pruning the misleading parts of the search tree. Results showed that the *One-Tree-ADF* drastically minimizes the total classification time while achieving high classification rates when compared to exact and approximate GED algorithms.

Our future work is two-fold. First, learning the order of the training graphs set. Second, transforming *One-Tree-ADF* into a parallel algorithm. These propositions can reduce the computation time of *One-Tree-ADF*.

References

1. Abu-Aisheh, Z., Raveaux, R., Ramel, J.-Y.: A graph database repository and performance evaluation metrics for graph edit distance. In: Liu, C.-L., Luo, B., Kropatsch, W.G., Cheng, J. (eds.) GbRPR 2015. LNCS, vol. 9069, pp. 138–147. Springer, Cham (2015). doi:10.1007/978-3-319-18224-7_14
2. Abu-Aisheh, Z., Raveaux, R., Ramel, J.-Y.: Anytime graph matching. Pattern Recogn. Lett. **84**, 215–224 (2016)
3. Abu-Aisheh, Z., Raveaux, R., Ramel, J.-Y., Martineau, P.: An exact graph edit distance algorithm for solving pattern recognition problems. In: ICPRAM, pp. 271–278 (2015)
4. Bougleux, S., Brun, L., Carletti, V., Foggia, P., Gaüzère, B., Vento, M.: Graph edit distance as a quadratic assignment problem. Pattern Recogn. Lett. **87**, 38–46 (2017)
5. Carletti, V., Gaüzère, B., Brun, L., Vento, M.: Approximate graph edit distance computation combining bipartite matching and exact neighborhood substructure distance. In: Liu, C.-L., Luo, B., Kropatsch, W.G., Cheng, J. (eds.) GbRPR 2015. LNCS, vol. 9069, pp. 188–197. Springer, Cham (2015). doi:10.1007/978-3-319-18224-7_19
6. Chang, C.L.: Finding prototypes for nearest neighbor classifiers. IEEE Trans. Comput. **23**(11), 1179–1184 (1974)
7. Cover, T., Hart, P.: Nearest neighbor pattern classification. IEEE Trans. Inf. Theor. **13**(1), 21–27 (2006)
8. Dreiseitl, S., Ohno-Machado, L.: Logistic regression and artificial neural network classification models: a methodology review. J. Biomed. Inform. **35**, 352–359 (2002)
9. Ferrer, M., Serratosa, F., Riesen, K.: A first step towards exact graph edit distance using bipartite graph matching. In: Liu, C.-L., Luo, B., Kropatsch, W.G., Cheng, J. (eds.) GbRPR 2015. LNCS, vol. 9069, pp. 77–86. Springer, Cham (2015). doi:10.1007/978-3-319-18224-7_8
10. Fukunaga, K., Narendra, P.M.: A branch and bound algorithm for computing k-nearest neighbors. IEEE Trans. Comput. **24**(7), 750–753 (1975)
11. Gates, G.W.: The reduced nearest neighbor rule. IEEE Trans. Inform. Theory **18**(5), 431–433 (1972)
12. Riesen, K., Neuhaus, M., Bunke, H.: Fast suboptimal algorithms for the computation of graph edit distance. SSPR **28**, 163–172 (2006)
13. Raphael, B., Hart, P., Nilsson, N.: A formal basis for the heuristic determination of minimum cost paths. IEEE TSSC **28**, 100–107 (2004)
14. Raveaux, R., Adam, S., Héroux, P., Trupin, É.: Learning graph prototypes for shape recognition. CVIU **115**(7), 905–918 (2011)
15. Riesen, K., Bunke, H.: IAM graph database repository for graph based pattern recognition and machine learning. In: da Vitoria, L.N., et al. (eds.) SSPR/SPR 2008. LNCS, vol. 5342, pp. 287–297. Springer, Heidelberg (2008)
16. Bunke, H., Riesen, K.: Approximate graph edit distance computation by means of bipartite graph matching. Image Vis. Comput. **28**, 950–959 (2009)
17. Riesen, K.: Structural Pattern Recognition with Graph Edit Distance. ACVPR. Springer, Cham (2015)
18. Riesen, K., Fischer, A., Bunke, H.: Combining bipartite graph matching and beam search for graph edit distance approximation. In: Gayar, N., Schwenker, F., Suen, C. (eds.) ANNPR 2014. LNCS, vol. 8774, pp. 117–128. Springer, Cham (2014). doi:10.1007/978-3-319-11656-3_11

19. Serratosa, F.: Fast computation of bipartite graph matching. Pattern Recogn. Lett. **45**, 244–250 (2014)
20. Aso, H., Omachi, S.: A fast algorithm for a k-nn classifier based on the branch and bound method and computational quantity estimation. Syst. Comput. Jpn. **31**(6), 1–9 (2000)
21. Vento, M.: A long trip in the charming world of graphs for pattern recognition. Pattern Recogn. **48**(2), 291–301 (2015)

Graph Edit Distance

Exact Computation of Graph Edit Distance for Uniform and Non-uniform Metric Edit Costs

David B. Blumenthal$^{(\boxtimes)}$ and Johann Gamper

Faculty of Computer Science, Free University of Bolzano,
Piazza Dominicani 3, 39100 Bolzano, Italy
{david.blumenthal,gamper}@inf.unibz.it

Abstract. The graph edit distance is a well-established and widely used distance measure for labelled, undirected graphs. However, since its exact computation is *NP*-hard, research has mainly focused on devising approximative heuristics and only few exact algorithms have been proposed. The standard approach A*-GED, a node-based best-first search that works for both uniform and non-uniform metric edit costs, suffers from huge runtime and memory requirements. Recently, two better performing algorithms have been proposed: DF-GED, a node-based depth-first search that works for uniform and non-uniform metric edit costs, and CSI_GED, an edge-based depth-first search that works only for uniform edit costs. Our paper contains two contributions: First, we propose a speed-up DF-GED^{u} of DF-GED for uniform edit costs. Second, we develop a generalisation CSI_GED^{nu} of CSI_GED that also covers non-uniform metric edit cost. We empirically evaluate the proposed algorithms. The experiments show, i.a., that our speed-up DF-GED^{u} clearly outperforms DF-GED and that our generalisation CSI_GED^{nu} is the most versatile algorithm.

Keywords: Graph matching · Graph similarity · Graph edit distance · Branch and bound

1 Introduction

Labelled, undirected graphs can be used for modelling various kinds of objects, such as social networks, molecular structures, and many more. Because of this, labelled graphs have received increasing attention over the past years. One task researchers have focused on is the following: Given a database $\mathcal{G}$ that contains labelled graphs, find all graphs $G \in \mathcal{G}$ that are sufficiently similar to a query graph H or to find the k graphs from $\mathcal{G}$ that are most similar to H. For approaching this task, a distance measure between undirected, labelled graphs G and H has to be defined. One of the most commonly used measures is the graph edit distance. Formally, a *labelled, undirected graph* G is a 4-tuple $G = \langle V^G, E^G, \ell_V^G, \ell_E^G \rangle$, where V^G is a set of nodes, E^G is a set of undirected edges, and $\ell_V^G : V^G \to \Sigma_V$ and $\ell_E^G : E^G \to \Sigma_E$ are labelling functions that assign nodes an edges to labels from alphabets Σ_V and Σ_E. Both Σ_V and Σ_E contain a special label ε reserved for dummy nodes and dummy edges. The *graph edit distance* $\lambda(G, H)$ between

© Springer International Publishing AG 2017
P. Foggia et al. (Eds.): GbRPR 2017, LNCS 10310, pp. 211–221, 2017.
DOI: 10.1007/978-3-319-58961-9_19

graphs G and H on common label alphabets Σ_V and Σ_E is defined as the minimum cost of an edit path between G and H. An *edit path* is a sequence of labelled graphs starting with G and ending at a graph that is isomorphic to H. Each graph along the path can be obtained from its predecessor by applying one of the following *edit operations*: Deleting or inserting an α-labelled edge, deleting or inserting an isolated α-labelled node, changing a node's or an edge's label from α to $\beta \neq \alpha$. Edit operations on nodes and edges come with associated *edit costs* $c_V : \Sigma_V \times \Sigma_V \to \mathbb{R}$ and $c_E : \Sigma_E \times \Sigma_E \to \mathbb{R}$, respectively. The cost of an edit path is defined as the sum of the costs of its edit operations. If the cost of each edit operation equals 1, we say that the edit costs are *uniform*. In many scenarios, it is natural to consider *non-uniform metric edit costs*. For instance, if the graphs model spacial objects and the node labels are Euclidean coordinates, the cost $c_V(\alpha, \beta)$ one has to pay for changing a node's label from α to β should probably be defined as the Euclidean distance between α and β.

It has been shown that, even for uniform edit costs, it is *NP*-hard to exactly compute the graph edit distance [14]. Exact algorithms that, if applied to large graphs, terminate within an acceptable amount of time are hence out of reach. Consequently, a substantial part of research on both uniform [14–16] and non-uniform [2–4, 6, 10, 12, 13] graph edit distance has focused on the task of devising heuristics that compute lower and/or upper bounds for $\lambda(G, H)$. Nonetheless, efficient exact algorithms are still important. This is because some of the objects that are readily modelled by labelled, undirected graphs — for instance, some molecular compounds — induce graphs with very few nodes [9]. For these graphs, queries of the kind "find all $G \in \mathcal{G}$ with $\lambda(G, H) \leq \tau$" can in principle be answered. Of course, one would first use efficiently computable upper and lower bounds in order to filter out candidates from $\mathcal{G}$. However, for the surviving candidates, $\lambda(G, H) \leq \tau$ has to be verified by means of an exact algorithm.

The standard approach A⋆-GED [11] for exactly computing $\lambda(G, H)$ carries out a node-based best-first search in order to find the optimal edit path. It is very slow and has huge memory requirements. Recently, three better performing algorithms BLP-GED [8], DF-GED [1], and CSI_GED [5] have been proposed. BLP-GED formulates the problem of computing $\lambda(G, H)$ as a binary linear program which is solved by calling the commercial solver CPLEX. It has been found to be faster and more memory-efficient than A⋆-GED. DF-GED carries out a node-based depth-first search for finding the cheapest edit path. It has been found to be much more memory-efficient and slightly faster than A⋆-GED. In contrast, CSI_GED carries out an edge-based depth-first search. It also has been found to be both faster and much more memory-efficient than A⋆-GED. While A⋆-GED, BLP-GED, and DF-GED cover non-uniform metric edit costs, CSI_GED only works for uniform edit costs. A direct comparison between BLP-GED, DF-GED, and CSI_GED is lacking.

Our paper contains the following contributions: In Sect. 2, we present a speed-up DF-GED^u of DF-GEDfor uniform edit costs. DF-GED^u exploits the fact that, in the uniform case, a subroutine that DF-GED employs at each node of its search tree can be implemented to run in linear rather than cubic time. In Sect. 3, we propose a generalisation CSI_GED^{nu} of CSI_GED that also covers non-uniform metric edit costs. This generalisation comes at the, price of a slightly increased run-time. However, this increase is very moderate, as the computational complexity is increased only at the initialisation of CSI_GED^{nu} and at the leafs of its search tree. In Sect. 4, we experimentally evaluate the performance of the newly proposed algorithms. The experiments show that, for uniform edit costs, our speed-up DF-GED^u clearly outperforms DF-GED, while CSI_GED and our generalisation CSI_GED^{nu} perform similarly. They also indicate that, neither for uniform nor for non-uniform edit costs, there is a clear winner between DF-GED^u and DF-GED, on the one side, and CSI_GED and CSI_GED^{nu}, on the other side. Finally, the experiments suggest that CSI_GED^{nu} is the most versatile algorithm: It covers both uniform and non-uniform edit costs and runs very stable even on datasets where other algorithms perform better. Section 5 concludes the paper.

2   DF-GED^u: Fast DF-GED for Uniform Edit Costs

In this section, we show how to speed-up the node-based depth-first search DF-GED for uniform edit costs. We first summarise DF-GED and then describe our speed-up DF-GED^u.

The Baseline Approach. DF-GED builds upon the following observation: If edit costs are metric, then $\lambda(G, H)$ can be defined equivalently as the minimum cost of an edit path that is induced by a node map [6]. Let $V^{G+|H|}$ and $V^{G+|H|}$ be the sets that are obtained from V^G and V^H by adding $|V^H|$ respectively $|V^G|$ isolated dummy nodes. A *node map* is an injective partial function $\pi : V^{G+|H|} \rightarrow V^{H+|G|}$, whose domain contains V^G and whose image contains V^H. For a given node map π, its *induced edit path* is defined as follows: If π maps a real node $i \in V^G$ to a dummy node j_ε, i is deleted. Conversely, if a dummy node i_ε is mapped to a real node $k \in V^H$, k is inserted. If a real node $i \in V^G$ is mapped to a real node $k \in V^H$, i's label is changed from $\ell_V^G(i)$ to $\ell_V^H(k)$. If $ij \in E^G$ but $\pi(i)\pi(j) \notin E^H$, the edge ij is deleted. If $kl \in E^H$ but $\pi^{-1}(k)\pi^{-1}(l) \notin E^G$, the edge kl is inserted. Finally, if an edge $ij \in E^G$ is mapped to an edge $kl \in E^H$, ij's label is changed from $\ell_E^G(ij)$ to $\ell_E^H(kl)$. The cost of the edit path induced by π is denoted by $g(\pi)$.

DF-GED performs a depth-first search on the set of all partial node maps between $V^{G+|H|}$ and $V^{H+|G|}$ starting with the empty node map. The tree's leafs correspond to complete node maps and its inner nodes correspond to incomplete node maps. DF-GED starts with sorting the nodes of V^G such that evident nodes will be processed first [3]. It also initialises an upper bound UB for $\lambda(G, H)$, using a fast sub-optimal heuristic [10]. For each visited node π of the search tree, values $g(\pi)$ and $h(\pi)$ are maintained. The value $g(\pi)$ denotes the cost of the corresponding incomplete induced edit path, and $h(\pi)$ is a lower bound for the cost from π to a leaf, i.e., complete node map, in π's down-shadow. Assume that all nodes in V^G up to node i have already been assigned by π. If i is the last node in V^G, π is extended to a complete node map by assigning a dummy node to each of the yet unassigned nodes $j \in V^G$, and UB is updated to $g(\pi)$ if

$g(\pi) < UB$. Otherwise, π's children $\pi' \in \{\pi \cup (i+1,j) : j \in V^H$ unassigned by $\pi\} \cup \{\pi \cup (i+1,j_\varepsilon)\}$ are considered in order of non-decreasing $g(\pi') + h(\pi')$. If $g(\pi') + h(\pi') < UB$, π is updated to π' and the process iterates. Otherwise, the branch rooted at π' is pruned. At termination, UB is returned.

Note that, for each visited partial node map π, the lower bound $h(\pi)$ has to be recomputed. DF-GED computes $h(\pi)$ as follows: For a given partial node map π, let $V^{G+|H|-\pi}$ and $V^{H+|G|-\pi}$ be the sets of unassigned nodes and $E^{G-\pi}$ and $E^{H-\pi}$ be the sets of unassigned edges filled up with dummy edges to ensure $|E^{G-\pi}| = |E^{H-\pi}|$. Furthermore, let $\ell_V^G(V^{G+|H|-\pi})$, $\ell_V^H(V^{H+|G|-\pi})$, $\ell_E^G(E^{G-\pi})$, and $\ell_E^H(E^{H-\pi})$ denote the multisets of labels of the unassigned nodes or edges contained in these sets. Then $h(\pi)$ is defined as $h(\pi) = h_V(\pi) + h_E(\pi)$, where $h_V(\pi)$ is the minimum cost of a linear assignment between $\ell_V^G(V^{G+|H|-\pi})$ and $\ell_V^H(V^{H+|G|-\pi})$ with assignment costs c_V, and $h_E(\pi)$ is the minimum cost of a linear assignment between $\ell_E^G(E^{G-\pi})$ and $\ell_E^H(E^{H-\pi})$ with assignment costs c_E. Since a minimum linear assignment can be computed in cubic time, e.g., by using the Hungarian Algorithm [7], the runtime complexity of computing $h(\pi)$ is thus cubic in n and m, where $n = |V^G| + |V^H|$ and $m = \max\{|E^G|, |E^H|\}$.

Our Speed-Up for Uniform Edit Costs. Our speed-up DF-GED^u builds upon the observation that, for uniform edit cost, $h(\pi)$ can be computed in linear time. For showing this, we need the following lemma:

Lemma 1. *Let A and B be two equally sized multisets and $c : A \times B \to \mathbb{R}$ be uniform in the sense that $c(a,b)$ equals 1 if $a \neq b$ and 0 otherwise. Then the cost of a minimum linear assignment between A and B for the assignment cost c equals $|A| - |A \cap B|$.*

Proof. Let $(a_i)_{i=1}^{|A|}$ and $(b_i)_{i=1}^{|B|}$ be orderings of A and B such that, for all $i \leq |A \cap B|$, it holds that $a_i = b_i$. Note that this implies $a_i \neq b_i$ for all $i > |A \cap B|$. We define $f : A \to B$ as $f(a_i) = b_i$. It is easy to see that f is a minimum linear assignment between A and B for the uniform assignment cost c. Its cost is $\sum_{a \in A} c(a, f(a)) = \sum_{i=1}^{|A \cap B|} c(a_i, b_i) + \sum_{i=|A \cap B|+1}^{|A|} c(a_i, b_i) = |A| - |A \cap B|$. $\square$

It has been shown that, if A and B are *sorted* multisets, the size of their intersection can be computed in linear time [14]. Together with Lemma 1, this immediately implies that, if c_V and c_E are uniform, $h_V(\pi)$ and $h_E(\pi)$ can be computed in $\mathcal{O}(n \log n)$ and $\mathcal{O}(m \log m)$ time, respectively: We first sort the labels of the nodes and the edges that have not been assigned by π in $\mathcal{O}(n \log n)$ and $\mathcal{O}(m \log m)$ time, respectively. Then, we compute the intersection sizes of the resulting sorted multisets in linear time. In order to further reduce the complexity of the computation of $h_V(\pi)$ and $h_E(\pi)$, we proceed as follows. When initialising DF-GED, we *once* sort $\ell_V^G(V^{G+|H|})$, $\ell_V^H(V^{H+|G|})$, $\ell_E^G(E^G)$, and $\ell_E^H(E^H)$, i.e., the multisets containing the labels of *all* nodes and edges. For each L of the resulting sorted multisets and each partial node map π, we maintain a boolean vector that indicates if the node or edge with label L_i is still unassigned by π. This vector can be updated in constant additional time when updating the cost $g(\pi)$ of the partial edit path induced by π. For each partial node map π, $h_V(\pi)$ and $h_E(\pi)$ can then be computed in linear time by using a variation of the algorithm for multiset intersection presented in [14].

3   CSI_GED^{nu}: CSI_GED for Non-uniform Metric Edit Costs

In this section, we show how to generalise the edge-based depth-first search CSI_GED to non-uniform metric edit costs. We first summarise CSI_GED and then describe our generalisation CSI_GED^{nu}.

The Baseline Approach. While DF-GED enumerates the space of all node maps, CSI_GED considers *valid edge maps* $\phi : \overrightarrow{E^G} \rightarrow \overleftrightarrow{E^H} \cup \{e_\varepsilon\}$. The set $\overrightarrow{E^G}$ contains one arbitrarily oriented edge (i,j) for each undirected edge $ij \in E^G$, $\overleftrightarrow{E^H}$ contains two directed edges (k,l) and (l,k) for each $kl \in E^H$, and e_ε denotes a dummy edge. An edge map ϕ induces a relation π_ϕ on $V^G \times V^H$: If $\phi(i,j) = (k,l)$, then $(i,k) \in \pi_\phi$ and $(j,l) \in \pi_\phi$. Since nodes cannot be assigned twice, ϕ is called *valid* if and only if π_ϕ is a partial injective function. A valid edge map ϕ also induces a partial edit path between G and H: If $\phi(i,j) = (k,l)$, ij's label is changed from $\ell_E^G(ij)$ to $\ell_E^H(kl)$. If $\phi(i,j) = e_\varepsilon$, the edge ij is deleted. If $\phi^{-1}[\{(k,l),(l,k)\}] = \emptyset$ holds for an edge $kl \in E^H$, kl is inserted. And if $\pi_\phi(i) = k$, i's label is changed from $\ell_V^G(i)$ to $\ell_V^H(k)$. The cost of the partial edit path induced by ϕ is denoted by $g(\phi)$. In general, ϕ's induced edit path is incomplete, since the sets $V^{G-\pi_\phi} \subseteq V^G$ and $V^{H-\pi_\phi} \subseteq V^H$ containing the nodes that are left unassigned by π_ϕ are in general non-empty. The following theorem constitutes the backbone of CSI_GED:

Theorem 1 (Cf. Theorem 1 in [5]). *If the edit costs c_V and c_E are uniform, then, for each node map $\pi : V^{G+|H|} \rightarrow V^{H+|G|}$, there is a valid edge map $\phi : \overrightarrow{E^G} \rightarrow \overleftrightarrow{E^H} \cup \{e_\varepsilon\}$ with $g(\pi) \geq g(\phi) + \Gamma(V^{G-\pi_\phi}, V^{H-\pi_\phi})$, where $\Gamma(V^{G-\pi_\phi}, V^{H-\pi_\phi}) = \max\{|V^{G-\pi_\phi}|, |V^{H-\pi_\phi}|\} - |\ell_V^G(V^{G-\pi_\phi}) \cap \ell_V^H(V^{H-\pi_\phi})|$. Moreover, $g(\phi) + \Gamma(V^{G-\pi_\phi}, V^{H-\pi_\phi}) \geq \lambda(G,H)$ holds for each valid each map ϕ.*

Theorem 1 implies that, for uniform edit costs, one can compute the graph edit distance by enumerating the space of all valid edge maps. To this purpose, CSI_GED carries out a depth-first search on the set of all valid partial edge maps starting with the empty edge map. CSI_GED maintains an upper bound for the graph edit distance, which is initialised as $UB = \infty$, and considers the edges $e_r \in \overrightarrow{E^G}$ in an arbitrary but fixed order. For each visited incomplete edge map ϕ, the current induced cost $g(\phi)$ and a lower bound $g'(\phi)$ for the induced cost of a complete edge map in ϕ's down-shadow are maintained. Assume that all edges in $\overrightarrow{E^G}$ up to e_r have already been assigned by ϕ. If e_r is the last edge in $\overrightarrow{E^G}$, ϕ is a complete valid edge map, and UB is updated to $g(\phi) + \Gamma(V^{G-\pi_\phi}, V^{H-\pi_\phi})$ if $g(\phi) + \Gamma(V^{G-\pi_\phi}, V^{H-\pi_\phi}) < UB$. Otherwise, ϕ's children $\phi' \in \{\phi \cup (e_{r+1}, e) : e \in \overleftrightarrow{E^H}$ unassigned by ϕ and $\phi \cup (e_{r+1}, e)$ valid$\} \cup \{\phi \cup (e_{r+1}, e_\varepsilon)\}$ are considered in order of non-decreasing $\mathcal{C}(e_{r+1}, e)$. $\mathcal{C}(e_{r+1}, e)$ is an estimate of the graph edit distance under the constraint that the edge e_{r+1} is mapped to e. Note that the estimated cost matrix $\mathcal{C}$ only has to be computed once at initialisation. If $g'(\phi') < UB$, ϕ is updated to ϕ' and the process iterates. Otherwise, the branch rooted at ϕ' is pruned. At termination, UB is returned.

Our Generalisation to Non-uniform Metric Edit Costs. The key-ingredient of our extension CSI_GED^{nu} is the following generalised version of Theorem 1:

Theorem 2. *If the edit costs c_V and c_E are metric, then, for each node map $\pi : V^{G+|H|} \to V^{H+|G|}$, there is a valid edge map $\phi : \overrightarrow{E^G} \to \overleftarrow{E^H} \cup \{e_\varepsilon\}$ with $g(\pi) \geq g(\phi) + \Gamma^{\mathrm{nu}}(V^{G-\pi_\phi}, V^{H-\pi_\phi})$, where $\Gamma^{\mathrm{nu}}(V^{G-\pi_\phi}, V^{H-\pi_\phi})$ is defined as the cost of a minimum linear assignment between $\ell_V^G(V^{G+|H|-\pi_\phi})$ and $\ell_V^H(V^{H+|G|-\pi_\phi})$ for the assignment cost c_V. Moreover, $g(\phi) + \Gamma^{\mathrm{nu}}(V^{G-\pi_\phi}, V^{H-\pi_\phi}) \geq \lambda(G, H)$ holds for each valid edge map ϕ.*

Proof. Given a node map π, we construct a valid edge map ϕ as follows: Let $(i, j) \in \overrightarrow{E^G}$. If the corresponding undirected edge ij is preserved under π, i.e., if $\pi(i)\pi(j) \in E^H$, we define $\phi(i, j) = (\pi(i), \pi(j))$. Otherwise, we set $\phi(i, j) = e_\varepsilon$. By construction, π_ϕ equals the restriction of π to those real nodes $i \in V^G$ that are incident with an edge that is preserved under π. This implies that ϕ is valid. Next, we compare the complete edit path P_π that is induced by π and the partial edit path P_ϕ that is induced by ϕ. We observe that P_ϕ contains all edge-deletions, -insertions, and -relabelings that appear in P_π, as well as all relabelings of nodes that are incident with a preserved edge. Apart from these edit operations, P_π also contains deletions and relabelings of nodes that are not incident with a preserved edge, as well as node-insertions. These latter operations can be viewed as a linear assignment between $\ell_V^G(V^{G+|H|-\pi_\phi})$ and $\ell_V^H(V^{H+|G|-\pi_\phi})$ for the assignment cost c_V, which, together with the observation above, implies $g(\pi) \geq g(\phi) + \Gamma^{\mathrm{nu}}(V^{G-\pi_\phi}, V^{H-\pi_\phi})$. For showing the second part of the theorem, we fix a valid edge map ϕ. Let π'_ϕ be a minimum linear assignment between $\ell_V^G(V^{G+|H|-\pi_\phi})$ and $\ell_V^H(V^{H+|G|-\pi_\phi})$ for the assignment cost c_V. Then $\pi = \pi_\phi \cup \pi'_\phi$ is a complete node map. By construction, we have $g(\phi) + \Gamma^{\mathrm{nu}}(V^{G-\pi_\phi}, V^{H-\pi_\phi}) \geq g(\pi) \geq \lambda(G, H)$, where the last inequality follows from the fact that, for metric edit costs, the graph edit distance can be defined as the minimum cost of an edit path that is induced by a node map. $\qquad\square$

Theorem 2 indicates how to extend CSI_GED to non-uniform metric edit costs: We just have to replace all occurrences of Γ by Γ^{nu}. As presented above, during the depth-first search carried out by CSI_GED, Γ has to be computed at the leafs of the search tree. At initialisation, further computations of Γ are required for computing the estimated cost matrix $\mathcal{C}$ and a constant that is required for the computation of g' (cf. [5] for these details of CSI_GED). Note that computing Γ requires linear time, whereas computing Γ^{nu} needs cubic time (cf. Sect. 2). This implies that our generalisation leads to an increased runtime of CSI_GED.

However, the increase is very moderate, as Γ^{nu} does not need to be computed at the inner nodes of the (exponentially large) search tree.

4   Empirical Evaluation

The aim of our experiments is to compare the performance of the algorithms CSI_GED, CSI_GED^{nu}, DF-GED, and DF-GED^{u} for both uniform and non-uniform metric edit costs. We implemented all algorithms in C++ making them employ the same data structures and subroutines. All tests were carried out on a machine with two Intel Xeon E5-2667 v3 processors with 8 cores each and 98 GB of main memory running GNU/Linux. We conducted tests on the datasets AIDS and FINGERPRINTS [9], which are widely used in the research community [5, 10–16]. Both datasets contain graphs with both node and edge labels for which non-uniform metric relabelling costs c_V and c_E are naturally induced by the domain [10]. For defining non-uniform metric edit costs, we thus only had to specify the deletion/insertion costs $c_V(\alpha, \varepsilon)$ and $c_E(\alpha, \varepsilon)$. This was done by setting $c_V(\alpha, \varepsilon) = \max\{c_V(\beta, \gamma) \mid \beta, \gamma \in \Sigma_V\}$ for all $\alpha \in \Sigma_V \smallsetminus \{\varepsilon\}$, and $c_E(\alpha, \varepsilon) = \max\{c_E(\beta, \gamma) \mid \beta, \gamma \in \Sigma_E\}$ for all $\alpha \in \Sigma_E \smallsetminus \{\varepsilon\}$, i.e., deleting and inserting nodes and edges was defined to be as expensive as the most expensive relabelling operations. Since both DF-GED and CSI_GED fail to compute the exact graph edit distance for graphs with more than 25 nodes within reasonable time [1, 5], we excluded larger graphs from AIDS. FINGERPRINTS only contains small graphs, anyway. We then used the experimental setup suggested in [5]: For both considered datasets $\mathcal{D}$ and all $i \in \{3, 6, \ldots, \max_{G \in \mathcal{D}} |G|\}$, we defined a size-constrained test-group $\mathcal{G}_i$ that contains four randomly selected graphs $G \in \mathcal{D}$ satisfying $|V^G| = i \pm 1$. For each tested algorithm ALG and each test-group $\mathcal{G}_i$, all six pairwise comparisons between graphs contained in $\mathcal{G}_i$ were carried out. We set a time limit of 1000 s and recorded the metrics *timeouts*, t, and *dev*. Since pretesting showed that the main memory demand of all tested algorithms is negligible, we did not record memory usage.

- *timeouts*(ALG, i): The number of timeouts on $\mathcal{G}_i$, i.e., of pairwise comparisons between graphs in $\mathcal{G}_i$ where ALG did not finish within 1000 s.
- t(ALG, i): ALG's average runtime across all six pairwise comparisons between graphs in $\mathcal{G}_i$.
- *dev*(ALG, i): ALG's average percentual deviation from the best tested algorithm as introduced in [1], i.e., the average of $100 \cdot [UB(\mathrm{ALG}) - UB^\star]/UB^\star$ across all six pairwise comparisons between graphs in $\mathcal{G}_i$. $UB(\mathrm{ALG})$ denotes the value of the upper bound UB maintained by ALG after 1000 s and $UB^\star$ is defined as $UB^\star = \min\{UB(\mathrm{ALG}') \mid \mathrm{ALG}'$ is tested algorithm$\}$.

Figure 1 shows the outcomes of our experiments for uniform edit costs. We observe that, on both datasets, our speed-up DF-GED^{u} outperforms DF-GED in terms of all recorded metrics, while CSI_GED and our generalisation CSI_GED^{nu} perform similarly. For instance, on FINGERPRINTS, DF-GED^{u} is on average 4.75 times faster than DF-GED, while avg$_i$ $t($CSI_GED$^{\mathrm{nu}}, i)/t($CSI_GED$, i) \approx 1.32$. On AIDS, we have avg$_i$ $t($DF-GED$, i)/t($DF-GED$^{\mathrm{u}}, i) \approx 2.05$ and avg$_i$ $t($CSI_GED$^{\mathrm{nu}}, i)/$

$t(\texttt{CSI\_GED}, i) \approx 1.31$. These results are readily explained by the fact that DF-GED has to carry out the cubic computation of the lower bound h at each node of its search tree, whereas $\texttt{CSI\_GED}^{nu}$ has to carry out the cubic computation of Γ^{nu} only at the leafs and at initialisation. Secondly, we see that, on FINGER-PRINTS (cf. Fig. 1a), the node-based approaches $\texttt{DF-GED}^{u}$ and DF-GED outperform the edge-based algorithms CSI_GED and $\texttt{CSI\_GED}^{nu}$, while, on AIDS (cf. Fig. 1b), the opposite is the case. Finally, we note that the edge-based algorithms are more stable: While their deviation never exceeds 2%, the node-based approaches' deviation explodes on comparisons between large graphs contained in AIDS.

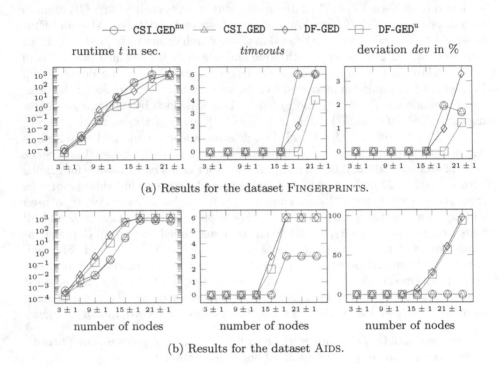

(a) Results for the dataset FINGERPRINTS.

(b) Results for the dataset AIDS.

Fig. 1. Results for uniform edit costs.

The results for non-uniform edit metric costs are displayed in Fig. 2. Note that the algorithms $\texttt{DF-GED}^{u}$ and CSI_GED do not appear in the evaluation, as they are designed only for uniform edit costs. The first observation is that, just like for uniform edit costs, the node-based approach DF-GED performs better on FINGERPRINTS (cf. Fig. 2a), while our edge-based generalisation $\texttt{CSI\_GED}^{nu}$ performs better on AIDS (cf. Fig. 2b). Secondly, we again note that the edge-based algorithm runs much more stable than the node-based approach: On FINGERPRINTS, i.e., the dataset where DF-GED performs better, we have $\max_i dev(\texttt{CSI\_GED}^{nu}, i) \approx 1.48$; whereas on AIDS, i.e., the dataset where $\texttt{CSI\_GED}^{nu}$ performs better, we observe $\max_i dev(\texttt{DF-GED}, i) \approx 47.97$.

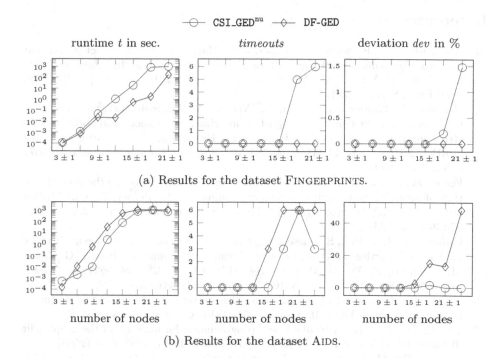

(a) Results for the dataset FINGERPRINTS.

(b) Results for the dataset AIDS.

Fig. 2. Results for non-uniform metric edit costs.

5   Conclusions and Future Work

Our experiments show that, for uniform edit costs, our speed-up DF-GED^u always outperforms DF-GED, while CSI_GED and our generalisation CSI_GED^{nu} perform similarly. We also observed that, neither for uniform nor for non-uniform metric edit costs, there is a clear winner between the node-based approaches DF-GED^u and DF-GED, on the one side, and the edge-based algorithms CSI_GED and CSI_GED^{nu}, on the other side. On FINGERPRINTS, the former two algorithms outperformed the latter in terms of runtime and timeouts, while on AIDS, the opposite outcome was observed. However, CSI_GED^{nu} and CSI_GED turned out to be more stable than DF-GED and DF-GED^u: While CSI_GED^{nu}'s and CSI_GED's deviation is small across all test-runs, DF-GED's and DF-GED^u's deviation explodes for comparisons between large graphs contained in the AIDS dataset. A global assessment of these observations indicates that, if there is no prior knowledge about the dataset and the graph edit distance has to be computed for both uniform and non-uniform metric edit costs, our generalisation CSI_GED^{nu} is the algorithm of choice. For future research, it might be interesting to individuate graph-properties that indicate if the node-based approaches DF-GED^u and DF-GED or the edge-based algorithms CSI_GED and CSI_GED^{nu} perform better. A meta-algorithm could then first compute these properties and select node-based or edge-based algorithms accordingly.

References

1. Abu-Aisheh, Z., Raveaux, R., Ramel, J.Y., Martineau, P.: An exact graph edit distance algorithm for solving pattern recognition problems. In: Marsico, M.D., Figueiredo, M.A.T., Fred, A.L.N. (eds.) ICPRAM 2015, vol. 1, pp. 271–278. SciTePress, Setúbal (2015)
2. Carletti, V., Gaüzère, B., Brun, L., Vento, M.: Approximate graph edit distance computation combining bipartite matching and exact neighborhood substructure distance. In: Liu, C.-L., Luo, B., Kropatsch, W.G., Cheng, J. (eds.) GbRPR 2015. LNCS, vol. 9069, pp. 188–197. Springer, Cham (2015). doi:10.1007/978-3-319-18224-7_19
3. Ferrer, M., Serratosa, F., Riesen, K.: Learning heuristics to reduce the overestimation of bipartite graph edit distance approximation. In: Perner, P. (ed.) MLDM 2015. LNCS (LNAI), vol. 9166, pp. 17–31. Springer, Cham (2015). doi:10.1007/978-3-319-21024-7_2
4. Gaüzère, B., Bougleux, S., Riesen, K., Brun, L.: Approximate graph edit distance guided by bipartite matching of bags of walks. In: Fränti, P., Brown, G., Loog, M., Escolano, F., Pelillo, M. (eds.) S+SSPR 2014. LNCS, vol. 8621, pp. 73–82. Springer, Heidelberg (2014). doi:10.1007/978-3-662-44415-3_8
5. Gouda, K., Hassaan, M.: CSI_GED: an efficient approach for graph edit similarity computation. In: ICDE 2016, pp. 265–276. IEEE Computer Society (2016)
6. Justice, D., Hero, A.: A binary linear programming formulation of the graph edit distance. IEEE Trans. Pattern Anal. Mach. Intell. **28**(8), 1200–1214 (2006)
7. Kuhn, H.W.: The hungarian method for the assignment problem. Nav. Res. Logist. Q. **2**(1–2), 83–97 (1955)
8. Lerouge, J., Abu-Aisheh, Z., Raveaux, R., Héroux, P., Adam, S.: Exact graph edit distance computation using a binary linear program. In: Robles-Kelly, A., Loog, M., Biggio, B., Escolano, F., Wilson, R. (eds.) S+SSPR 2016. LNCS, vol. 10029, pp. 485–495. Springer, Cham (2016). doi:10.1007/978-3-319-49055-7_43
9. Riesen, K., Bunke, H.: IAM graph database repository for graph based pattern recognition and machine learning. In: da Vitoria Lobo, N., Kasparis, T., Roli, F., Kwok, J.T., Georgiopoulos, M., Anagnostopoulos, G.C., Loog, M. (eds.) S+SSPR 2008. LNCS, vol. 5342, pp. 287–297. Springer, Heidelberg (2008)
10. Riesen, K., Bunke, H.: Approximate graph edit distance computation by means of bipartite graph matching. Image Vis. Comput. **27**(7), 950–959 (2009)
11. Fankhauser, S., Riesen, K., Bunke, H.: Speeding up graph edit distance computation through fast bipartite matching. In: Jiang, X., Ferrer, M., Torsello, A. (eds.) GbRPR 2011. LNCS, vol. 6658, pp. 102–111. Springer, Heidelberg (2011). doi:10.1007/978-3-642-20844-7_11
12. Riesen, K., Fischer, A., Bunke, H.: Computing upper and lower bounds of graph edit distance in cubic time. In: Gayar, N., Schwenker, F., Suen, C. (eds.) ANNPR 2014. LNCS (LNAI), vol. 8774, pp. 129–140. Springer, Cham (2014). doi:10.1007/978-3-319-11656-3_12
13. Serratosa, F.: Fast computation of bipartite graph matching. Pattern Recogn. Lett. **45**, 244–250 (2014)
14. Zeng, Z., Tung, A.K.H., Wang, J., Feng, J., Zhou, L.: Comparing stars: on approximating graph edit distance. PVLDB **2**(1), 25–36 (2009)
15. Zhao, X., Xiao, C., Lin, X., Wang, W.: Efficient graph similarity joins with edit distance constraints. In: Kementsietsidis, A., Salles, M.A.V. (eds.) ICDE 2012, pp. 834–845. IEEE Computer Society (2012)

16. Zheng, W., Zou, L., Lian, X., Wang, D., Zhao, D.: Graph similarity search with edit distance constraint in large graph databases. In: He, Q., Iyengar, A., Nejdl, W., Pei, J., Rastogi, R. (eds.) CIKM 2013, pp. 1595–1600. ACM (2013)

Improved Graph Edit Distance Approximation with Simulated Annealing

Kaspar Riesen[1,3]([⊠]), Andreas Fischer[2], and Horst Bunke[3]

[1] Institute for Information Systems, University of Applied Sciences FHNW,
Riggenbachstrasse 16, 4600 Olten, Switzerland
kaspar.riesen@fhnw.ch
[2] Department of Informatics,
University of Fribourg and HES-SO, 1700 Fribourg, Switzerland
andreas.fischer@unifr.ch
[3] Institute of Computer Science and Applied Mathematics,
University of Bern, Neubrückstrasse 10, 3012 Bern, Switzerland
bunke@iam.ch

Abstract. The present paper is concerned with graph edit distance, which is widely accepted as one of the most flexible graph dissimilarity measures available. A recent algorithmic framework for approximating the graph edit distance overcomes the major drawback of this distance model, viz. its exponential time complexity. Yet, this particular approximation suffers from an overestimation of the true edit distance in general. Overall aim of the present paper is to improve the distance quality of this approximation by means of a post-processing search procedure. The employed search procedure is based on the idea of simulated annealing, which turns out to be particularly suitable for complex optimization problems. In an experimental evaluation on several graph data sets the benefit of this extension is empirically confirmed.

1 Introduction

Due to their power and flexibility, graphs have found widespread application in pattern recognition and related fields [1,2]. Prominent examples of a classes of patterns, which can be formally represented in a more suitable and natural way by means of graphs rather than with feature vectors, are chemical compounds [3], binary executables [4], or networks [5].

The problem of computing graph dissimilarity is commonly solved via a particular *graph matching* algorithm. Graph matching has been the topic of numerous studies in pattern recognition over the last decades [1,2], resulting in powerful methods such as, for instance, *spectral methods* [6] or *graph kernels* [3]. *Graph edit distance* [7], introduced about 30 years ago, is still one of the most flexible graph distance models available. Yet, the run time of exact graph edit distance computation is exponential in the number of nodes of the involved graphs, which limits its applicability to rather small graphs.

In [8] the authors of the present paper introduced an algorithmic framework for the approximation of graph edit distance in cubic time. Yet, one of the major

© Springer International Publishing AG 2017
P. Foggia et al. (Eds.): GbRPR 2017, LNCS 10310, pp. 222–231, 2017.
DOI: 10.1007/978-3-319-58961-9_20

problems of this particular approximation framework is that it overestimates the true edit distance quite often. The present paper is concerned with an extension of this approximation that aims at making the distance approximation more accurate. The idea of this extension is based on a post-processing procedure that takes the result of the original approximation as a starting point for a (non-exhaustive) search process.

Note that in [9] several search procedures for the improvement of the approximation accuracy have already been proposed (amongst others, greedy forward search procedures). The novelty of the present paper is twofold. First, it presents a search strategy which takes into account both a lower and an upper bound on the true edit distance (rather than only the upper bound as proposed in [9]). Second, we make use of a different search method which is based on *simulated annealing* [10,11].

The basic idea of simulated annealing is to explore the search space in a random fashion and accepting solutions as long as they are getting better than the previous solution. Yet, in contrast with pure greedy algorithms, simulated annealing also accepts worse solutions with a certain probability (which slowly decreases during run time). The property of accepting worse solutions is fundamental as this allows to escape local minima during the search process.

The remainder of this paper is organized as follows. Next, in Sect. 2, the approximation framework for graph edit distance is reviewed. In Sect. 3, the novel search procedure based on simulated annealing is described in detail. Eventually, in Sect. 4, we empirically confirm the benefit of this extension on three graph data sets. Finally, in Sect. 5, we conclude the paper.

2   Graph Edit Distance (GED)

2.1   Basic Definition of GED

A graph g is a four-tuple $g = (V, E, \mu, \nu)$, where V is the finite set of nodes, $E \subseteq V \times V$ is the set of edges, $\mu : V \to L_V$ is the node labeling function, and $\nu : E \to L_E$ is the edge labeling function. The labels for both nodes and edges can be given by the set of integers $L = \{1, 2, 3, \ldots\}$, the vector space $L = \mathbb{R}^n$, a set of symbolic labels $L = \{\alpha, \beta, \gamma, \ldots\}$, or a combination of various label alphabets from different domains. Unlabeled graphs are obtained by assigning the same (empty) label $\varnothing$ to all nodes and edges, i.e. $L_V = L_E = \{\varnothing\}$.

Given two graphs, $g_1 = (V_1, E_1, \mu_1, \nu_1)$ and $g_2 = (V_2, E_2, \mu_2, \nu_2)$, the basic idea of *graph edit distance* (*GED*) [7] is to transform g_1 into g_2 using edit operations, viz. *insertions*, *deletions*, and *substitutions* of both nodes and edges. The substitution of two nodes u and v is denoted by $(u \to v)$, the deletion of node u by $(u \to \varepsilon)$, and the insertion of node v by $(\varepsilon \to v)$[1]. A set of edit operations $\lambda(g_1, g_2) = \{e_1, \ldots, e_k\}$ that completely transform g_1 into g_2 is called an edit path between g_1 and g_2.

---

[1] A similar notation is used for edges.

Let $\Upsilon(g_1, g_2)$ denote the set of all admissible edit paths between two graphs g_1 and g_2. To find the most suitable edit path out of $\Upsilon(g_1, g_2)$, one introduces a cost $c(e_i)$ for every edit operation e_i, measuring the strength of the corresponding operation. The idea of such a cost is to define whether or not an edit operation represents a strong modification of the graph. The *graph edit distance* $d_{\lambda_{\min}}$ between two graphs $g_1 = (V_1, E_1, \mu_1, \nu_1)$ and $g_2 = (V_2, E_2, \mu_2, \nu_2)$ is then defined by

$$d_{\lambda_{\min}}(g_1, g_2) = \min_{\lambda \in \Upsilon(g_1, g_2)} \sum_{e_i \in \lambda} c(e_i).$$

2.2   Approximate Computation of GED

The problem of minimizing the graph edit distance can be reformulated as an instance of a *Quadratic Assignment Problem (QAP)* which in turn belong to the class of $\mathcal{NP}$-*complete* problems. QAPs basically consist of a linear and a quadratic term which have to be simultaneously optimized. In case of graph edit distance, the linear term of QAPs can be used to model the sum of node edit costs, while the latter is commonly used to represent the sum of edge edit costs (see [12] for further details).

The graph edit distance approximation framework introduced in [8] reduces the QAP of graph edit distance computation to an instance of a *Linear Sum Assignment Problem (LSAP)*. Similar to QAPs, LSAPs deal with the question how the entities of two sets can be optimally assigned to each other. We formally represent assignments by means of permutations $(\varphi_1, \ldots, \varphi_n)$ of the integers $(1, 2, \ldots, n)$. Such a permutation refers to the assignment where the i-th entity of the first set is mapped to the entity at position φ_i in the second set $(i = 1, \ldots, n)$.

For solving LSAPs, which cope with a linear term only, a large number of efficient algorithms exist (see [13] for an exhaustive survey). The time complexity of the best performing exact algorithms for LSAPs is cubic in the size of the problem. Hence, LSAPs can be – in contrast with QAPs – quite efficiently solved.

In order to reformulate the graph edit distance problem to an instance of an LSAP, the use of a square $(n + m) \times (n + m)$ cost matrix $\mathbf{C}$ has been proposed in [8]. This particular cost matrix represents the costs of all possible node substitutions as well as all possible node deletions and node insertions. The framework proposed in [8] optimizes the linear term of the LSAP stated on $\mathbf{C}$.

By omitting the quadratic term during the assignment process, we neglect the structural relationships between the nodes (i.e. the edges between the nodes). In order to integrate knowledge about the graph structure, to each entry $c_{ij} \in \mathbf{C}$, i.e. to each cost of a node edit operation $(u_i \rightarrow v_j)$, the minimum sum of edge edit operation costs, implied by the corresponding node operation, is added. This particular encoding of the minimum matching cost arising from the local edge structure enables the LSAP to consider information about the local, yet not global, edge structure of a graph.

A minimum cost permutation $(\varphi_1, \ldots, \varphi_{n+m})$ derived on $\mathbf{C} = (c_{ij})$ via LSAP solving algorithm corresponds to the assignment of all nodes of g_1 to all nodes of g_2. Assignment ψ includes edit operations of the form $(u_i \rightarrow v_j)$, $(u_i \rightarrow \varepsilon)$, and

$(\varepsilon \to v_j)^2$. Two different distance approximations can now be instantly derived from this node assignment, viz. an upper and a lower bound on the true graph edit distance.

$$\psi = ((u_1 \to v_{\varphi_1}), (u_2 \to v_{\varphi_2}), \ldots, (u_{m+n} \to v_{\varphi_{m+n}}))$$

For the upper bound we observe that edit operations on edges are uniquely defined by the edit operations on their adjacent nodes. That is, whether an edge (u, v) is substituted with an existing edge from the other graph, deleted, or inserted actually depends on the operations performed on both adjacent nodes u and v (and whether or not there is an edge between the matching nodes of the other graph). Hence, we can use the node assignment ψ to infer the complete set of globally consistent edge edit operations. The sum of costs of the node edit operations plus the costs of the implied edge operations gives us a first approximation value for the graph edit distance. Note that this approximation generally overestimates the true edit distance and actually builds an upper bound on the exact distance [14]. Thus, we denote this approximation with $d_{up}(g_1, g_2)$, or d_{up} for short.

The second approximation, which actually provides a lower bound d_{low} on the true edit distance [14], can be additionally inferred from the optimal assignment $(\varphi_1, \ldots, \varphi_n)$. Remember that every entry $c_{ij} \in \mathbf{C}$ reflects the cost of the corresponding node edit operation $(u_i \to v_j)$ plus the minimal cost of editing the incident edges of u_i to the incident edges of v_j. Hence, given an optimal permutation $(\varphi_1, \ldots, \varphi_{(n+m)})$, the minimal sum $\sum_{i=1}^{(n+m)} c_{i\varphi_i}$ can be subdivided into costs for node edit operations and costs for edge edit operations. Since every edge (u_i, u_j) is adjacent with two individual nodes u_i and u_j, every edge is considered twice in two independent entries in the optimal sum $\sum_{i=1}^{(n+m)} c_{i\varphi_i}$ (viz. once in entry $c_{i\varphi_i}$ and once in entry $c_{j\varphi_j}$). In order to derive a suitable approximation for the true edit distance, the cost of edge edit operations encoded in the sum $\sum_{i=1}^{(n+m)} c_{i\varphi_i}$ has thus to be multiplied by $\frac{1}{2}$. In summary, we obtain a lower bound on the true edit distance by summing up the cost of all node and half the cost of all edge edit operations, given the optimal assignment.

It is important to note that the permutation $(\varphi_1, \ldots, \varphi_{n+m})$ can be arbitrarily permuted and the resulting approximation d_{up} remains an admissible upper bound on the true edit distance. Yet, this does not account for the lower bound as defined above. That is, d_{low} constitutes a lower bound on the exact edit distance, if, and only if, the underlying permutation $(\varphi_1, \ldots, \varphi_{n+m})$ refers to the optimal solution of the LSAP stated on $\mathbf{C}$.

3   Improving the Accuracy with Simulated Annealing

It has been observed that both bounds d_{up} and d_{low} might introduce a (substantial) approximation error compared to the exact edit distance $d_{\lambda_{\min}}$.

---

[2] Edit operations of the form $(\varepsilon \to \varepsilon)$ can be dismissed, of course.

The present work aims at improving the overall distance quality of the approximation by means of a post processing procedure which searches within the interval $[d_{up}, d_{low}]$. The proposed search procedure is based on *simulated annealing*, which emulates a phenomenon in material science, viz. the annealing of solids. Simulated annealing has been originally proposed to obtain a state of minimum energy of a multiparticle physical system [10] and has later been adopted to solve difficult optimization problems [11].

The basic idea of solving optimization problems with simulated annealing is to start with a (random) initial solution and then randomly disturb it. As long as the resulting solution is better than the previous one, it is accepted and used in the following step. If the resulting solution is worse than the previous one, it may still be accepted with a certain probability. This probability is typically reciprocally proportional to the quality difference of the current and the previous solution and proportional to the current temperature. Usually, one starts with a high temperature in order to rather frequently allow deteriorations in the first iterations. Yet, during the running process the temperature is gradually decreased, and thus the probability that a worse solution is accepted becomes smaller. This reflects the idea of initially sampling the search space in larger steps and then gradually focusing on smaller, promising areas for the final solution.

The detailed algorithmic procedure for the improvement of the distance accuracy is given in Algorithm 1. As input parameters the algorithm takes the cost matrix $\mathbf{C}$, the upper and lower bound of the true edit distance d_{up} and d_{low}, the maximum number of iterations N, the starting temperature T, as well as the temperature decrease factor F.

On line 1 of Algorithm 1 two counters ($counter_1$ and $counter_2$) are initialized with zero. The former controls the number of iterations, while the latter is used to compute the probability of resetting the current search to a new random starting point (details follow below). Next, on line 2 and 3, a list with the first $(n + m)$ integers is initialized (in ascending order) and d_{min} as well as $d_{current}$ are initialized with the original upper bound d_{up}.

On line 4 the main loop of the search procedure starts. In every iteration we aim at improving, i.e. decreasing, the current upper bound $d_{current}$ by means of slightly changing the assignment ψ. In any case $d_{current}$ remains a valid upper bound on the exact edit distance. However, remember that the lower bound d_{low} cannot be improved, i.e. increased, during the proposed search process.

The main loop of Algorithm 1 is repeated until the current upper bound $d_{current}$ becomes equal to d_{low}. In this case we have found the optimal edit distance and can stop the procedure. Yet, this can only occur when the lower bound is equal to the true edit distance, of course (i.e. when $d_{low} = d_{\lambda_{min}}$). Otherwise, the maximum number of iterations N have to be carried out. In either case, d_{min}, which corresponds to the minimal upper bound that has been found during the search process, is finally returned by the algorithm.

In every iteration of the main loop a new candidate for the upper bound is generated by means of the sub-procedure *Candidate-Generator*, which takes $order_{current}$ and $\mathbf{C}$ as parameters (see line 6). This sub-procedure, outlined in

Algorithm 1. Compute-Improvement($\mathbf{C}, d_{up}, d_{low}, N, T, F$)

```
 1: counter₁=0 and counter₂=0
 2: order_current = (1, 2, . . . , (n + m))
 3: d_min = d_up and d_current = d_up
 4: while ((d_min − d_low) > 0 and counter₁ < N) do
 5:     counter₁++
 6:     (d_cand, order_cand) = Candidate-Generator(order_current, C)
 7:     Δ = |d_cand − d_current|
 8:     select random number r from [0, 1]
 9:     if (d_cand < d_current) or (r < exp (−Δ/(Δ_avg×T))) then
10:         d_current = d_cand
11:         order_current = order_cand
12:     end if
13:     if (d_current < d_min) then
14:         d_min = d_current
15:         counter₂=0
16:     else
17:         counter₂++
18:     end if
19:     select random number r from [0, 1]
20:     if (r < counter₂/N) then
21:         order_current = random permutation of (1, 2, . . . , (n + m))
22:     end if
23:     T = F × T
24: end while

25: return  d_min
```

Algorithm 2, randomly changes the current order on one position. Formally, the integer at position r in $order_{current}$ is moved to the head of the current list (the remaining parts remain unaltered). Next, the LSAP stated on $\mathbf{C}$ is solved with a suboptimal assignment algorithm in $O((n + m)^2)$ time [15]. This algorithm iterates through the rows of $\mathbf{C}$ and assigns every node to the minimum unused node in the respective row in a greedy manner. By removing column φ_i in $\mathbf{C}$ it is ensured that every column of the cost matrix is considered exactly once (i.e. $\forall j$ refers to available columns in $\mathbf{C}$). This assignment procedure crucially depends on the order in which the rows are processed (actually defined in $order_{current}$). Due to the (slight) change of the processing order introduced at the beginning of Algorithm 2, an alternative assignment ψ and thus an alternative upper bound can be expected. Finally, we return both the candidate processing order $order_{cand}$ and the corresponding distance approximation d_{cand} to the main procedure.

Both $order_{cand}$ and d_{cand} are accepted when d_{cand} is lower than $d_{current}$ (i.e. we observe an improvement of the current upper bound) – see line 9 to 12 of Algorithm 1. If the distance approximation d_{cand} is greater than (or equal to) the current upper bound $d_{current}$, it may still be accepted with probability

$$P = \exp \left(\frac{-\Delta}{\Delta_{avg} \times T} \right),$$

where Δ refers to the absolute difference between d_{cand} and $d_{current}$, the normalizing factor Δ_{avg} corresponds to the running average of all values of Δ at that time, and T is the current temperature. Note the influence of Δ and T on the probability P. The greater the deterioration Δ, the smaller is P. Vice versa, the greater the current temperature T, the greater is P (yet, note that temperature T is gradually lowered by factor F at the end of every iteration – see line 23).

On line 13 to line 18 we verify whether the current distance $d_{current}$ is smaller than the minimal upper bound d_{min} that has been found so far. Whenever a new minimal distance has been found, $counter_2$ is reset to zero, otherwise $counter_2$ is increased by one (i.e. we count the number of iterations without improvements of the minimal upper bound). This counter is eventually used to control whether or not the current solution is reset to a new random starting point. Formally, the probability that the current processing order ($order_{current}$) is randomly disturbed on all positions increases with $counter_2$ (see line 20 to 22). The rationale behind this resetting is that whenever the number of iterations without improvements exceeds a certain limit, a restart of the search procedure from another point in the search domain might be beneficial.

Algorithm 2. Candidate-Generator($order = (i_{(1)}, i_{(2)}, \ldots, i_{((n+m)})), \mathbf{C}$)

1: select random integer r from $[0, (n + m)]$
2: $order = (i_{(r)}, i_{(1)}, i_{(2)}, \ldots, i_{(r-1)}, i_{(r+1)}, \ldots, i_{((n+m)}))$
3: $\psi = \{\}$
4: **for** $i \in order$ **do**
5:     $\varphi_i = \arg\min_{\forall j} c_{ij}$
6:     Remove column φ_i from $\mathbf{C}$
7:     $\psi = \psi \cup \{(u_i \to v_{\varphi_i})\}$
8: **end for**
9: **return** $(d_\psi, order)$

---

4   Experimental Evaluation

4.1   Experimental Setup

The experimental evaluation aims at investigating the benefit of the post processing search procedure proposed in the present paper in a graph matching scenario. In particular, we measure the approximation error and the computation time on three different real world data sets from the IAM graph database repository [16][3]. The first graph data set involves graphs that represent molecular compounds (AIDS). The graphs from the second and third data set represent images of fingerprints (FP) and images of symbols from architectural and electronic drawings (GREC). For details on the graph extraction methods and the graph characteristics we refer to [16]. From all data sets, subsets of 1,000 graphs are randomly selected on which 1,000,000 pairwise graph edit distance computations are conducted.

Rather than choosing an appropriate starting temperature T it might be more intuitive to define the probabilities P_s and P_e of accepting a worse solution at the

---

[3] www.iam.unibe.ch/fki/databases/iam-graph-database.

beginning and at the end of the optimization process, respectively. Eventually, one can define the starting and end temperature according to

$$T_s = \frac{1}{\ln(P_s)} \quad \text{and} \quad T_e = \frac{1}{\ln(P_e)}.$$

The end temperature T_e is merely used for a proper definition of the decrease factor F for the temperature $T_{(n+1)}$ in iteration $(n+1)$ with respect to the current temperature T_n. Formally, given the the total number of iterations N, the decrease factor F can be defined as

$$F = \left(\frac{T_e}{T_s}\right)^{1/(N-1)}.$$

In our evaluation we set the starting and end probability to $P_s = 0.8$ and $P_e = 0.01$ and we test the novel algorithm with $N = 1000$ and $N = 10,000$ iterations (referred to as BP-SA(1) and BP-SA(10), respectively).

4.2   Empirical Investigation

In Table 1 the mean computation time per graph pair (t) as well as the approximation error, i.e. the degree of overestimation (o), is indicated for the different edit distance algorithms on all data sets. Exact-GED and BP-GED refer to an exact computation via tree search algorithm and the original approximation framework presented in [8] (these two algorithms are the reference systems). The first reference system is mainly used to control whether our novel method's computation time remains below the computation time of an exact algorithm, while the second reference system is mainly used to investigate the impact of the novel method on the approximation quality.

We first focus on the computation time. We note that BP-GED needs some fractions of a millisecond on average for one graph matching. With the proposed extension we observe an increase of the mean computation time to 1–3 ms and 10–25 ms on average with BP-SA(1) and BP-SA(10), respectively. Yet, comparing these matching times with the matching times of the exact algorithm (which takes 3–5 s per matching on average), the increase of the run time seems to be acceptable.

Taking the sum of distances of BP-GED as reference point for the overestimation (i.e. we take the sum of distances returned by BP-GED as 100%), we observe reductions of the approximation error of approximately 77%, 98%, and 85% on the three data sets (using BP-SA(1)). The approximation error can be further reduced by increasing the number of iterations from $N = 1000$ to $N = 10,000$. That is, with 10,000 iterations we can report reductions of the approximation error of approximately 89%, 99%, and 93%.

The substantial improvement of the approximation accuracy can be also observed in the scatter plots in Fig. 1 (on the GREC data set[4]). These scatter

---

[4] On the other data sets very similar plots can be observed.

Table 1. The mean run time for one matching (t) and overestimation error (o) using a specific graph edit distance algorithm.

Data Set	Algorithm							
	Exact-GED		BP-GED		BP-SA(1)		BP-SA(10)	
	t	o	t	o	t	o	t	o
AIDS	5.63 s	0.00	0.07 ms	100.00	2.95 ms	23.42	25.44 ms	10.82
FP	5.00 s	0.00	0.29 ms	100.00	1.24 ms	1.91	9.46 ms	0.44
GREC	3.10 s	0.00	0.20 ms	100.00	2.21 ms	14.23	17.84 ms	6.40

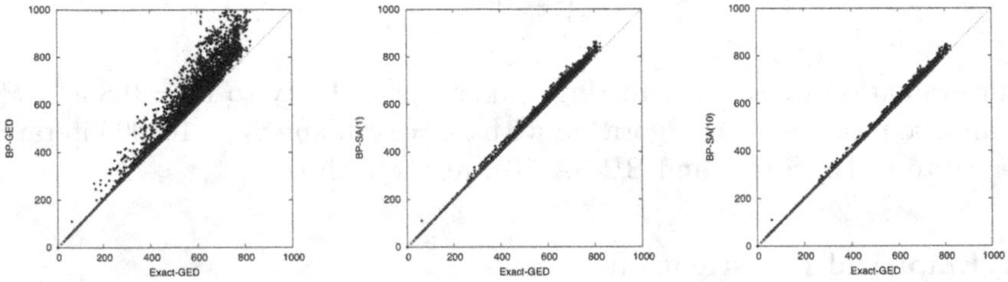

(a) BP-GED   (b) BP-SA(1)   (c) BP-SA(10)

Fig. 1. Exact (x-axis) vs. approximate (y-axis) graph edit distance on the GREC data computed with (a) original framework BP-GED, (b) BP-SA(1), and (c) BP-SA(10).

plots give us a visual representation of the accuracy of our approximations. We plot for each pair of graphs its exact (horizontal axis) and approximate (vertical axis) distance value. The reduction of the overestimation using our proposed extension is clearly observable and illustrates the power of BP-SA.

5   Conclusions

In the present paper we propose to improve the graph edit distance quality of a recent approximation framework by means of simulated annealing. The basic idea of this search process is to start with the upper- and lower bound on the true edit distance and then randomly search in the neighborhood of the current solution. As long as we improve the current distance, the new solution is accepted. Yet, also a deterioration of the solution might be accepted by the algorithm with a certain probability (that depends on both the level of deterioration and the search progress). This allows the search procedure to overcome local minima and possibly find the globally optimal solution. With an empirical investigation on three data sets we observe that substantial improvements of the approximation quality can be made with our novel extension.

Acknowledgements. This work has been supported by the *Hasler Foundation* Switzerland.

References

1. Conte, D., Foggia, P., Sansone, C., Vento, M.: Thirty years of graph matching in pattern recognition. Int. J. Pattern Recognit. Artif. Intell. **18**(3), 265–298 (2004)
2. Foggia, P., Percannella, G., Vento, M.: Graph matching and learning in pattern recognition in the last 10 years. Int. J. Pattern Recognit. Artif. Intell. **28**(1), 1450001 (2014)
3. Gaüzère, B., Brun, L., Villemin, D.: Two new graphs kernels in chemoinformatics. Pattern Recognit. Lett. **33**(15), 2038–2047 (2012)
4. Kinable, J., Kostakis, O.: Malware classification based on call graph clustering. J. Comput. Virol. **7**(4), 233–245 (2011)
5. Dickinson, P.J., Bunke, H., Dadej, A., Kraetzl, M.: Matching graphs with unique node labels. Pattern Anal. Appl. **7**(3), 243–254 (2004)
6. Wilson, R.C., Hancock, E.R., Luo, B.: Pattern vectors from algebraic graph theory. IEEE Trans. Pattern Anal. Mach. Intell. **27**(7), 1112–1124 (2005)
7. Bunke, H., Allermann, G.: Inexact graph matching for structural pattern recognition. Pattern Recognit. Lett. **1**, 245–253 (1983)
8. Riesen, K., Bunke, H.: Approximate graph edit distance computation by means of bipartite graph matching. Image Vis. Comput. **27**(4), 950–959 (2009)
9. Riesen, K., Bunke, H.: Improving bipartite graph edit distance approximation using various search strategies. Pattern Recognit. **48**(4), 1349–1363 (2015)
10. Metropolis, N., Rosenbluth, A.W., Rosenbluth, M.N., Teller, A.H., Teller, E.: Equation of state calculations by fast computing machines. J. Chem. Phys. **21**(6), 1087 (1953)
11. Kirkpatrick, S., Gelatt, C.D., Vecchi, M.P.: Optimization by simulated annealing. Science **4598**, 671–680 (1983)
12. Riesen, K.: Structural Pattern Recognition with Graph Edit Distance: Approximation Algorithms and Applications. Advances in Computer Vision and Pattern Recognition. Springer, Heidelberg (2015)
13. Burkard, R., Dell'Amico, M., Martello, S.: Assignment Problems. Society for Industrial and Applied Mathematics, Philadelphia (2009)
14. Riesen, K., Fischer, A., Bunke, H.: Estimating graph edit distance using lower and upper bounds of bipartite approximations. Int. J. Pattern Recognit. Artif. Intell. **29**(2), 1550011 (2015)
15. Riesen, K., Ferrer, M., Dornberger, R., Bunke, H.: Greedy graph edit distance. In: Perner, P. (ed.) MLDM 2015. LNCS, vol. 9166, pp. 3–16. Springer, Cham (2015). doi:10.1007/978-3-319-21024-7_1
16. Riesen, K., Bunke, H.: IAM graph database repository for graph based pattern recognition and machine learning. In: da Vitoria Lobo, N., Kasparis, T., Roli, F., Kwok, J.T., Georgiopoulos, M., Anagnostopoulos, G.C., Loog, M. (eds.) Structural, Syntactic, and Statistical Pattern Recognition. LNCS, vol. 5342, pp. 287–297. Springer, Heidelberg (2008)

An Edit Distance Between Graph Correspondences

Carlos Francisco Moreno-García[1], Francesc Serratosa[2(✉)],
and Xiaoyi Jiang[3]

[1] School of Computer Science and Digital Media,
The Robert Gordon University, Aberdeen, UK
c.moreno-garcia@rgu.ac.uk
[2] Department of Computer Science and Mathematics,
Universitat Rovira i Virgili, Tarragona, Spain
francesc.serratosa@urv.cat
[3] Department of Mathematics and Computer Science,
University of Münster, Münster, Germany
xjiang@uni-muenster.de

Abstract. The Hamming Distance has been largely used to calculate the dissimilarity of a pair of correspondences (also known as labellings or matchings) between two structures (i.e. sets of points, strings or graphs). Although it has the advantage of being simple in computation, it does not consider the structures that the correspondences relate. In this paper, we propose a new distance between a pair of graph correspondences based on the concept of the edit distance, called Correspondence Edit Distance. This distance takes into consideration not only the mapped elements of the correspondences, but also the attributes on the nodes and edges of the graphs being mapped. In addition to its definition, we also present an efficient procedure for computing the correspondence edit distance in a special case. In the experimental validation, the results delivered using the Correspondence Edit Distance are contrasted against the ones of the Hamming Distance in a case of finding the weighted means between a pair of graph correspondences.

Keywords: Graph correspondence · Hamming distance · Edit distance · Weighted mean

1 Introduction

A graph correspondence (or simply referred as a correspondence) is defined as a bijective function which designates a set of element-to-element mappings between the nodes of a pair of graphs. It can be generated either manually or automatically, with the purpose of finding the similarity between these two graphs. In the case that a

This research is supported by projectsTIN2016-77836-C2-1-R, ColRobTransp MINECO DPI2016-78957-R AEI/FEDER EU and by Consejo Nacional de Ciencia y Tecnologías (CONACyT) México).

© Springer International Publishing AG 2017
P. Foggia et al. (Eds.): GbRPR 2017, LNCS 10310, pp. 232–241, 2017.
DOI: 10.1007/978-3-319-58961-9_21

correspondence is obtained through an automatic method; the process is most commonly done through an optimisation process called error-tolerant graph matching. Several graph matching methods have been proposed in recent years [1–3] and therefore, it is possible to generate more than one correspondence between a single pair of graphs. In these scenarios, it may be interesting to know how different the generated correspondences are with respect to a ground truth correspondence, or also to analyse how different two correspondences are, and thus the requirement of a specifically designed distance between correspondences. So far in literature, the most commonly used distance between correspondences is the Hamming Distance (HD), which measures the number of mappings that are different between two correspondences. This distance has been used either to measure the accuracy of graph matching algorithms [4, 5] or to perform classification [6]. Nonetheless, the HD falls short on truly representing the dissimilarity between a pair of correspondences.

To justify this claim, consider the following toy example. Assume that three separate parties (human experts or automatic systems) deduce respectively three correspondences f^1, f^2 and f^3 between two graphs G and G' as shown in Fig. 1 (numbers in nodes represent their attribute). Notice that if the HD is used to calculate the dissimilarity between these correspondences, the result is $HD(f^1, f^2) = 2$ and $HD(f^1, f^3) = 2$, implying that both f^2 and f^3 are equally dissimilar with respect to f^1. Nonetheless, if we consider the cost of matching nodes on G and G' as the Euclidean distance between the attributes, then it can be seen that $Cost(f^1) = 1 + 0 + 1 + 1 = 3$, $Cost(f^2) = 1 + 0 + 1 + 3 = 5$ and $Cost(f^3) = 6 + 5 + 1 + 1 = 13$. Notice that the HD fails at reflecting that the cost difference between f^1 and f^3 is larger than between f^1 and f^2.

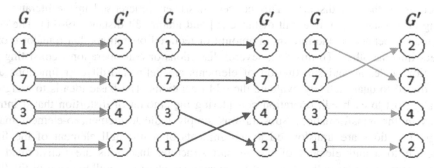

Fig. 1. A first example of two correspondences f^1 and f^2 between two graphs.

The rest of the paper is structured as follows. The next section briefly introduces the basic definitions. In Sect. 3, we present the newly proposed distance between a pair of correspondences. In Sect. 4, we contrast the new distance against the Hamming distance in the case of finding the weighted mean correspondences. Finally, Sect. 5 is reserved for conclusions and further work.

2 Basic Definitions

Let us represent an attributed graph as a four-tuple $G = (V, E, \gamma, \mu)$, where elements $v_i \in \Sigma$ represent the set of nodes, elements $e_i \in E$ represent the set of edges, and γ and μ are functions that assign a set of attributes to each node or edge respectively. Such graph may contain a specific kind of nodes called "null nodes", which are an additional set of nodes which have differentiated attributes (i.e. distinct values to the range of the original attribute values). Moreover, given a pair of graphs $G = (V, E, \gamma, \mu)$, and $G' = (V', E', \gamma', \mu')$, of the same order n (naturally or due to the presence of null nodes), we define the set T of all possible correspondences, such that each correspondence in T maps all nodes of G to nodes in G', $f : V \rightarrow V'$ in a bijective manner. Let f^1 and f^2 denote two arbitrarily selected correspondences in T. We can calculate how similar these two correspondences are through the Hamming distance (HD) between f^1 and f^2

$$HD\left(f^1, f^2\right) = \sum_{i=1}^{n}\left(1 - \partial\left(v'_a, v'_b\right)\right) \tag{1}$$

Where a and b are defined such that $f^1(v_i) = v'_a$ and $f^2(v_i) = v'_b$, and ∂ is the well-known Kronecker Delta function

$$\partial(x, y) = \begin{cases} 0 \ if \ x \neq y \\ 1 \ if \ x = y \end{cases} \tag{2}$$

One of the most widely used frameworks to evaluate the distance between two data structures is the edit distance. This concept has been concretised in the literature as string edit distance [7], tree edit distance [8] and graph edit distance [9–11]. The edit distance is defined as the minimum amount of required operations that transform one object into the other. To this end, several distortions or edit operations, consisting of insertion, deletion and substitution of elements are defined. Edit cost functions are introduced to quantitatively evaluate the edit operations. The basic idea is to assign a penalty cost to each edit operation considering the amount of distortion that it introduces in the transformation. Substitutions simply indicate element-to-element mappings. Deletions are transformed to assignments of a non-null element of the first structure to a null element of the second structure. Insertions are transformed to assignments of a non-null element of the second structure to a null element of the first structure. Given two graphs G and G' and a correspondence f between them, the edit cost would be

$$Graph\_EditCost\left(G, G', f\right) = \sum_{v_i \in V} DV\left(v_i, v'_a\right) + \sum_{e_{ij} \in E} DE\left(e_{ij}, e'_{ab}\right) \tag{3}$$

where $f(v_i) = v'_a$, $f(v_j) = v'_b$, and DV and DE the distances between nodes and edges respectively. In the case that one of the nodes is a null node, then $DV\left(v_i, v'_a\right) = K_v$, which is the assigned penalty cost for nodes. Similarly for edges, $DE\left(e_{ij}, e'_{ab}\right) = K_e$ in case that one of the edges is a "null edge" (i.e. non-existing edge). If both nodes and adjacent edges are null, these functions return a zero. In the case that both nodes or both edges are non-null, these functions are application dependent. For instance, if the attributes of the nodes and edges are in $\mathbb{R}^n$, it is usual to apply the Euclidean distance or the weighted Euclidean distance.

Thus, the graph edit distance (GED) is defined as the minimum cost under any bijection in T

$$GED(G, G') = \min_{f \in T}\{Graph\_EditCost(G, G', f)\} \qquad (4)$$

Several algorithms have been presented in the literature to compute the GED in an exact or an approximate From this vast pool of options, one of the most widely used algorithms to calculate the GED based on the local substructures [12–14] of the graphs is the bipartite graph matching(BP) framework [15–19].

3 Correspondence Edit Distance

In this section, we present a first step towards a concretisation of an edit distance for correspondences, which we have called Correspondence Edit Distance (CED). In contrast to the HD, the CED aims to consider both the attributes and the local sub-structure of the nodes mapped by the correspondences. Given G and G' and two correspondences f^1 and f^2 between them, the elements to be considered by the CED must be the elements within the correspondence (mappings) within f^1 and f^2. To that aim, correspondences f^1 and f^2 are defined as sets of mappings $f^1 = \{m_1^1, \ldots, m_i^1, \ldots, m_n^1\}$ and $f^2 = \{m_1^2, \ldots, m_a^2, \ldots, m_n^2\}$, where $m_i^1 = (v_i, f^1(v_i))$ and $m_a^2 = (v_a, f^2(v_a))$. This means that we do not intend to compute the distance between G and G', but rather the distance between f^1 and f^2 while also considering the attributes of graphs G and G'.

Figure 2 (left) shows an illustrative example of our proposal using two graphs with no edges, four nodes each (in both graphs, the fourth node is a null node marked as ϕ and ϕ') and two correspondences between them: f^1 (blue) composed of m_1^1, m_2^1, m_3^1 and m_4^1, and f^2 (red) composed of m_1^2, m_2^2, m_3^2 and m_4^2. Notice that m_4^1 and m_4^2 map the null node of G, and thus will be onwards referred as "null mappings". Figure 2 (right) shows a bijective function $h = \{h_1, h_2, h_3, h_4\}$ (green) between f^1 and f^2. Then, the cost of h is calculated as the sum of distances between all mapping-to-mapping relations in h. For this example, the cost yielded by the mappings in h_1 is zero, given the two mappings are the same. For the rest of cases, depending on the attributes and the penalty costs K_v, K_e, the substitution costs would be calculated for the mappings involved.

Notice that for the CED it is important to first define a bijective function $h \in H$ between mappings, where H is the set of all possible bijections between a pair of correspondences. Given such a bijective function h, the edit cost function $Corr\_EditCost$ is defined in terms of the distances between mappings

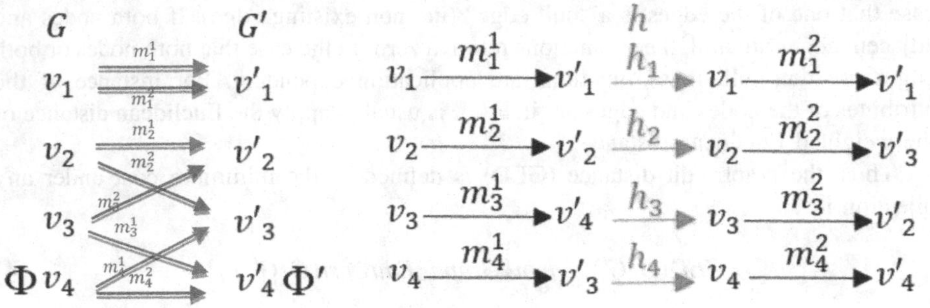

Fig. 2. Left: Two graphs G and G' and two correspondences f^1 and f^2 between them. Right: A bijective function h between f^1 and f^2. (Color figure online)

$$Corr\_EditCost\left(G, G', f^1, f^2, h\right) = \sum_{m_i^1 \in f^1} DM\left(G, G', m_i^1, h(m_i^1)\right) \qquad (5)$$

where DM is the distance (cost) between two mappings related by h. Then, the CED is defined in a similar way as the GED, that is

$$CED\left(G, G', f^1, f^2\right) = \min_{h \in H}\left\{Corr\_EditCost\left(G, G', f^1, f^2, h\right)\right\} \qquad (6)$$

Due to the combinatorial nature, the computation of CED is not easy in general. In the following we thus consider a special case which enables an efficient CED computation. If the aim of defining h is to relate the mappings which may resemble the most, then the most straightforward solution is to set all mapping-to-mapping relations in h as $h_j : m_j^1 \rightarrow m_j^2$. Figure 2 shows an example of this solution. In this case, the DM (Eq. 5) becomes the distance between the local substructures DS of the nodes being mapped, that is

$$DM\left(G, G', m_i^1, m_i^2\right) = DS\left(G', f^1(v_i), f^2(v_i)\right) \qquad (7)$$

Notice that a key difference between *Graph_EditCost* (Eq. 3) and *Corr_EditCost* (Eq. 5) is that in the first case, the distance functions DV and DE are defined between the nodes and adjacent edges of G and G', while in the second case, the distance between local substructures DS is obtained between nodes and adjacent edges on the same graph G'. In other words, to compute DS it is only necessary to compute the distance (cost) between the local substructure being mapped by f^1 in G' and the local substructures being mapped by f^2 in the same G'.

For this special case, the computation of the CED is presented in Algorithm 1. If the ith pair of mappings of f^1 and f^2 is equal, then it is excluded from the CED calculation. Moreover, the exclusion also prevails for the cases that two null mappings are paired, or that the two mappings refer to a null node (ϕ) in G'.

Algorithm 1. *Correspondence Edit Distance*
Input:G', f^1, f^2
Output: CED
Begin
CED=0
for $j = 1:n$
if $[f^1(i) \neq f^2(i)] \wedge \neg \{[f^1(i)) = \phi \wedge f^2(i) = \phi] \vee [f^1(i) = \phi \wedge f^2(i) = \phi]\}$
  CED $= CED + DS(G', f^1(i), f^2(i))$
  end if
end for
End Algorithm

4   Validation

To demonstrate in the most practical way that the use of either the HD or the CED produces different outcomes, we propose to use the scenario of calculating the weighted mean between a pair of correspondences. The concept of the weighted mean between two elements x and y has been largely used on data structures such as strings [20], graphs [21] and data clusters [22] to find an element z such that

$$Dist(x, y) = Dist(x, z) + Dist(y, z) \tag{8}$$

In practice, the weighted mean is used to implement methods that approximate towards the generalised median [23] of a set of strings [24–26], graphs [27], data clusters [28] or correspondences [29], as well as to define frameworks such as the

Fig. 3. Correspondences f^1 (top) and f^2 (bottom) between the graphs. (Color figure online)

"consensus" calculation between a set of correspondences, where the aim is to find the most accurate representative prototype from a pool of set-of-points correspondences or graph correspondences [30–33].

Using the first two images of the "BOAT" sequence in the "Tarragona Rotation Zoom" database [6], we randomly select 7 out of the 50 original nodes provided. A node represents a salient point in the image and the normalised SURF features [34] are its attribute. Afterwards, a graph is constructed using these nodes with edges conformed through the Delaunay triangulation. Two correspondences f^1 and f^2 are generated using two different matching algorithms. Notice that since graphs have been enlarged with a null node each to create mutually bijective correspondences, both have a total of eight mappings, with 7 of them being different one from the other (green lines) and one being equal (red line). The result of this process is shown in Fig. 3.

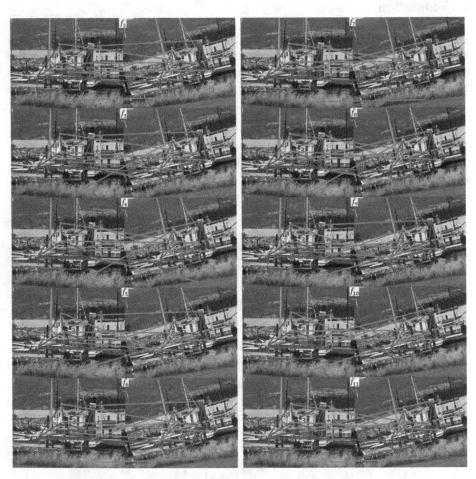

Fig. 4. All weighted means between f^1 and f^2 excluding the first and last one.

To find all weighted mean correspondences, we have implemented an A^* search algorithm which generates all possible correspondences between the two graphs and selects the ones that hold

$$Dist(f^1, f^2) = Dist(f^1, \bar{f}) + Dist(f^2, \bar{f}) \tag{9}$$

Using either HD or CED, the algorithm obtains the same weighted means, but with a different numerical value. For this test, the algorithm found the set of correspondences $\mathcal{W} = \bar{f}_1, \ldots \bar{f}_{12}$, as weighted means, where two of them are the original f_1 and f_2, thus $\bar{f}_1 = f^1$ and $\bar{f}_{12} = f^2$. Figure 4 shows the correspondences $\bar{f}_2, \ldots \bar{f}_{11}$, in $\mathcal{W}$.

Figure 5 shows the distance value using HD (+) or CED (O) ($K_V = K_E = 0.2$) between each of the 12 weighted means towards f_1, normalised by the distance between f_1 and f_2, that is

$$\alpha_i = \frac{Dist(f_1, \bar{f}_i)}{Dist(f_1, f_2)}, \ 1 \leq i \leq 12 \tag{10}$$

Notice that using the HD for the weighted means in $\mathcal{W}$ achieves seven different distance values, with repetitions such as $\alpha_3 = \alpha_4 = \alpha_5 = 0.\bar{3}$ and $\alpha_8 = \alpha_9 = \alpha_{10} = 0.\bar{6}$. Conversely, all weighted means in $\mathcal{W}$ deliver different distance values when the CED is used. The main conclusion drawn from this validation is that CED can deliver more diverse distance values than HD since it considers the attributes of the nodes and edges of the graphs being mapped. This characteristic allows to find better distributed weighted means when intending to use algorithms that aim at approximating towards the generalised median.

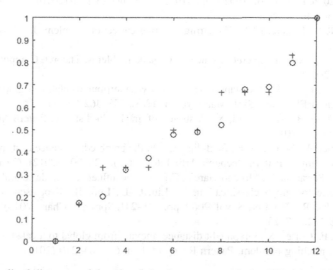

Fig. 5. Normalised distances of the 12 weighted means considering HD (+) and CED (O). The horizontal axis represents the different weighted means $\bar{f}_i$; $1 \leq i \leq 12$.

5   Conclusion and Further Work

In this paper, we present a first approach towards a new distance between a pair of correspondences called Correspondence Edit Distance (CED), based on the well-known concept of the edit distance. In contrast to the classic HD, CED is defined through the attributes of the nodes and their local substructure from the graphs being mapped. This characteristic allows more flexibility and versatility in cases such as obtaining the weighted mean correspondences for their use in algorithms that approach towards the generalised median or the consensus correspondence. In a near future, we intend to present an algorithm to calculate the generalised median correspondence through the use of the CED.

References

1. Conte, D., Foggia, P., Sansone, C., Vento, M.: Thirty years of graph matching. Int. J. Pattern Recognit. Artif. Intell. **18**(3), 265–298 (2004)
2. Foggia, P., Percannella, G., Vento, M.: Graph matching and learning in pattern recognition on the last ten years. Int. J. Pattern Recognit. Artif. Intell. **28**(1), 1450001 (2014). (40 pages)
3. Vento, M.: A long trip in the charming world of graphs for pattern recognition. Pattern Recognit. **48**(2), 291–301 (2015)
4. Caetano, T.S., McAuley, J.J., Cheng, L., Le, Q.V., Smola, A.J.: Learning graph matching. IEEE Trans. Pattern Anal. Mach. Intell. **31**(6), 1048–1058 (2009)
5. Zhou, F., De la Torre, F.: Factorized graph matching. IEEE Trans. Pattern Anal. Mach. Intell. **38**(9), 1774–1789 (2016)
6. Moreno-García, C.F., Cortés, X., Serratosa, F.: A graph repository for learning error-tolerant graph matching. In: Robles-Kelly, A., Loog, M., Biggio, B., Escolano, F., Wilson, R. (eds.) S+SSPR 2016. LNCS, vol. 10029, pp. 519–529. Springer, Cham (2016). doi:10.1007/978-3-319-49055-7_46
7. Wagner, R.A., Fischer, M.J.: The string-to-string correction problem. J. ACM **21**(1), 168–173 (1974)
8. Bille, P.: A survey on tree edit distance and related problems. Theoret. Comput. Sci. **337**(9), 217–239 (2005)
9. Sanfeliu, A., Fu, K.S.: A distance measure between attributed relational graphs for pattern recognition. IEEE Trans. Syst. Man Cybern. **13**(3), 353–362 (1983)
10. Gao, X., Xiao, B., Tao, D., Li, X.: A survey of graph edit distance. Pattern Anal. Appl. **13**(1), 113–129 (2010)
11. Solé-Ribalta, A., Serratosa, F., Sanfeliu, A.: On the graph edit distance cost: properties and applications. Int. J. Pattern Recognit Artif Intell. **26**(5), 1260004 (2012). (24 pages)
12. Cortés, X., Serratosa, F., Moreno-García, C.F.: On the influence of node centralities on graph edit distance for graph classification. In: Liu, C.-L., Luo, B., Kropatsch, W.G., Cheng, J. (eds.) GbRPR 2015. LNCS, vol. 9069, pp. 231–241. Springer, Cham (2015). doi:10.1007/978-3-319-18224-7_23
13. Serratosa, F., Cortés, X.: Graph edit distance: moving from global to local structure to solve the graph-matching problem. Pattern Recognit. Lett. **65**, 204–210 (2015)

14. Cortés, X., Serratosa, F., Riesen, K.: On the relevance of local neighbourhoods for greedy graph edit distance. In: Robles-Kelly, A., Loog, M., Biggio, B., Escolano, F., Wilson, R. (eds.) S+SSPR 2016. LNCS, vol. 10029, pp. 121–131. Springer, Cham (2016). doi:10.1007/978-3-319-49055-7_11

15. Riesen, K., Bunke, H.: Approximate graph edit distance computation by means of bipartite graph matching. Image Vis. Comput. **27**(7), 950–959 (2009)

16. Serratosa, F.: Fast computation of bipartite graph matching. Pattern Recognit. Lett. **45**, 244–250 (2014)

17. Serratosa, F.: Computation of graph edit distance: reasoning about optimality and speed-up. Image Vis. Comput. **40**, 38–48 (2015)

18. Serratosa, F.: Speeding up fast bipartite graph matching trough a new cost matrix. Int. J. Pattern Recognit. Artif. Intell. **29**(2), 1550010 (2015). (17 pages)

19. Sanroma, G., Penate-Sanchez, A., Alquezar, R., Serratosa, F., Moreno-Noguer, F., Andrade-Cetto, J., Gonzalez, M.A.: MSClique: multiple structure discovery through the maximum weighted clique problem. PLoS ONE **11**(1), e0145846 (2016)

20. Bunke, H., Jiang, X., Abegglen, K., Kandel, A.: On the weighted mean of a pair of strings. Pattern Anal. Appl. **5**(1), 23–30 (2002)

21. Bunke, H., Günter, S.: Weighted mean of a pair of graphs. Computing **67**(3), 209–224 (2001)

22. Franek, L., Jiang, X., He, C.: Weighted mean of a pair of clusterings. Pattern Anal. Appl. **17**(1), 153–166 (2014)

23. Jiang, X., Münger, A., Bunke, H.: On median graphs: properties, algorithms, and applications. IEEE Trans. Pattern Anal. Mach. Intell. **23**(10), 1144–1151 (2001)

24. Jiang, X., Abegglen, K., Bunke, H., Csirik, J.: Dynamic computation of generalised median strings. Pattern Anal. Appl. **6**(3), 185–193 (2003)

25. Jiang, X., Wentker, J., Ferrer, M.: Generalized median string computation by means of string embedding in vector spaces. Pattern Recognit. Lett. **33**(7), 842–852 (2012)

26. Franek, L., Jiang, X.: Evolutionary weighted mean based framework for generalized median computation with application to strings. In: Gimel'farb, G., et al. (eds.) S+SSPR 2012. LNCS, vol. 7626, pp. 70–78. Springer, Heidelberg (2012). doi:10.1007/978-3-642-34166-3_8

27. Ferrer, M., Valveny, E., Serratosa, F., Riesen, K., Bunke, H.: Generalized median graph computation by means of graph embedding in vector spaces. Pattern Recognit. **43**(4), 1642–1655 (2010)

28. Franek, L., Jiang, X.: Ensemble clustering by means of clustering embedding in vector spaces. Pattern Recognit. **47**(2), 833–842 (2014)

29. Moreno-García, C.F., Serratosa, F., Cortés, X.: Generalised median of a set of correspondences based on the hamming distance. In: Robles-Kelly, A., Loog, M., Biggio, B., Escolano, F., Wilson, R. (eds.) S+SSPR 2016. LNCS, vol. 10029, pp. 507–518. Springer, Cham (2016). doi:10.1007/978-3-319-49055-7_45

30. Moreno-García, C.F., Serratosa, F.: Correspondence consensus of two sets of correspondences through optimisation functions. Pattern Anal. Appl. **20**(1), 201–213 (2015)

31. Moreno-García, C.F., Serratosa, F.: Online learning the consensus of multiple correspondences between sets. Knowl. Based Syst. **90**, 49–57 (2015)

32. Moreno-García, C.F., Serratosa, F.: Consensus of multiple correspondences between sets of elements. Comput. Vis. Image Underst. **142**, 50–64 (2016)

33. Moreno-García, C.F., Serratosa, F.: Obtaining the consensus of multiple correspondences between graphs through online learning. Pattern Recognit. Lett. (2016)

34. Bay, H., Ess, A., Tuytelaars, T., Van Gool, L.: Speeded-up robust features (SURF). Comput. Vis. Image Underst. **110**(3), 346–359 (2008)

A Survey on Applications of Bipartite Graph Edit Distance

Michael Stauffer[1,4(✉)], Thomas Tschachtli[1], Andreas Fischer[2,3],
and Kaspar Riesen[1]

[1] Institute for Information Systems, University of Applied Sciences
and Arts Northwestern Switzerland, Riggenbachstr. 16, 4600 Olten, Switzerland
{michael.stauffer,kaspar.riesen}@fhnw.ch

[2] Department of Informatics, University of Fribourg, 1700 Fribourg, Switzerland
andreas.fischer@unifr.ch

[3] Institute for Complex Systems, University of Applied Sciences and Arts Western
Switzerland, 1705 Fribourg, Switzerland

[4] Department of Informatics, University of Pretoria, Pretoria, South Africa

Abstract. About ten years ago, a novel graph edit distance framework based on bipartite graph matching has been introduced. This particular framework allows the approximation of graph edit distance in cubic time. This, in turn, makes the concept of graph edit distance also applicable to larger graphs. In the last decade the corresponding paper has been cited more than 360 times. Besides various extensions from the methodological point of view, we also observe a great variety of applications that make use of the bipartite graph matching framework. The present paper aims at giving a first survey on these applications stemming from six different categories (which range from document analysis, over biometrics to malware detection).

Keywords: Applications of bipartite graph matching · Graph-based pattern representations

1 Introduction

Most pattern recognition applications are either based on statistical (i.e. vectorial) or structural data structures (i.e. strings, trees, or graphs). Graphs, in contrast to feature vectors, are able to represent both entities and binary relationships that might exist between subparts of these entities. Moreover, graphs can adapt their size and complexity to the size and complexity of the actual pattern to be modelled. Due to their representational power and flexibility, graphs have found widespread application in pattern recognition and related fields. Prominent examples of classes of patterns, which can be formally represented in a more suitable and natural way by means of graphs rather than with feature vectors, are chemical compounds [1], documents [2], proteins [3], and networks [4] (see [5] for an early survey on applications of graphs in pattern recognition).

© Springer International Publishing AG 2017
P. Foggia et al. (Eds.): GbRPR 2017, LNCS 10310, pp. 242–252, 2017.
DOI: 10.1007/978-3-319-58961-9_22

The availability of a dissimilarity or similarity measure is a basic requirement for pattern recognition and analysis. For graph dissimilarity computation,
commonly solved via a particular *graph matching* algorithm, no standard model
has been established to date. For an excellent and exhaustive review on graph
matching methods emerged during the last forty years, the reader is referred
to [6,7].

The present paper is concerned with the graph matching paradigm of *graph
edit distance* [8,9]. In fact, the concept of graph edit distance is considered as one
of the most flexible and versatile graph matching models available. Yet, the major
drawback of graph edit distance is its computational complexity that restricts
its applicability to graphs of rather small size. Graph edit distance belongs to
the family of *Quadratic Assignment Problems (QAPs)*, which in turn belong to
the class of $\mathcal{NP}$-*complete* problems. That is, an exact and efficient algorithm for
the graph edit distance problem can not be developed unless $\mathcal{P} = \mathcal{NP}$.

About ten years ago, an algorithmic framework, which allows the approximate computation of graph edit distance in a substantially faster way than traditional methods on general graphs, has been introduced [10,11]. The basic idea
of this approach, termed *Bipartite Graph Edit Distance (BP)*, is to reduce the
difficult QAP of graph edit distance computation to a *Linear Sum Assignment
Problem (LSAP)*. LSAPs basically constitute the problem of finding an optimal
assignment between two independent sets of entities. For LSAPs quite an arsenal
of efficient (i.e. polynomial) algorithms exist (see [12] for an exhaustive survey
on LSAP algorithms).

The graph dissimilarity framework BP presented in [10,11] resolves several
major issues that appear when graph edit distance is reformulated to an instance
of an LSAP. In a first step the graphs to be matched are subdivided into individual nodes including local structural information. Next, in step 2, an algorithm
solving the LSAP is employed in order to find an optimal assignment of the nodes
(plus local structures) of both graphs. Finally, in step 3, an approximate graph
edit distance, which is globally consistent with the underlying edge structures of
both graphs, is derived from the assignment of step 2.

The time complexity of this matching framework is cubic with respect to the
number of nodes of the involved graphs. Hence, BP is also applicable to larger
graphs. Due to this benefit, the underlying methodology has been employed in
a great variety of applications. The contribution of the present paper is to give
a first survey on these application fields and the corresponding methods that
actually use the BP framework.

2 Applications

In the last decade, the original paper (that describes BP for the first time) [10] as
well as its extended version [11] have been cited more than 360 times. Regarding
these citing papers we observe two main categories. The first category is concerned with methodological extensions of BP. There are, for instance, papers that
use another basic cost model than proposed in the original framework [13,14],

or works that aim at making the approximation faster [15,16], or more accurate [17,18]. The second category of citing papers is concerned with different applications of the approximate graph matching framework BP. The main focus of the present paper is to review and categorise the papers of this second category.

A taxonomy of the application fields and the corresponding papers (reviewed in the following subsections) is given in Fig. 1. In all of these applications, graphs are used to represent real-word (or abstract) objects or patterns, such as for instance images, proteins, or business processes (to mention just a few examples). Eventually, the BP framework is used to measure the (dis)similarity between pairs of graph-based representations.

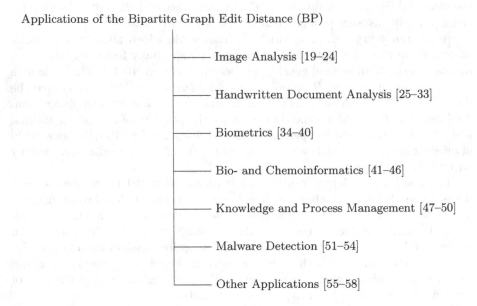

Applications of the Bipartite Graph Edit Distance (BP)

Image Analysis [19–24]

Handwritten Document Analysis [25–33]

Biometrics [34–40]

Bio- and Chemoinformatics [41–46]

Knowledge and Process Management [47–50]

Malware Detection [51–54]

Other Applications [55–58]

Fig. 1. Taxonomy of the reviewed application fields and papers that use the framework for Bipartite Graph Edit Distance (BP).

2.1   Image Analysis

Image analysis is often based on measuring the similarity of objects represented by 2D- or 3D-images. In the present scenario, graphs are used to represent these images, while the dissimilarity between pairs of images is measured by BP. In fact, many of the reviewed application fields presented below can be seen as special case of image analysis. In the present section, however, we present applications that are not part of any of the following subsections.

In [19], for instance, graphs are used to represent envelope images. That is, segmented regions are represented by nodes, while edges are inserted between specific pairs of nodes. The dissimilarities returned by BP are finally used to build a retrieval system. Another image analysis application is presented in [20]. In this case lunar surface images are formalised with graphs, where nodes represent SIFT-keypoints, while edges represent a Delaunay triangulation of the nodes. The BP distances are eventually used for localisation tasks. In [21] graphs are used to represent thinned images of archaeological structures (so called *Kites*) in order to find similar structures in large aerial image databases. Finally, in [22] the BP framework has been employed for shoe print classification, while in [23] the BP algorithm is used for the computation of similarities between petroglyphs.

Graphs are also used for 3D-images analysis. In [24], for instance, graphs are used to represent topological building structures by a so called *Room Connectivity Graph*. To this end, nodes represent rooms and are labelled by three-dimensional characteristics of the room. The edges are used to represent the connectivity between rooms labelled by two-dimensional features (i.e. width and height). BP is then employed in a retrieval scenario.

2.2 Handwritten Document Analysis

In recent years, handwritten (historical) documents have become increasingly digital available. However, the accessibility to these digitised documents with respect to browsing and searching is often limited. A first approach to bridge this gap is presented in [25], which aims at the recognition of unconstrained handwriting images. In this approach, nodes represent keypoints on the skeletonised word images, while edges represent strokes between keypoints. The BP framework (which has been extended in this particular case) is eventually used to define graph similarity features.

Keyword spotting allows to retrieve arbitrary keywords in handwritten documents. In case of graph-based keyword spotting, graphs commonly represent (parts of) segmented word images. The nodes of these graphs are, for instance, based on keypoints [26–29] or prototype strokes [30,31], while edges are commonly used to represent the connectivity between pairs of keypoints or prototype strokes. The actual spotting for certain words in a document is then based on dissimilarity computations between the query graph and document graphs using BP.

The BP dissimilarity framework has not only been applied for spotting keywords in historical documents, but also for clustering ancient ornamental initials (so called lettrines) [32]. In this particular case, each lettrine is represented by a *Region Adjacency Graph (RAG)*, where nodes are used to represent homogenous regions. Finally, edges are inserted based on the adjacency of regions.

In [33] a graph database for ancient handwritten documents is proposed and evaluated by means of a word classification experiment using BP. In particular, on the basis of segmented word images, six different graph representation formalisms are proposed and compared with each other using the BP dissimilarity model.

2.3   Biometrics

Biometrics are often used to verify or identify an individual based on certain biometrical characteristics (e.g. iris, fingerprint, or signature). In [34,35], retina vessels are used as a biometric trait for user verification. Formally, nodes are used to represent keypoints in skeletonised vessel images, while edges represent the vessels between selected keypoints. Finally, BP is used to match the retina vessel graphs [34] or to derive different graph similarity measures for building a multiple classifier system [35]. In [36,37] a very similar approach is applied on palm veins rather than retina vessels.

Moreover, graphs are also used to for fingerprint identification as introduced in [38]. In particular, fingerprint images are segmented into core areas (i.e. areas with same ridge direction), which are in turn represented by nodes, while edges are inserted between adjacent areas. The resulting fingerprint graphs are then classified using the distances derived from the BP framework.

In recent years, a trend towards high coverage of video surveillance can be observed. Thus, person re-identification over serval camera scenes evolved to a crucial task. In [39] a graph-based approach is presented for this task. To this end, segmented camera images are represented by means of a RAG [32]. The BP framework is then used in conjunction with a *Laplacian*-kernel in order to re-identify persons.

Last but not least, BP is also used for on-line signature verification [40]. First, the signatures are divided into segments which are in turn represented by graphs. That is, nodes represent the sample points of a segment, while edges are inserted between specific pairs of nodes. Finally, two signatures are compared with each other by measuring a sum of BP dissimilarities between pairs of graphs.

2.4   Bio- and Chemoinformatics

Bio- and chemoinformatics combine approaches and techniques of a broad field to analyse and interpret biological (i.e. DNA, protein sequences) or chemical structures (i.e. molecules), respectively. An important application in the field of bioinformatics is the analysis of deviations in biological structures to detect cancer. In [41], for instance, graphs are used to represent tissue image of biological cells. In this case nodes are used to represent tissue components, while their spatial relationship is represented by edges. Subsequently, the BP framework is used to classify graphs representing normal, low-grade and high-grade cancerous tissue images. In [42], a similar approach is introduced to detect irregularities in blood vessels rather than biological cells.

Chemoinformatics has become a well established field of research. Chemoinformatics is mainly concerned with the prediction of molecular properties by means of informational techniques. The assumption that two similar molecules should have similar activities and properties, is one of the major principles in this particular field. Clearly, molecules can be readily described with labelled graphs, where atoms are represented by nodes, while bonds between atoms (e.g. *single*, *double*, *triple*, or *aromatic*) are represented by edges.

In [43, 44] the approximation of graph edit distance by means of BP is used to build a novel graph kernel for activity predictions of molecular compounds. In [45] various graph embeddings methods and graph kernels, which are in part built upon the BP framework, are evaluated in diverse chemoinformatics applications. Finally, in [46] an algorithm to compute single summary graphs from a collection of molecule graphs has been proposed. The formulation of the cost of a matching, which is actually used in this methodology, is based on BP.

2.5 Knowledge and Process Management

In the last decades, a trend towards digitalisation of business models can be observed throughout most industries. Knowledge and process management ensures a thorough information flow, which is actually needed to manage both physical and intellectual resources. Nowadays, business processes are often supported (or completely created) by means of web services. Thus, the rediscoverability of composite web services (described by means of an OWL-S process) is of high relevance and issued in [47]. To this end, a graph is used to represent a composite process. Nodes represent process states and atomic services, while directed edges are used to represent the control flow. By a similar principle, business (sub)-processes rather than web services are retrieved in [48]. In particular, business process activities represent nodes, while directed edges are used to represent the business process flow. Finally, the BP framework is used to find similarities between business (sub)-processes.

Another application in this field is presented in [49], where semantical enriched documents (so called *Resource Description Framework (RDF)* ontologies) are represented by graphs. To this end, document key concepts (e.g. db:Bob_Dylan, db:Folk_Music) are represented by nodes, while directed edges are used to represent semantic relations (e.g. dbp:genre) and labelled by their importance. Finally, the similarity of documents is computed by an adapted BP matching framework.

Based on (similar) RDF ontologies, an approach to estimate the execution time of SPARQL (the RDF query language) queries is presented in [50]. In this scenario the BP distances of an unknown query to a set of training queries are used as query features.

2.6 Malware Detection

Anti-virus companies receive huge amounts of samples of potentially harmful executables. This growing amount of data makes robust and automatic detection of malware necessary.

The differentiation between malicious and original binary executables is actually another field where the framework BP has been extensively used. In [51–53], for instance, malware detection based on comparisons of call graphs has been proposed. In particular, the authors propose to represent malware samples as call graphs such that certain variations of the malware can be generalised. This approach enables the detection of structural similarities between samples in a robust way. For pairwise comparisons of these call graphs the approximation of BP is employed.

In [54] a similar approach has been pursued for the detection of malware by using weighted contextual API dependency graphs in conjunction with an extended version of BP for graph comparison. Finally, in [51] BP has been employed for the development of a polynomial time algorithm for calculating the differences between two binaries.

2.7   Other Applications

A further application where the BP matching algorithm has been applied, is for instance, the retrieval of stories (for storytelling) [55]. In this scenario, nodes represent goals and actions, while edges represent time and order. Thus, similar stories can be retrieved by means of the BP framework. In [56] a similar approach is introduced to retrieve sketches used to define the building behaviour of non-player characters in computer games. Finally, the BP framework is also used to detect plagiarism. In particular, [57] and [58] BP is used to detect plagiarism in Haskell programs and in textual documents, respectively.

3   Conclusion

In this paper we have reviewed about 40 different papers applying the BP matching framework. The reviewed applications stem from different fields like image analysis, handwritten document analysis, biometrics, bio- and chemoinformatics, knowledge and process management, malware detection, and others. In future work, we plan to extend this survey by integrating not only applications, but also methodological extensions of the BP matching algorithm.

Acknowledgments. This work has been supported by the Hasler Foundation Switzerland.

References

1. Mahé, P., Ueda, N., Akutsu, T., Perret, J.L., Vert, J.P.: Graph kernels for molecular structure-activity relationship analysis with support vector machines. J. Chem. Inf. Model. **45**(4), 939–951 (2005)
2. Schenker, A.: Graph-Theoretic Techniques for Web Content Mining, vol. 62. World Scientific, Singapore (2005)
3. Borgwardt, K.M., Kriegel, H.P., Vishwanathan, S.V.N., Schraudolph, N.N.: Graph kernels for disease outcome prediction from protein-protein interaction networks. In: Pacific Symposium on Biocomputing, pp. 4–15 (2007)
4. Dickinson, P.J., Bunke, H., Dadej, A., Kraetzl, M.: Matching graphs with unique node labels. Pattern Anal. Appl. **7**(3), 243–254 (2004)
5. Conte, D., Foggia, P., Sansone, C., Vento, M.: Graph matching applications in pattern recognition and image processing. Int. Conf. Image Process. **3**, 21–24 (2003)

6. Conte, D., Foggia, P., Sansone, C., Vento, M.: Thirty years of graph matching in pattern recognition. Int. J. Pattern Recognit. Artif. Intell. **18**(03), 265–298 (2004)
7. Foggia, P., Percannella, G., Vento, M.: Graph matching and learning in pattern recognition in the last 10 Years. Int. J. Pattern Recognit. Artif. Intell. **28**(01), 1450001 (2014)
8. Bunke, H., Allermann, G.: Inexact graph matching for structural pattern recognition. Pattern Recognit. Lett. **1**(4), 245–253 (1983)
9. Sanfeliu, A., Sanfeliu, A., Fu, K.S.: A distance measure between attributed relational graphs for pattern recognition. IEEE Trans. Syst. Man. Cybern. **13**(3), 353–362 (1983)
10. Riesen, K., Neuhaus, M., Bunke, H.: Bipartite graph matching for computing the edit distance of graphs. In: Escolano, F., Vento, M. (eds.) GbRPR 2007. LNCS, vol. 4538, pp. 1–12. Springer, Heidelberg (2007). doi:10.1007/978-3-540-72903-7_1
11. Riesen, K., Bunke, H.: Approximate graph edit distance computation by means of bipartite graph matching. Image Vis. Comput. **27**(7), 950–959 (2009)
12. Burkard, R., Dell'Amico, M., Martello, S.: Assignment Problems. SIAM, Philadelphia (2009)
13. Serratosa, F.: Fast computation of bipartite graph matching. Pattern Recognit. Lett. **45**(1), 244–250 (2014)
14. Gaüzère, B., Bougleux, S., Riesen, K., Brun, L.: Approximate graph edit distance guided by bipartite matching of bags of walks. In: Fränti, P., Brown, G., Loog, M., Escolano, F., Pelillo, M. (eds.) S+SSPR 2014. LNCS, vol. 8621, pp. 73–82. Springer, Heidelberg (2014). doi:10.1007/978-3-662-44415-3_8
15. Riesen, K., Ferrer, M., Bunke, H.: Approximate graph edit distance in quadratic time. IEEE Trans. Comput. Biol. Bioinform. (99), 1 (2015). http://ieeexplore.ieee.org/document/7264987/
16. Fischer, A., Riesen, K., Bunke, H.: Improved quadratic time approximation of graph edit distance by combining Hausdorff matching and greedy assignment. Pattern Recognit. Lett. **87**, 55–62 (2017)
17. Riesen, K., Bunke, H.: Improving bipartite graph edit distance approximation using various search strategies. Pattern Recognit. **48**(4), 1349–1363 (2015)
18. Riesen, K., Fischer, A., Bunke, H.: Estimating graph edit distance using lower and upper bounds of bipartite approximations. Int. J. Pattern Recognit. Artif. Intell. **29**(02), 1550011 (2015)
19. Liu, L., Lu, Y., Suen, C.Y.: Retrieval of envelope images using graph matching. In: International Conference on Document Analysis and Recognition, pp. 99–103 (2011)
20. Zhang, Y., Yang, X., Qiao, H., Liu, Z., Liu, C., Wang, B.: A graph matching based key point correspondence method for lunar surface images. In: World Congress on Intelligent Control and Automation, pp. 1825–1830 (2016)
21. Madi, K., Seba, H., Kheddouci, H., Barge, O.: A graph-based approach for Kite recognition. Pattern Recognit. Lett. **87**, 186–194 (2017)
22. Hasegawa, M., Tabbone, S.: A local adaptation of the histogram radon transform descriptor: an application to a shoe print dataset. In: Gimel'farb, G., et al. (eds.) SSPR/SPR 2012. LNCS, vol. 7626, pp. 675–683, Springer, Heidelberg (2012). doi:10.1007/978-3-642-34166-3_74
23. Seidl, M., Wieser, E., Zeppelzauer, M., Pinz, A., Breiteneder, C.: Graph-based shape similarity of petroglyphs. In: Agapito, L., Bronstein, M.M., Rother, C. (eds.) ECCV 2014. LNCS, vol. 8925, pp. 133–148. Springer, Cham (2015). doi:10.1007/978-3-319-16178-5_9

24. Wessel, R., Blümel, I., Ochmann, S., Vock, R.: Efficient retrieval of 3D building models using embeddings of attributed subgraphs. In: ACM Conference on Information and Knowledge Management, pp. 2097–2100 (2011)

25. Fischer, A., Suen, C.Y., Frinken, V., Riesen, K., Bunke, H.: A fast matching algorithm for graph-based handwriting recognition. In: Kropatsch, W.G., Artner, N.M., Haxhimusa, Y., Jiang, X. (eds.) GbRPR 2013. LNCS, vol. 7877, pp. 194–203. Springer, Heidelberg (2013). doi:10.1007/978-3-642-38221-5_21

26. Riesen, K., Brodić, D., Milivojević, Z.N., Maluckov, Č.A.: Graph based keyword spotting in medieval slavic documents – a project outline. In: Ioannides, M., Magnenat-Thalmann, N., Fink, E., Žarnić, R., Yen, A.-Y., Quak, E. (eds.) EuroMed 2014. LNCS, vol. 8740, pp. 724–731. Springer, Cham (2014). doi:10.1007/978-3-319-13695-0_74

27. Wang, P., Eglin, V., Garcia, C., Largeron, C., Llados, J., Fornes, A.: A novel learning-free word spotting approach based on graph representation. In: International Workshop on Document Analysis Systems, pp. 207–211 (2014)

28. Wang, P., Eglin, V., Garcia, C., Largeron, C., Llados, J., Fornes, A.: A coarse-to-fine word spotting approach for historical handwritten documents based on graph embedding and graph edit distance. In: International Conference on Pattern Recognition, pp. 3074–3079 (2014)

29. Stauffer, M., Fischer, A., Riesen, K.: Graph-based keyword spotting in historical handwritten documents. In: Robles-Kelly, A., Loog, M., Biggio, B., Escolano, F., Wilson, R. (eds.) S+SSPR 2016. LNCS, vol. 10029, pp. 564–573. Springer, Cham (2016). doi:10.1007/978-3-319-49055-7_50

30. Bui, Q.A., Visani, M., Mullot, R.: Unsupervised word spotting using a graph representation based on invariants. In: International Conference on Document Analysis and Recognition, pp. 616–620 (2015)

31. Riba, P., Llados, J., Fornes, A.: Handwritten word spotting by inexact matching of grapheme graphs. In: International Conference on Document Analysis and Recognition, pp. 781–785 (2015)

32. Jouili, S., Coustaty, M., Tabbone, S., Ogier, J.M.: NAVIDOMASS: structural-based approaches towards handling historical documents. In: International Conference on Pattern Recognition, pp. 946–949 (2010)

33. Stauffer, M., Fischer, A., Riesen, K.: A Novel Graph Database for Handwritten Word Images. In: Robles-Kelly, A., Loog, M., Biggio, B., Escolano, F., Wilson, R. (eds.) S+SSPR 2016. LNCS, vol. 10029, pp. 553–563. Springer, Cham (2016). doi:10.1007/978-3-319-49055-7_49

34. Arakala, A., Davis, S.A., Horadam, K.J.: Retina features based on vessel graph substructures. In: International Joint Conference on Biometrics, pp. 1–6 (2011)

35. Lajevardi, S.M., Arakala, A., Davis, S.A., Horadam, K.J.: Retina verification system based on biometric graph matching. IEEE Trans. Image Process. **22**(9), 3625–3635 (2013)

36. Horadam, K.J., Arakala, A., Davis, S., Lajevardi, S.M.: Hand vein authentication using biometric graph matching. IET Biom. **3**(4), 302–313 (2014)

37. Arakala, A., Hao, H., Davis, S., Horadam, K.J.: The palm vein graph for biometric authentication. In: Camp, O., Weippl, E., Bidan, C., Aïmeur, E. (eds.) ICISSP 2015. CCIS, vol. 576, pp. 199–218. Springer, Cham (2015). doi:10.1007/978-3-319-27668-7_12

38. Choi, Y., Kim, G.: Graph-based fingerprint classification using orientation field in core area. IEICE Electron. Express **7**(17), 1303–1309 (2010)

39. Brun, L., Conte, D., Foggia, P., Vento, M.: A graph-kernel method for re-identification. In: Kamel, M., Campilho, A. (eds.) ICIAR 2011. LNCS, vol. 6753, pp. 173–182. Springer, Heidelberg (2011). doi:10.1007/978-3-642-21593-3_18

40. Wang, K., Wang, Y., Zhang, Z.: On-line signature verification using segment-to-segment graph matching. In: International Conference on Document Analysis and Recognition, pp. 804–808 (2011)

41. Ozdemir, E., Gunduz-Demir, C.: A hybrid classification model for digital pathology using structural and statistical pattern recognition. IEEE Trans. Med. Imaging 32(2), 474–483 (2013)

42. Núñez, J.M., Bernal, J., Ferrer, M., Vilariño, F.: Impact of keypoint detection on graph-based characterization of blood vessels in colonoscopy videos. In: Luo, X., Reichl, T., Mirota, D., Soper, T. (eds.) CARE 2014. LNCS, vol. 8899, pp. 22–33. Springer, Cham (2014). doi:10.1007/978-3-319-13410-9_3

43. Brun, L., Conte, D., Foggia, P., Vento, M., Villemin, D.: Symbolic learning vs. graph kernels: an experimental comparison in a chemical application. In: East-European Conference on Advances in Databases and Information Systems (2010)

44. Gaüzère, B., Brun, L., Villemin, D.: Two new graph kernels and applications to chemoinformatics. In: Jiang, X., Ferrer, M., Torsello, A. (eds.) GbRPR 2011. LNCS, vol. 6658, pp. 112–121. Springer, Heidelberg (2011). doi:10.1007/978-3-642-20844-7_12

45. Gaüzère, B., Hasegawa, M., Brun, L., Tabbone, S.: Implicit and explicit graph embedding: comparison of both approaches on chemoinformatics applications. In: Gimel'farb, G., et al. (eds.) SSPR/SPR 2012. LNCS, vol. 7626, pp. 510–512. Springer, Heidelberg (2012). doi:10.1007/978-3-642-34166-3_56

46. Koop, D., Freire, J., Silva, C.T.: Visual summaries for graph collections. In: IEEE Pacific Visualization Symposium, pp. 57–64 (2013)

47. Cuzzocrea, A., Coi, J.L., Fisichella, M., Skoutas, D.: Graph-based matching of composite OWL-S services. In: Xu, J., Yu, G., Zhou, S., Unland, R. (eds.) DAS-FAA 2011. LNCS, vol. 6637, pp. 28–39. Springer, Heidelberg (2011). doi:10.1007/978-3-642-20244-5_4

48. Niedermann, F.: Deep business optimization: concepts and architecture for an analytical business process optimization platform. Ph.D. thesis, University of Stuttgart (2015)

49. Schuhmacher, M., Ponzetto, S.P.: Knowledge-based graph document modeling. In: ACM International Conference on Web Search and Data Mining, New York, pp. 543–552 (2014)

50. Hasan, R., Gandon, F.: A machine learning approach to SPARQL query performance prediction. In: IEEE/WIC/ACM International Joint Conferences on Web Intelligence (WI) and Intelligent Agent Technologies (IAT), pp. 266–273 (2014)

51. Bourquin, M., King, A., Robbins, E.: BinSlayer: accurate comparison of binary executables. In: ACM SIGPLAN on Program Protection and Reverse Engineering, New York, pp.1–10 (2013)

52. Elhadi, A.A.E., Maarof, M.A., Osman, A.H.: Malware detection based on hybrid signature behaviour application programming interface call graph. Am. J. Appl. Sci. 9(3), 283–288 (2012)

53. Kostakis, O., Kinable, J., Mahmoudi, H., Mustonen, K.: Improved call graph comparison using simulated annealing. In: ACM Symposium on Applied Computing, New York, pp. 1516–1523 (2011)

54. Zhang, M., Duan, Y., Yin, H., Zhao, Z.: Semantics-aware android malware classification using weighted contextual api dependency graphs. In: ACM SIGSAC Conference on Computer and Communications Security, New York, pp.1105–1116 (2014)
55. Paul, S.: Exploring story similarities using graph edit distance algorithms (2013)
56. Flórez-Puga, G., González-Calero, P.A., Jiménez-Díaz, G., Díaz-Agudo, B.: Supporting sketch-based retrieval from a library of reusable behaviours. Expert Syst. Appl. **40**(2), 531–542 (2013)
57. Kammer, M., Bodlaender, H., Hage, J.: Plagiarism detection in Haskell programs using call graph matching (2011)
58. Røkenes, H.D.: Graph-based natural language processing: graph edit distance applied to the task of detecting plagiarism (2012)

Graphs and Information Theory

Minimising Entropy Changes in Dynamic Network Evolution

Jianjia Wang$^{(\boxtimes)}$, Richard C. Wilson, and Edwin R. Hancock

Department of Computer Science,
University of York, York YO10 5DD, UK
jw1157@york.ac.uk

Abstract. The modelling of time-varying network evolution is critical to understanding the function of complex systems. The key to such models is a variational principle. In this paper we explore how to use the Euler-Lagrange equation to investigate the variation of entropy in time evolving networks. We commence from recent work where the von Neumman entropy can be approximated using simple degree statistics, and show that the changes in entropy in a network between different time epochs are determined by correlations in the changes in degree statistics of nodes connected by edges. Our variational principle is that the evolution of the structure of the network minimises the change in entropy with time. Using the Euler-Lagrange equation we develop a dynamic model for the evolution of node degrees. We apply our model to a time sequence of networks representing the evolution of stock prices on the New York Stock Exchange (NYSE). Our model allows us to understand periods of stability and instability in stock prices, and to predict how the degree distribution evolves with time. We show that the framework presented here provides allows accurate simulation of the time variation of degree statistics, and also captures the topological variations that take place when the structure of a network changes violently.

Keywords: Dynamic networks · Financial markets · Euler-Lagrange equation

1 Introduction

One of the challenges presented by networks is how to model and predict their evolution with time [1]. This problem can be addressed either at the local or the global level. At the local level the aim is to model how the detailed connectivity structure changes with time [2,6], while at the global level the aim is to model the evolution of characteristics which capture the structure of network, and allow different types of network to be distinguished from one to another. Methods falling into the former category include generative models which allow the detailed edge connectivity structure to be estimated from noisy or uncertain input data [5]. The latter includes models of the degree distribution, and other global properties such as edge density or structural community indices [3,4,8].

© Springer International Publishing AG 2017
P. Foggia et al. (Eds.): GbRPR 2017, LNCS 10310, pp. 255–265, 2017.
DOI: 10.1007/978-3-319-58961-9_23

However, in both cases a model is required to describe how vertices interact through edges and how this interaction evolves with time. In the case of a generative model it is captured probabilistically and can be described by various forms of regressive or autoregressive models [5,7]. In the latter case it is easier to couch the model in terms of how different node degrees co-occur on the edges connecting them [2]. Both models require a means fitting to data or of learning their parameters.

In recent work we have addressed the former problem by detailing a generative model of graph-structure [5] and have shown how it can be applied to network time series using an autoregressive model [7,9]. One of the key elements for this model is a means of approximating the von Neumann entropy of both directed and undirected graphs [13]. The von Neumann entropy is the Shannon entropy associated with the re-scaled eigenvalues of the normalised Laplacian matrix. We have shown how the Shannon entropy can be approximated using a quadratic entropy, and this leads to simple expressions for the von Neumann entropy in terms of the degrees of nodes forming edges. The fitting of the generative model to both statics and time vary data (in its autoregressive form) involves the use of a description length criterion [1,7]. The criterion is two-part, with a data likelihood terms modelling goodness of fit and the approximate von Neumann entropy controlling the complexity of the fitted structure.

The aim in this paper is to explore whether our model of network entropy can be extended to model the way in which the node degree distribution evolves with time, taking into account the effect of degree correlations caused by the degree structure of edges. According to our model of the von Neumann entropy, those edges that connect high degree vertices have the lowest entropy, while those connecting low degree vertices have the highest entropy [2]. Moreover the change in entropy of an edge between different epochs depends on the product of the degree of one vertex, and the degree change of the second vertex. In other words the change in entropy depends on the structure of the degree change correlations.

We exploit this property by modelling the evolution if network structure using the Euler-Lagrange equations. Our variational principle is that the evolution minimises changes in entropy. Using our approximation of the von Neumann entropy this leads to update equations for the node degree which include the effects of correlations induced by the edges of the network. It is effectively a type of diffusion process that models how the degree distribution propagates across the network. In fact, it has elements similar to preferential attachment [11], since it favours edges that connect high degree nodes [12].

The remainder of the paper is organized as follows. In Sect. 2, we provide a detailed analysis of entropy changes in dynamic networks using the Euler-Lagrange equation. In Sect. 3, we apply the resulting characterization to the real-world time-varying networks, i.e. New York Stock Exchange (NYSE) data. Finally, we conclude the paper and make suggestions for future work.

2   Variational Principle on Graphs

2.1   Preliminaries

Let $G(V, E)$ be an undirected graph with node set V and edge set $E \subseteq V \times V$, and let $|V|$ represent the total number of nodes on graph $G(V, E)$. The adjacency matrix A of a graph is defined as

$$A = \begin{cases} 0 & \text{if } (u, v) \in E \\ 1 & \text{otherwise.} \end{cases} \tag{1}$$

Then the degree of node u is $d_u = \sum_{v \in V} A_{uv}$.

The normalized Laplacian matrix $\tilde{L}$ of the graph G is defined as $\tilde{L} = D^{-\frac{1}{2}} L D^{\frac{1}{2}}$, where $L = D - A$ is the Laplacian matrix and D denotes the degree diagonal matrix whose elements are given by $D(u, u) = d_u$ and zeros elsewhere. The element-wise expression of $\tilde{L}$ is

$$\tilde{L}_{uv} = \begin{cases} 1 & \text{if } u = v \text{ and } d_u \neq 0 \\ -\frac{1}{\sqrt{d_u d_v}} & \text{if } u \neq v \text{ and } (u, v) \in E \\ 0 & \text{otherwise.} \end{cases} \tag{2}$$

2.2   Network Entropy

Severini et al. [14] import ideas from quantum mechanics to the graph domain to define the density matrix with normalized Laplacian. A density matrix can be obtained by scaling the combinatorial Laplacian matrix by the reciprocal of the number of nodes in the graph, i.e. $\rho = \frac{\tilde{L}}{|V|}$. This opens up the possibility of computing the von Neumann entropy to characterize a graph. The von Neumann entropy is defined as the Shannon entropy associated with the density matrix eigenvalues. It is given in terms of the eigenvalues $\lambda_1, \dots, \lambda_V$ of the density matrix ρ,

$$S = -\mathrm{Tr}(\rho \log \rho) = -\sum_{i=1}^{|V|} \frac{\hat{\lambda}_i}{|V|} \log \frac{\hat{\lambda}_i}{|V|} \tag{3}$$

Recently, Han et al. [13] have shown how to approximate the calculation of von Neumann entropy in terms of simple degree statistics. Their approximation allows the cubic complexity of computing the von Neumann entropy from the Laplacian spectrum, to be reduced to one of quadratic complexity. The idea is to approximate the Shannon entropy $\rho \log \rho$ by the quadratic entropy $\rho(1 - \rho)$. The resulting approximation to the von Neumann entropy is

$$S = 1 - \frac{1}{|V|} - \frac{1}{|V|^2} \sum_{(u,v) \in E} \frac{1}{d_u d_v} \tag{4}$$

This expression for the von Neumann entropy allows the approximate entropy of the network to be efficiently computed. It has been shown to be an effective tool for characterizing structural properties of networks, with extremal values for cycle and fully connected graphs.

Suppose that two undirected graphs $G_t = (V_t, E_t)$ and $G_{t+1} = (V_{t+1}, E_{t+1})$ represent the structure of a time-varying complex network at two consecutive epochs t and $t+1$ respectively. Then the change of von Neumann entropy between two undirected graphs can be written

$$\Delta S = S(G_{t+1}) - S(G_t) = \frac{1}{|V|^2} \sum_{(u,v) \in E, E'} \frac{d_u \Delta_v + d_v \Delta_u + \Delta_u \Delta_v}{d_u(d_u + \Delta_u)d_v(d_v + \Delta_v)} \quad (5)$$

where Δ_u is the change of degree for node u, i.e., $\Delta_u = d_u^{t+1} - d_u^t$; Δ_v is similarly defined as the change of degree for node v, i.e., $\Delta_v = d_v^{t+1} - d_v^t$. The entropy change is sensitive to degree correlations for pairs of nodes connected by an edge.

2.3 Euler-Lagrange Equation

We would like to understand the dynamics of a network which evolves so as to minimise this entropy change between different sequential epochs. To do this we cast the evolution process into a variational setting of the Euler-Lagrange equation, and consider the system which optimises the functional

$$\mathcal{E}(q) = \int_{t_1}^{t_2} \mathcal{G}\left[t, q(t), \dot{q}(t)\right] dt. \quad (6)$$

where t is time, $q(t)$ is the variable of the system as a function of time, and $\dot{q}(t)$ is the time derivative of $q(t)$. Then, the Euler-Lagrange equation is given by

$$\frac{\partial \mathcal{G}}{\partial q}\left[t, q(t), \dot{q}(t)\right] - \frac{d}{dt}\frac{\partial \mathcal{G}}{\partial \dot{q}}\left[t, q(t), \dot{q}(t)\right] = 0 \quad (7)$$

Here we consider an evolution which changes just the edge connectivity structure of the vertices, and does not change the number of vertices in the graph. As a result, the factors $1 - \frac{1}{|V|}$ and $\frac{1}{|V|^2}$ are constants and do not affect the solution of the Euler-Lagrange equation. We aim to study evolutions that minimise the entropy, i.e. minimise the entropy change between time intervals. Then we apply the Euler-Lagrange equation with $\mathcal{G} = \Delta S$ the entropy change in Eq. (5) to obtain

$$\mathcal{G}\left[t, u(t), \Delta_u(t), v(t), \Delta_v(t)\right] = \frac{d_u \Delta_v + d_v \Delta_u + \Delta_u \Delta_v}{(d_u + \Delta_u)(d_v + \Delta_v)} \quad (8)$$

For the vertex indexed u with degree d_u the Euler-Lagrange equation in Eq. (7) gives,

$$\frac{\partial \mathcal{G}}{\partial d_u} - \frac{d}{dt}\frac{\partial \mathcal{G}}{\partial \Delta_u} = 0 \quad (9)$$

First, solving for the partial derivative of the degree d_u, we find

$$\frac{\partial \mathcal{G}}{\partial d_u} = -\frac{d_v \Delta u}{(d_u + \Delta u)^2 (d_v + \Delta v)} \tag{10}$$

Then computing the partial time derivative to the first order degree difference Δ_u, we obtain

$$\frac{\partial \mathcal{G}}{\partial \Delta_u} = \frac{d_v d_u}{(d_u + \Delta u)^2 (d_v + \Delta v)} \tag{11}$$

Substituting Eqs. (10) and (11) into Eq. (9), the solution for Euler-Lagrange equation in terms of node degree difference is

$$d_u \Delta_v = -2 d_v \Delta_u \tag{12}$$

As a result, solving the Euler-Lagrange equation which minimises the change in entropy over time gives a relationship between the degree changes of nodes connected by an edge. Since we are concerned with understanding how network structure changes with time, the solution of the Euler-Lagrange equation provides a way of modelling the effects of these structural change on the degree distribution across nodes in the network. The update equation for the node degree is at time epochs t and $t + 1$ is

$$d_u^{t+1} = d_u^t + \sum_{v \sim u} \dot{\Delta}_v \Delta_t = d_u^t + \sum_{v \sim u} \left(\frac{\Delta_u}{\Delta_t}\right)_v \Delta_t \tag{13}$$

In other words by summing over all edges connected to node u, we increment the degree at node u due to changes associated with the degree correlations on the set of connecting edges. We then leverage the solution of the Lagrange equation to simplify the degree update equation, to give the result

$$d_u^{t+1} = d_u^t - \sum_{v \sim u} \left(\frac{d_u}{2 d_v}\right) \Delta_v \tag{14}$$

This can be viewed as a type of diffusion process, which updates edge degree so as to satisfy constraints on degree change correlation so as to minimise the entropy change between time epochs. Specifically, the update of degree reflects the effects of correlated degree changes between nodes connected by an edge.

3   Experimental Evaluation

3.1   Data Set

We test our method on data provided by the New York Stock Exchange. This dataset consists of the daily prices of 3,799 stocks traded continuously on the New York Stock Exchange over 6000 trading days. The stock prices were obtained from the Yahoo! financial database [15]. A total of 347 stock were selected from this set, for which historical stock prices from January 1986 to February 2011 are available. In our network representation the nodes correspond to stock and the edges indicate that there is a statistical similarity between the time series associated to the stock closing prices [10].

To establish the edge-structure of the network we use a time window of 20 days is to compute the cross-correlation coefficients between the time-series for each pair of stock. Connections are created between a pair of stock if the cross-correlation exceeds an empirically determined threshold. In our experiments we set the correlation co-efficient threshold to the value to $\xi = 0.85$. This yields a time-varying stock market network with a fixed number of 347 nodes and varying edge structure for each of 6,000 trading days. The edges of the network therefore represent how the closing prices of the stock follow each other.

3.2   Network Dynamics

Using the degree update equation derived from the principle of minimum entropy change and the Euler-Lagrange equation, we aim to simulate the behavior of the financial market networks. Here we focus on how the degree distribution evolves with time. We compare the simulated structure and the observed network properties, and provide a way to identify the consequence of structural variations in time-evolving networks.

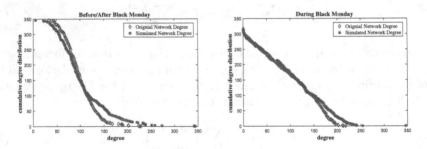

Fig. 1. Degree distribution of original networks and simulated networks before/after Black Monday (left) and during Black Monday (right). Around the Black Monday, the network is highly connected with large number of nodes having high degree. During Black Monday, the network is becomes disconnected and most vertices are disjoint, which results in the degree distribution following the power-law.

Our procedure is as follows. We select a network at a particular epoch from the time series, and simulate its evolution using the degree update equation in Eq. 14. Then we compare the degree distributions for the real network sampled at a subsequent time and the simulation of the degree distribution after an identical elapsed time. One of the most salient events in the NYSE is Black Monday. This event occurred on October 19, 1987, during which the world stock markets crashed, dropping in value in a very short time. The crash began in

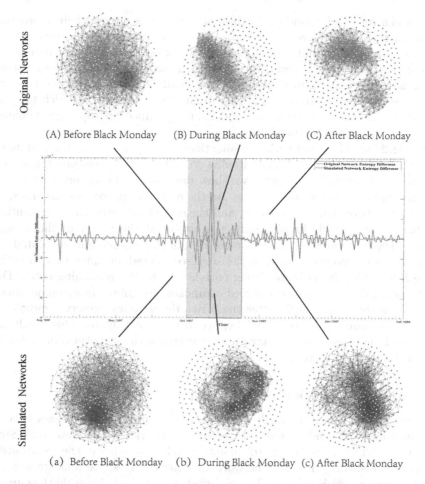

Fig. 2. The first order von Neumann entropy difference of the NYSE networks during and around the Black Monday. We show a visualization of the original and simulated networks at three specific days in Black Monday financial crisis. The red line corresponds to the entropy difference for the original networks (A-C), and the blue line represents the simulated networks with Euler-Lagrange equation (a-c). (Color figure online)

Hong Kong and spread west to Europe, hitting the United States after other markets had already declined by a significant margin.

We compare the prediction of consecutive time steps at different epochs, before/after and during the Black Monday crisis. The results are shown in Fig. 1. The most obvious feature is that the degree distribution for the networks before and after Black Monday is quite different to that during the crisis period. During the Black Monday crisis, large number of vertices in the network are disconnected. This results in a power-law degree distribution. However, for time epochs before and after Black Monday, the disconnected nodes recover their interactions to one another. This increases the number of connections among vertices, and causes departures from the power law distribution. This phenomenon is also observed in the networks simulated networks using our degree update equation. This is an important result that shows empirically that the simulated networks reflect the structural properties of the original networks from which they are generated. Moreover, our dynamic model can reproduce the topological changes that occur during the financial crisis.

In Fig. 2, we show network visualizations corresponding to three different instants of time around the Black Monday crisis. In order to compare the original and simulated network structures, we show the connected components (community structures) at three time epochs. As the network approaches the crisis, the network structure changes violently, and the community structure substantially vanishes. Only a single highly connected cluster at the center of the network persists. These features can be observed in both the original and simulated networks. At the crisis epoch, most stocks are disconnected, meaning that the prices evolve independently without strong correlations to the remaining stock. During the crisis, the persistent connected component exhibits a more homogeneous structure as shown in Fig. 2. After the crisis, the network preserves most of its existing community structure and begins to reconnect again. This result also agrees with findings in other literatures concerning the structural organization of financial market networks [10].

3.3 Anomaly Detection

We now validate our framework by analyzing the entropy differences between simulated networks and actual stock market networks in the New York Stock Exchange (NYSE). In order to quantitatively investigate the relationship between a financial crisis and network entropy changes, we analyze a set of well documented crisis periods. These periods are marked alongside the curve of the first order entropy difference in Fig. 3, for all business days in our dataset.

Figure 4 plots the entropy difference for simulated networks with two different values of the time interval. We commence simulation with start time t_1 and continue to the end time t_2. In the study reported in Sect. 3.2 this time interval is one day, i.e. $\Delta t = t_1 - t_2 = 1$. In other words we use the network on the previous day to predict the structure for the next day. We now consider the effects of varying parameter to longer intervals, i.e. $\Delta t = 5$ days and $\Delta t = 10$ days respectively, and investigate the effect on the performance of our method. Increasing the value of the simulation interval results in significant fluctuations, and reduces the amplitude of the entropy change associated with the different crises. So our method is capable of modelling local trends which gives rise to structural changes in the network, but it is less effective when used to predict changes in network structure over an extended time interval.

Overall, the most striking observation is that the largest peaks of entropy difference can be used to identify the corresponding financial crisis both in the original and simulated networks. This shows that the entropy difference is a

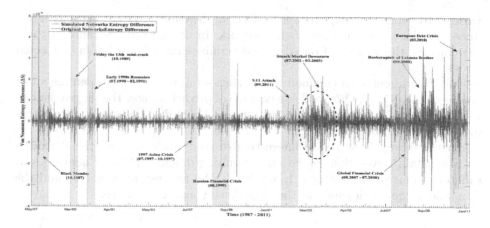

Fig. 3. The von Neumann entropy difference in NYSE (1987–2011) for original financial networks and simulated networks. Critical financial events, i.e., Black Monday, Friday the 13th mini-crash, Early 1990s Recession, 1997 Asian Crisis, 9.11 Attacks, Downturn of 2002–2003, 2007 Financial Crisis, the Bankruptcy of Lehman Brothers and the European Debt Crisis, are associated with large entropy differences.

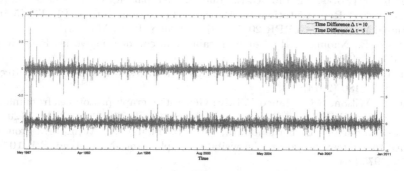

Fig. 4. The performance of time difference for simulated networks in NYSE (1987–2011). The smaller number of time intervals in simulation, the better identifying the financial crisis.

sensitive to significant structural changes in networks. The financial crises are characterized by significant entropy changes, whereas outside these critical periods the entropy difference remains stable.

4   Conclusion

In this paper, we explore how to model the time evolution of networks using a variational principle. We use the Euler-Lagrange equations to model the evolution of networks that undergo changes in structure by minimising the change in von Neumann entropy. This treatment leads to model of how the node degree varies with time, and captures the effects of degree change correlations introduced by the edge-structure of the network. In other words, because of these correlations, the variety of one degree determines the translation in connected nodes.

We experiment with the model on a time-serial networks representing stock trades on the NYSE. Our model is capable of predicting how the degree distribution evolves with time. Moreover, it can also be used to detect abrupt changes in network structure.

In the future, it would be interesting to study different variational models for the network evolution, based on minimising different physical quantities or different forms of the entropy. It would also be interesting to understand the dynamics of quantities such as the edge density and its variance.

References

1. Wolstenholme, R.J., Walden, A.T.: An efficient approach to graphical modeling of time series. IEEE Trans. Sig. Process. **63**, 3266–3276 (2015). ISSN: 1053–587X
2. Ye, C., Torsello, A., Wilson, R.C.: Hancock, ER 2015, Thermodynamics of time evolving networks. In: Liu, C.-L., Luo, B., Kropatsch, W.G., Cheng, J. (eds.) Graph-Based Representations in Pattern Recognition: 10th IAPR-TC-15 International Workshop, GbRPR 2015, Beijing, China, 13–15 May 2015
3. Ernesto, E., Naomichi, H.: Communicability in complex networks. Phys. Rev. E **77**, 036111 (2008)
4. Bai, L., Hancock, E.: Depth-based complexity traces of graphs. Pattern Reçogn. **47**, 1172–1186 (2014)
5. Han, L., Wilson, R.C., Hancock, E.R.: Generative graph prototypes from information theory. IEEE Trans. Pattern Anal. Mach. Intell. **37**(10), 2013–2027 (2015)
6. Lucas, L., Bartolo, L., Fernando, B., Jordi, L., Juan, C.N., Author, A.: From time series to complex networks: the visibility graph. Proc. Natl. Acad. Sci. **105**(13), 4972–4975 (2008)
7. Andreas, L., Simonetto, A., Leus, G.: Distributed autoregressive moving average graph filters. IEEE Sig. Process. Lett. **22**(11), 1931–1935 (2015)
8. Bai, L., Hancock, E.R., Han, L., Ren, P.: Graph clustering using graph entropy complexity traces. In: 21st International Conference on Pattern Recognition (ICPR) 2012, Tsukuba, Japan, 11–15 November 2012
9. Ye, C., Wilson, R.C., Hancock, E.R.: Correlation network evolution using mean reversion autoregression. In: Robles-Kelly, A., Loog, M., Biggio, B., Escolano, F., Wilson, R. (eds.) S+SSPR 2016. LNCS, vol. 10029, pp. 163–173. Springer, Cham (2016). doi:10.1007/978-3-319-49055-7_15
10. Silva, F.N., Comin, C.H., Peron, T.K.D., Rodrigues, F.A., Ye, C., Wilson, R.C., Hancock, E., Costa, L.F.: Modular Dynamics of Financial Market Networks, pp. 1–13 (2015)
11. Barabsi, A.L., Albert, R.: Emergence of scaling in random networks. Science **286**(5439), 509–512 (1999)
12. Wang, J., Wilson, R., Hancock, E.R.: Network entropy analysis using the Maxwell-Boltzmann partition function. In: The 23rd International Conference on Pattern Recognition (ICPR), pp. 1–6 (2016)

13. Han, L., Escolano, F., Hancock, E.R., Wilson, R.C.: Graph characterizations from von Neumann entropy. Pattern Recogn. Lett. **33**, 1958–1967 (2012)
14. Passerini, F., Severini, S.: The von Neumann entropy of networks. Int. J. Agent Technol. Syst. **1**(4), 58–67 (2008)
15. Yahoo! Finance. http://finance.yahoo.com

Synchronization Over the Birkhoff Polytope for Multi-graph Matching

Michele Schiavinato and Andrea Torsello[✉]

Dipartimento di Scienze Ambientali, Informatica e Statistica,
Universitá Ca' Foscari Venezia, via Torino 155, 30172 Mestre, VE, Italy
andrea.torsello@unive.it

Abstract. In this paper we address the problem of simultaneously matching multiple graphs imposing cyclic or transitive consistency among the correspondences. This is obtained through a synchronization process that projects doubly-stochastic matrices onto a consistent set. We overcome the lack of group structure of the Birkhoff polytope, *i.e.*, the space of doubly-stochastic matrices, by making use the Birkhoff-Von Neumann theorem stating that any doubly-stochastic matrix can be seen as the expectation of a distribution over the permutation matrices, and then cast the synchronization problem as one over the underlying permutations. This allows us to transform any graph-matching algorithm working on the Birkhoff polytope into a multi-graph matching algorithm. We evaluate the performance of two classic graph matching algorithms in their synchronized and un-synchronized versions with a state-of-the-art multi-graph matching approach, showing that synchronization can yield better and more robust matches.

Keywords: Transformation synchronization · Doubly-stochastic matrix · Birkhoff polytope · Graph matching

1 Introduction

Graph-based representations have found widespread application in several domains due to their ability to characterize complex systems in terms of parts and relations, capturing the fundamental state of the system in a way that is invariant to transformations that are irrelevant to the classification task at hand. Concrete examples include the use of graphs to represent shapes [13], metabolic networks [7], protein structure [6], and road maps [8]. However, this enhanced expressive power comes at the cost of the inability to utilize most of the pattern analysis toolset directly and in general in the requirement of using approaches that are computationally more demanding.

Structural pattern recognition in its first 30 years of research has mainly focused its attention to the graph matching problem as the fundamental means of dealing with structural representation and assessing their similarity [3]. In fact, with correspondences at hand, standard similarity-based recognition and

© Springer International Publishing AG 2017
P. Foggia et al. (Eds.): GbRPR 2017, LNCS 10310, pp. 266–275, 2017.
DOI: 10.1007/978-3-319-58961-9_24

classification techniques can be imported to the structural domain. However, graph matching is in general very computationally demanding and can introduce bias in the inference process [15].

Alternatively, graphs can be embedded in a low-dimensional pattern space using either multidimensional scaling, non-linear manifold leaning techniques, or by adopting the famous kernel trick through the definition of graph kernels [1,5,9,12]. One drawback of these approaches is that they neglect the locational information for the substructures in a graph, thus limiting the precision of the resulting similarity measures.

More recently, in an attempt to increase matching performance and reducing the bias in the inference process, some researchers have started studying the problem of simultaneously extracting correspondence information from whole sets of graphs, rather than limiting the analysis to each pair. In this multi-graph matching setting, we aim at improve correspondence estimation by incorporating transitivity constraints among the matches. Namely, if node u in graph A matches node v in graph B and, in turn, the latter node v matches node w in graph C, then node u in A must match node w in C.

Williams et al. [16], impose the transitive vertex-matching constraint in a softened Bayesian manner, favoring inference triangles through fuzzy compositions of pairwise matching functions. Sole-Ribalta and Serratosa [14] extended the Graduated Assignment algorithm [4] to the multi-graph scenario by raising the assignment matrices associated to pair of graphs to assignment hypercube, or tensors, between all the graphs. For computational efficiency, the hypercube is constructed via sequential local pair matching, but still result in a potentially exponential expansion of the state space. More recently, Yan et al. [19,20] proposed a new framework explicitly extending the Integer Quadratic Programming (IQP) formulation of pairwise matching to the multi-graph matching scenario. The resulting IQP is then solved through alternating optimization approach. Yan et al. [17,18] introduced a method to iteratively approximating the global-optimal affinity matching score in a pool of graphs using the consistency between all the pairwise matching as a regularizer for the whole process. Conversely Zhou et al. [22] avoided the semi-definite programming formulation (SDP) proposing a method for multi-image matching as a low-rank matrix recovery problem based on the nuclear-norm relaxation. Pachauri et al. [10] and Schiavinato et al. [11] on the other hand, start from given pairwise correspondence estimations, and synchronize them, that is fin the set of correspondences that satisfy the transitivity constraint that are closer to the given ones in the least square sense.

The advantage of this permutation synchronization approach is that it can be paierd with any given graph-matching algorithm in the literature it does not require any additional memory other than what is required to store the original $\binom{n}{2}$ correspondences among n graphs. However, it offers only an *ex post* correction through a relaxation process and cannot be fully integrated with an iterative matching process to direct its convergence to a better solution.

1.1 Contribution

In this paper we aim at extending the synchronization approach in such a way that it can be used within well-known graph-matching approaches transforming them into multi-graph matching algorithms. In particular, we aim at defining a synchronization process for doubly-stochastic matrices, a probabilistic relaxation of correspondence matrices commonly used as a state space in several iterative matching processes [4,21].

The problem with defining a synchronization process over the doubly-stochastic matrices, is that, contrary to the permutation or orthogonal groups used in other approaches, the Birkhoff polytope does not have a group structure necessary even for defining the notion of transitivity.

Here we use the Birkhoff-Von Neumann theorem stating that any doubly-stochastic matrix can be seen as the expectation of a distribution over the permutation matrices, and synchronize the doubly-stochastic matrices by implicitly constructing a low entropy distribution over synchronized permutations that fit the given observations in a least square sense.

2 Synchronization Over the Birkhoff Polytope

The Birkhoff-Von Neumann theorem states that any doubly stochastic matrix can be constructed as the convex linear combination of a set of permutation matrices. This implies that, given a probability distribution $\mathbf{q}$ over the group σ_n of $n \times n$ permutation matrices, the expected value of such distribution

$$\mathbf{O} = \langle \mathbf{q}P \rangle = \sum_k = q_k \mathbf{P}^k$$

is a doubly-stochastic matrix and that any doubly-stochastic matrix can be constructed this way. Unfortunately this construction is not unique and several distributions lead to the same expected value. In general, however, we are interested in sparse, low entropy distributions.

We exploit this property to lower the definition of transitivity to that over the permutation group Σ_n and then raise it back to the Brikhoff polytope.

Let $\mathbf{P}_{ij}$, $i, j = 1, \ldots, N$ be a set of permutation matrices. We say that they satisfy the *transitivity* property if

$$\forall i, j, k = 1, \ldots, N \qquad \overline{P}_{ij}\overline{P}_{jk} = \overline{P}_{ik}. \tag{1}$$

It can be shown [11] that if the matrices $\mathbf{P}_{ij}$ are transitive, then there exist a reference canonical ordering of the vertices and a set $Q_i \in \Sigma_n$ $i = 1, \ldots, N$ of alignment matrices that map vertices in G_i to the reference order, such that

$$\forall i, j = 1, \ldots, N \overline{P}_{ij} = Q_i Q_j^T. \tag{2}$$

Let $\mathbf{P}_{ij}^k = Q_i^k (Q_j^k)^T$ be a sequence of transitive permutation matrices, where k is the sequence index, while i, j span over the set of graphs. Further, let q_k be a distribution over the sequence, then

$$\forall i, j = 1, \ldots, N \quad O_{ij} = \sum_k q_k P_{ij}^k \tag{3}$$

forms a set of doubly-stochastic matrices over the given graphs that are composed as expectation of permutations that satisfy the transitivity property. We say that any doubly-stochastic matrix thus constructed is *transitive*. The problem of synchronization over the Birckhoff polytope can thus reduced to that of finding the transitive set of doubly-stochastic matrices closest to a given set in a least square sense. However, the search space is huge, $O(n!^N)$ where N is the number of graphs and n is the number of nodes in each graph. We solve this by looking for a sparse distribution q over the set of transitive permutations. This is achieved through the introduction of an entropic regularizer over q:

$$\underset{q,\bar{Q}}{\operatorname{argmin}} \sum_{i,j=1}^N \| O_{ij} - \sum_k^{n!^N} q_k Q_i^k Q_j^{kT} \|_2^2 + \lambda H(q) \tag{4}$$

where $\lambda \in \mathbb{R}$ is a free parameter and $H(q) = -\sum_k q_k \ln(q_k)$ denotes the entropy.

Assuming the sparsity of the resulting q, we find an approximate solution to (4) through *Matching Pursuit*.

Let $R_{ij}^{(t)} = \sum_k^{n!^N} q_k^{(t)} Q_i^k Q_j^{kT}$ with $i = 1, \ldots, N$ be the set of synchronized doubly stochastic matrices at iteration t, we can write the solution at the next iteration as

$$R_{ij}^{(t+1)} = (1 - \alpha) R_{ij}^{(t)} + \alpha Q_i^{\hat{k}(t+1)} Q_j^{\hat{k}(t+1) T} \tag{5}$$

where $\hat{k}^{(t+1)}$ is the index which denotes the optimal residual alignment and α is a value in $[0,1]$. Moreover, under the sparsity assumption, we can assume that we only bring in new entries over the distribution q, so the update step for the probability distribution q becomes

$$q^{(t+1)} = (1 - \alpha) q^{(t)} + \alpha e_{\hat{k}(t+1)} \tag{6}$$

where $e_{\hat{k}(t+1)}$ is a vector of zeros where the unique one is placed in position $\hat{k}^{(t+1)}$. This assumption on q allows us to ignore the entropic term $\lambda H(q)$ from (4).

With this formulation, the matching pursuit iteration is computed by solving

$$\underset{\hat{k},\alpha}{\min} \sum_{i,j=1}^N \| O_{ij} - (1 - \alpha) R_{ij}^{(t)} - \alpha Q_i^{\hat{k}} Q_j^{\hat{k} T} \|_2^2 + \lambda H(q^{(t+1)}) \tag{7}$$

We can observe that $\hat{k}$ does not depend by α, thus we can iteratively solve for $\hat{k}$, and then for α given the current set of correspondences introduced into the reconstruction of the doubly stochastic matrices.

2.1   Solving for $\hat{k}$

Let the matrix $\mathbf{M}_{ij}^{(t)} = \mathbf{O}_{ij} - (1-\alpha)\mathbf{R}_{ij}^{(t)}$ we can rewrite the objective function in the problem (7) as follows (without considering the entropic term $\lambda H(\mathbf{q}^{(t+1)})$):

$$\sum_{i,j=1}^{N} ||\mathbf{O}_{ij} - (1-\alpha)\mathbf{R}_{ij}^{(t)} - \alpha\mathbf{Q}_i^{\hat{k}}\mathbf{Q}_j^{\hat{k}T}||_2^2 = \sum_{i,j=1}^{N} ||\mathbf{M}_{ij}^{(t)} - \alpha\mathbf{Q}_i^{\hat{k}}\mathbf{Q}_j^{\hat{k}T}||_2^2$$

$$= \sum_{i,j=1}^{N} ||\mathbf{M}_{ij}^{(t)}||_2^2 + \alpha^2 n^2 - 2\alpha\,\mathrm{Tr}\left(\mathbf{Q}_i^{\hat{k}}\mathbf{Q}_j^{\hat{k}T}\mathbf{M}_{ij}^{(t)}\right) \quad (8)$$

Note that the optimization over the index $\hat{k}$ in the set of synchronized permutations can be substituted for the direct optimization over the synchronized permutations $\bar{\mathbf{Q}} = \{\mathbf{Q}_i\}$, $i = 1 \ldots, N$. Further, under the assumption that α is small, we can set $\bar{\mathbf{M}}_{ij}^{(t)} = \mathbf{O}_{ij} - \mathbf{R}_{ij}^{(t)}$ resulting in the optimization problem

$$\operatorname*{argmax}_{\bar{\mathbf{Q}}} \sum_{i,j=1}^{N} \mathrm{Tr}\left(\mathbf{Q}_i\mathbf{Q}_j^{T}\bar{\mathbf{M}}_{ij}^{(t)}\right) \quad (9)$$

which can be solved with any approach extracting synchronized permutations, such as [11].

2.2   Solving for α

The entropic term $H(\mathbf{q}^{(t+1)})$ can be written explicitly as follows:

$$H(\mathbf{q}^{(t+1)}) = -\sum_k \left((1-\alpha)q_k + \alpha\delta_{k\hat{k}}\right)\ln\left((1-\alpha)q_k + \alpha\delta_{k\hat{k}}\right) \quad (10)$$

where $\delta_{k\hat{k}}$ denotes the Kronecker delta operator. This can be re-written as:

$$H(\mathbf{q}^{(t+1)}) = -(1-\alpha)\sum_{k\neq\hat{k}} q_k\left(\ln q_k + \ln(1-\alpha)\right) - \alpha\ln\alpha$$

$$= (1-\alpha)H(\mathbf{q}^{(t)}) - (1-\alpha)\ln(1-\alpha) - \alpha\ln\alpha$$

$$= (1-\alpha)H(\mathbf{q}^{(t)}) + H(\alpha) \quad (11)$$

The problem (7) can be solved by gradient descent of the energy function:

$$E = \sum_{i,j=1}^{N} ||\mathbf{O}_{ij} - (1-\alpha)\mathbf{R}_{ij}^{(t)} - \alpha\mathbf{Q}_i^{\hat{k}(t+1)}\mathbf{Q}_j^{\hat{k}(t+1)T}||_2^2 + \lambda H(\mathbf{q}^{(t+1)}) \quad (12)$$

Differentiating E with respect to α yields:

$$\frac{dE}{d\alpha} = -\lambda\left(H(\mathbf{q}^{(t)}) + \ln\left(\frac{\alpha}{1-\alpha}\right)\right)$$

$$+ 2\sum_{i,j=1}^{N}\left[\alpha||\mathbf{R}_{ij}^{(t)} - \mathbf{Q}_i^{\hat{k}(t+1)}\mathbf{Q}_j^{\hat{k}(t+1)T}||_2^2 + \mathrm{Tr}\left((\mathbf{O}_{ij} - \mathbf{R}_{ij}^{(t)})^T(\mathbf{R}_{ij}^{(t)} - \mathbf{Q}_i^{\hat{k}(t+1)}\mathbf{Q}_j^{\hat{k}(t+1)T})\right)\right]$$

$$(13)$$

and with the derivative to hand we extract the optimal *alpha* $[0, 1]$ and recon-struct the new solution using (5).

$$\mathbf{R}_{ij}^{(t+1)} = (1 - \alpha)\mathbf{R}_{ij}^{(t)} + \alpha\mathbf{Q}_i^{\hat{k}(t+1)}\mathbf{Q}_j^{\hat{k}(t+1)\,T}$$

Note that, $\alpha = 0$, or more unlikely $\alpha = 1$, means that the basic pursuit step cannot reduce the entropy-regularized energy and we take that as a stopping criterion for our basis pursuit approach.

3   Synchronized Algorithms

In this section we describe our experimental setup and evaluation strategies for the method presented in this paper. To this end, we synchronize two well-known graph-matching algorithms working in the Birkhoff Polytope, namely Graduated Assignment [4] and Path Following Algorithm [21]. In particular, we included a synchronization step inside their main updating loops, maintaining the relaxed correspondences consistency among all the graphs throughout the execution of the algorithms.

Considering a single iteration of the main loop we perform the pairwise gradu-ated assignment/path following iteration for all the graphs computing the doubly stochastic permutation matrices $\mathcal{O}$. In either case there is a β parameter which governs the whole process, pushing it towards the vertices of the polytope, i.e., towards permutation matrices. After the pairwise computation of all the doubly stochastic permutation matrices in a given β-iteration, we perform our synchro-nization process resulting in synchronized doubly stochastic matrices. This phase is controlled by our parameter λ, which we set to be proportional to β since the goal of both parameters is to push the solution towards the vertices of the poly-tope. When the synchronized graduated assignment/path following algorithms converge, we discretize the solutions applying the typical maximum bipartite assignment problem through the well-know Hungarian algorithm to each doubly stochastic matrix.

4   Experimental Setup and Evaluation

We evaluated our work by comparing the two synchronized algorithms to their un-synchronized counterparts and to a state-of-the art multi-graph match-ing algorithm: the Consistency-driven Non-Factorized Alternating Optimization (CDAO) [20]. We summarized the main parameter setting for these 5 methods as follows:

GA, S-GA. For the Synchronized Graduated Assignment (S-GA) we initial-ized $\beta = n$ and all the doubly stochastic matrices are initialized as a random perturbation from the barycenter of the polytope. The entropy scale para-meter λ_s, i.e., the proportionality factor between β and λ, was set to 10^{-6}. Moreover, the growth rate and exit threshold for β were set to 1.075 and 200 respectively.

PF, S-PF. For the Synchronized Path Following algorithm (S-PF) we initialized the doubly stochastic matrices $\mathbf{O}_{pq}$ as int the original work [21] performing a convex quadratic optimization problem by Frank-Wolfe algorithm. The entropy scale parameter was set to 1 while the increasing rate for β was 0.15.

CDAO. For this method we respected the original setup for the Consistency-driven Non-Factorized Alternating Optimization algorithm in [20] initializing the max number of iteration $T_{max} = 2$ and using the Reweighted Random Walks [2] as pairwise graph matching solver.

We performed tests over several random synthetic graph datasets with different levels of distortion, variations in edge density and proportion of outlier nodes. This evaluation approach follows a widely adopted protocol [2,19,20]. The dataset is generated from a set of N root graphs G^r, $r = 1, \ldots, N$, with n_{in} inlier nodes randomly connected with edge density ρ. Edge attributes a_{ij}^r are randomly drawn form a uniform distribution in $[0, 1]$. From these root graphs, we generate several perturbed sets, by varying (a) edge attributes, adding Gaussian noise sampled from $N(0, \sigma^2)$ for increasing values of σ; (b) edge density ρ; and (c) adding a number of outlier nodes.

We introduced another synthetic test as [14] whose aim is to control the topological structure of the graphs. The construction of a synthetic dataset $\mathcal{G}$ in this experiment is based on the generation of an initial seed $P^r = \{([0, 1]; [0, 1])_i\}_{i=1,\ldots,n}$ of 2D points which are related to the n nodes. Each perturbed graph is generated through a random Gaussian perturbation of the points in P^r, from which we extract a Delaunay triangulation. The computation of the affinity matrix $\mathbf{M}_{pq} = (m_{ia,jb})$ for each pair of graphs (G^p, G^q) is defined as

$$m_{ia,jb} = \exp\left(-\frac{(a_{ij}^p - a_{ab}^q)^2}{\sigma^2}\right)$$

where σ^2 is a scale factor which we set to 0.15. No single-node weight is considered, so we set the unary affinity as $m_{ia,ia} = 0$.

We present our results in terms of vertex correspondences from the permutations given by the graph-matching methods. The evaluation strategy is based on the computation of a matching accuracy (MA) between the common n nodes of two graphs $G^p, G^q \in \mathcal{G}$, which is defined as the ratio between the number of correspondences found (C_{pq}^{ALG}) with respect to those of the ground truth (C_{pq}^{TRU}) and the total number of possible matching as follows:

$$\mathrm{MA}(G^p, G^q) = \frac{|C_{pq}^{ALG} \cap C_{pq}^{TRU}|}{n}$$

We underline that we only calculate the accuracy for common inlier nodes ignoring the matching results over outliers. Given a whole dataset $\mathcal{G}$ of N graphs, the agglomerated matching accuracy (MA) can be expressed as the mean measure:

$$\mathrm{MA}(\mathcal{G}) = \frac{\sum_{p=1}^{N-1} \sum_{q=p+1}^{N} \mathrm{MA}(G^p, G^q)}{N(N-1)/2}$$

In Fig. 1 we plot the final results of all the synthetic tests varying the para-meter of (a) deformation, (b) edge density, (c) number of outlier nodes and (d) topological noise. All these experiments are repeated over 10 trials, for which we plot average and standard error. Each synthetic dataset has $N = 10$ graphs with $n_{in} = 20$ nodes. For the deformation and edge density tests we set $n_{out} = 0$, for outlier and density tests we set the Gaussian deformation with standard deviation as $\sigma = 0.05$ and $\sigma = 0.2$ respectively.

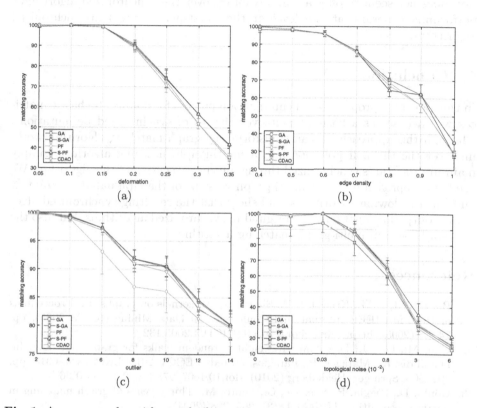

Fig. 1. Average results with standard error for synthetic test at varying of the levels of (a) deformation, (b) edge density, (c) number of outlier nodes and (d) topological noise performing Graduated Assignment (GA), Synchronized Graduated Assignment (S-GA), Path Following (PF), Synchronized Path Following (S-PF) and Consistency-driven Non-Factorized Alternating Optimization (CDAO) for Multi-graph Matching algorithms.

From Fig. 1 we can see that in general for high deformations the synchro-nized algorithms are the best performers regardless of the original algorithm chosen and they generally outperform CDAO as well. It is interesting to not that for deformation, edge density and outlier Graduated Assignment seems to be just as robust as the synchronized algorithms, while is exhibit high sensitiv-ity on the topological noise using Delaunay triangulations. On the other hand, the synchronized version of Graduated Assignment seems to under-perform for low topological noise, going back to very high precision for larger noise. On the other hand, the Synchronized Path Following algorithm is almost always the top performer, even when the original Path Following algorithm appears to be the worst-performing of the lot. This appears to point to the fact that path-following ans synchronization provide complementary information. Finally, CDAO, which was built as a multi-graph matching algorithm optimizing a global objective function does not seem to offer a real advantage over the synchronized algorithms, performing generally at the level of the worst-performing non-synchronized algorithms.

5   Conclusion

In this paper we proposed a synchronization process for doubly stochastic matrices which is set as a basis pursuit over the set of synchronized permutations. Through this approach we can transform any graph-matching algorithm working over the Birkhoff polytope into a multi-graph matching algorithm simply maintain the states synchronized throughout the execution of the algorithm. We used this approach to create multi-graph versions of the Graduated Assignment and Path Following Algorithms, and show that the resulting synchronized algorithms outperform not only the original unsynchronized algorithms, but also the state-of-the-art in multi-graph matching algorithms.

References

1. Borgwardt, K.M., Kriegel, H.P.: Shortest-path kernels on graphs. In: Proceedings of the Fifth IEEE International Conference on Data Mining (ICDM 2005), pp. 74–81 (2005). http://dx.doi.org/10.1109/ICDM.2005.132
2. Cho, M., Lee, J., Lee, K.M.: Reweighted random walks for graph matching. In: Daniilidis, K., Maragos, P., Paragios, N. (eds.) ECCV 2010. LNCS, vol. 6315, pp. 492–505. Springer, Heidelberg (2010). doi:10.1007/978-3-642-15555-0_36
3. Conte, D., Foggia, P., Sansone, C., Vento, M.: Thirty years of graph matching in pattern recognition. IJPRAI 18(3), 265–298 (2004)
4. Gold, S., Rangarajan, A.: A graduated assignment algorithm for graph matching. IEEE Trans. Pattern Anal. Mach. Intell. 18(4), 377–388 (1996)
5. Haussler, D.: Convolution kernels on discrete structures. Technical report UCS-CRL-99-10, University of California at Santa Cruz, Santa Cruz, CA, USA (1999). http://citeseer.ist.psu.edu/haussler99convolution.html
6. Ito, T., Chiba, T., Ozawa, R., Yoshida, M., Hattori, M., Sakaki, Y.: A comprehensive two-hybrid analysis to explore the yeast protein interactome. Proc. Natl. Acad. Sci. 98(8), 4569 (2001)
7. Jeong, H., Tombor, B., Albert, R., Oltvai, Z., Barabási, A.: The large-scale organization of metabolic networks. Nature 407(6804), 651–654 (2000)
8. Kalapala, V., Sanwalani, V., Moore, C.: The structure of the united states road network. Preprint, University of New Mexico (2003)
9. Kashima, H., Tsuda, K., Inokuchi, A.: Marginalized kernels between labeled graphs. In: ICML, pp. 321–328 (2003)

10. Pachauri, D., Kondor, R., Singh, V.: Solving the multi-way matching problem by permutation synchronization. Adv. NIPS **2013**, 1860–1868 (2013)
11. Schiavinato, M., Gasparetto, A., Torsello, A.: Transitive assignment kernels for structural classification. In: Feragen, A., Pelillo, M., Loog, M. (eds.) SIMBAD 2015. LNCS, vol. 9370, pp. 146–159. Springer, Cham (2015). doi:10.1007/978-3-319-24261-3_12
12. Shervashidze, N., Schweitzer, P., van Leeuwen, E.J., Mehlhorn, K., Borgwardt, K.M.: Weisfeiler-Lehman graph kernels. J. Mach. Learn. Res. **12**, 2539–2561 (2011). http://dblp.uni-trier.de/db/journals/jmlr/jmlr12.html#ShervashidzeSLMB11
13. Siddiqi, K., Shokoufandeh, A., Dickinson, S., Zucker, S.: Shock graphs and shape matching. Int. J. Comput. Vis. **35**(1), 13–32 (1999)
14. Solé-Ribalta, A., Serratosa, F.: Models and algorithms for computing the common labelling of a set of attributed graphs. Comput. Vis. Image Underst. **115**(7), 929–945 (2011)
15. Torsello, A.: An importance sampling approach to learning structural representations of shape. In: 2008 IEEE Computer Society Conference on Computer Vision and Pattern Recognition (CVPR 2008), 24–26 June 2008, Anchorage, Alaska, USA. IEEE Computer Society (2008). http://dx.doi.org/10.1109/CVPR.2008.4587639
16. Williams, M.L., Wilson, R.C., Hancock, E.R.: Multiple graph matching with Bayesian inference. Pattern Recogn. Lett. **18**, 080 (1997)
17. Yan, J., Cho, M., Zha, H., Yang, X., Chu, S.M.: A general multi-graph matching approach via graduated consistency-regularized boosting. CoRR abs/1502.05840 (2015). http://arxiv.org/abs/1502.05840
18. Yan, J., Li, Y., Liu, W., Zha, H., Yang, X., Chu, S.M.: Graduated consistency-regularized optimization for multi-graph matching. In: Fleet, D., Pajdla, T., Schiele, B., Tuytelaars, T. (eds.) ECCV 2014. LNCS, vol. 8689, pp. 407–422. Springer, Cham (2014). doi:10.1007/978-3-319-10590-1_27
19. Yan, J., Tian, Y., Zha, H., Yang, X., Zhang, Y., Chu, S.M.: Joint optimization for consistent multiple graph matching. In: Proceeding IEEE International Conference on Computer Vision, pp. 1649–1656. IEEE Computer Society (2013)
20. Yan, J., Wang, J., Zha, H., Yang, X., Chu, S.: Consistency-driven alternating optimization for multigraph matching: a unified approach. IEEE Trans. Image Process. **24**(3), 994–1009 (2015)
21. Zaslavskiy, M., Bach, F., Vert, J.-P.: A path following algorithm for graph matching. In: Elmoataz, A., Lezoray, O., Nouboud, F., Mammass, D. (eds.) ICISP 2008. LNCS, vol. 5099, pp. 329–337. Springer, Heidelberg (2008). doi:10.1007/978-3-540-69905-7_38
22. Zhou, X., Zhu, M., Daniilidis, K.: Multi-image matching via fast alternating minimization. CoRR abs/1505.04845 (2015). http://arxiv.org/abs/1505.04845

Adaptive Feature Selection Based on the Most Informative Graph-Based Features

Lixin Cui[1], Yuhang Jiao[1], Lu Bai[1(✉)], Luca Rossi[2(✉)], and Edwin R. Hancock[3]

[1] Central University of Finance and Economics, Beijing, China
bailucs@cufe.edu.cn
[2] Aston University, Birmingham, UK
l.rossi@aston.ac.uk
[3] University of York, York, UK

Abstract. In this paper, we propose a novel method to adaptively select the most informative and least redundant feature subset, which has strong discriminating power with respect to the target label. Unlike most traditional methods using vectorial features, our proposed approach is based on graph-based features and thus incorporates the relationships between feature samples into the feature selection process. To efficiently encapsulate the main characteristics of the graph-based features, we probe each graph structure using the steady state random walk and compute a probability distribution of the walk visiting the vertices. Furthermore, we propose a new information theoretic criterion to measure the joint relevance of different pairwise feature combinations with respect to the target feature, through the Jensen-Shannon divergence measure between the probability distributions from the random walk on different graphs. By solving a quadratic programming problem, we use the new measure to automatically locate the subset of the most informative features, that have both low redundancy and strong discriminating power. Unlike most existing state-of-the-art feature selection methods, the proposed information theoretic feature selection method can accommodate both continuous and discrete target features. Experiments on the problem of P2P lending platforms in China demonstrate the effectiveness of the proposed method.

1 Introduction

Many real-world applications, including image processing, bioinformatics analysis, face recognition, and P2P lending analysis [16], are represented by high dimensional data. However, only a small number of features are really significant to describe the target label [12]. One way to overcome this problem is to use feature selection.

Mutual information (MI) [8,9,15,18] is a well-known means of measuring the mutual dependency of two variables, and has received much attention for developing new feature selection methods. Typical examples include (1) the Information-based Feature Selection method (MIFS) [8], (2) the Maximum-Relevance Minimum-Redundancy Feature Selection method (MRMR) [18],

P. Foggia et al. (Eds.): GbRPR 2017, LNCS 10310, pp. 276–287, 2017.
DOI: 10.1007/978-3-319-58961-9_25

(3) the Joint-Information Feature Selection method (JMI) [20], and (4) the MIFS method under the assumption of a uniform distribution for input features (MIFS-U) [14]. Unfortunately, these methods suffer from two widely known drawbacks. First, these methods require the number of selected features in advance. Second, these methods mine subsets of the most informative features in a greedy manner [10]. To overcome the shortcomings, Liu et al. [15] have developed the Adaptive MI based Feature Selection method (AMIF) that can automatically determine the size of most informative feature subsect, by maximizing the average pairwise informativeness. Zhang and Hancock [21] have developed a Hypergraph based Information-Theoretic Feature Selection method (HITF) that can automatically determine the most informative feature subset through dominant hypergraph clustering [17].

Unfortunately, the aforementioned information theoretic feature selection methods cannot incorporate the relationship between pairwise samples of each feature dimension. More specifically, for a dataset with N features denoted as $\mathcal{X} = \{\mathbf{f}_1, \ldots, \mathbf{f}_i, \ldots, \mathbf{f}_N\}$, each feature $\mathbf{f}_i$ has M samples as $\mathbf{f}_i = \{f_{i1}, \ldots, f_{ia}, \ldots, f_{ib}, \ldots, f_{iM}\}^T$. Traditionally, existing information theoretic feature selection methods accommodate each feature $\mathbf{f}_i$ as a vector, and thus ignore the relationship between pairwise samples f_{ia} and f_{ib} in $\mathbf{f}_i$. This drawback limits the precise information theoretic measure between pairwise features. To address this shortcoming, Cui et al. [11] have proposed a new feature selection method in terms of graph-based features. They transform each vectorial feature into a graph structure that encapsulates the relationship between pairwise samples from the feature. The most relevant vectorial features are identified by selecting the graph-based features that are most similar to the graph-based target feature, in terms of the Jensen-Shannon divergence measure between graphs. Unfortunately, this method cannot adaptively determine the most relevant feature subset. It is fair to say that developing effective information theoretic feature selection method still remains a challenge.

This paper aims to overcome the shortcomings of existing information theoretic feature selection methods by developing a new algorithm that can incorporate the relationship between feature samples into the feature selection process. In summary, the main contributions are threefold. First, like Cui et al. [11], for the above dataset $\mathcal{X}$ having N features, we transform each vectorial feature $\mathbf{f}_i$ into a graph-based feature $\mathbf{G}_i$. Here, $\mathbf{G}_i$ is a complete weighted graph, where each vertex v_a represents a corresponding sample f_{ia} in $\mathbf{f}_i$ and each weighted edge $\{v_a, v_b\}$ represents the relationship between pairwise samples f_{ia} and f_{ib}. We use the Euclidean distance to measure the relationship between f_{ia} and f_{ib}. Similarly, for the target feature $\mathbf{Y}$ (e.g., the class labels), we also compute a target feature graph $\mathbf{G}_Y$. We argue that the graph-based features can reflect richer characteristics than the original vectorial features. Furthermore, for the feature graphs $\mathbf{G}_i$ and $\mathbf{G}_Y$, we probe each graph structure in terms of the steady state random walk (SSRW) [3] and compute a probability distribution of the walk visiting the vertices. Second, with the probability distributions of the feature graphs $\mathbf{G}_i$ and $\mathbf{G}_Y$ to hand, we propose a new information theoretic criterion to measure the joint relevance of different pairwise feature combinations with respect to the target feature, through the Jensen-Shannon divergence (JSD). Third, we use the new information theoretic measure to automatically locate the subset of the most informative and less redundant features by solving a quadratic program problem [17]. We show that, unlike most existing feature selection methods, the proposed feature selection method can accommodate both continuous and discrete target variables. Experimental results on the analysis of P2P lending platforms in China demonstrate the effectiveness of the proposed method.

2   Preliminary Concepts

2.1   The Steady State Random Walk (SSRW)

As mentioned in the previous section, we propose to use the SSRW to capture the main characteristics of the graph-based features. The main advantages of using SSRWs are twofold. First, SSRWs can accommodate weighted information residing on edges. Second, the computational complexity of probing a graph structure using SSRWs is quadratic in the number of vertices, i.e., SSRWs can be efficiently performed on graphs. As a result, SSRWs represent an elegant way of efficiently characterizing the graph-based features. Below, we review the main concepts underpinning SSRWs.

Let $G(V, E)$ be a weighted graph, V be the vertex set, and E be edge set. Assume $\omega : V \times V \to \mathbb{R}^+$ is a edge weight function. If $\omega(u, v) > 0$ ($\omega(u, v) = \omega(v, u)$), we say that (u, v) is an edge of G, i.e., the vertices $u \in V$ and $v \in V$ are adjacent. The vertex degree matrix of G is a diagonal matrix D whose elements are given by $D(v, v) = d(v) = \sum_{u \in V} \omega(v, u)$. Based on [3], the probability of the steady state random walk visiting each vertex v is $p(v) = d(v) / \sum_{u \in V} d(u)$. Furthermore, from the probability distribution $P = \{p(1), \ldots, p(v), \ldots, p(|V|)\}$, we can straightforwardly compute the Shannon entropy of G as

$$H_S(G) = - \sum_{v \in V} p(v) \log p(v). \tag{1}$$

2.2   The Jensen-Shannon Divergence

In information theory, the JSD is a dissimilarity measure between probability distributions. Let two (discrete) probability distributions be $\mathcal{P} = (p_1, \ldots, p_a, \ldots, p_A)$ and $\mathcal{Q} = (q_1, \ldots, q_b, \ldots, q_B)$, then the JSD between $\mathcal{P}$ and $\mathcal{Q}$ is defined as

$$I_D(\mathcal{P}, \mathcal{Q}) = H_S\left(\frac{\mathcal{P} + \mathcal{Q}}{2}\right) - \frac{1}{2} H_S(\mathcal{P}) - \frac{1}{2} H_S(\mathcal{Q}), \tag{2}$$

where $H_S(\mathcal{P}) = \sum_{a=1}^{A} p_a \log p_a$ is the Shannon entropy of the probability distribution $\mathcal{P}$. In [3], the JSD has been used as a means of measuring the information theoretic dissimilarity between graphs associated with their probability distributions. In this work, we are also concerned with the similarity measure between graph-based features. Therefore, we transform the JSD into its negative form and obtain the corresponding exponential function value to denote the information theoretic similarity measure I_S between probability distributions, i.e.,

$$I_S(\mathcal{P}, \mathcal{Q}) = \exp\{-I_D(\mathcal{P}, \mathcal{Q})\}. \tag{3}$$

3 Methodology of the Proposed Feature Selection Method

3.1 Graph-Based Features from Vectorial Features

In this subsection, we introduce how to transform each vectorial feature into a complete weighted graph. The advantages of using the graph-based representation are twofold. First, graph structures have stronger ability to encapsulate global topological information than vectors. Second, the graph-based features can incorporate the relationships between samples of each original vectorial feature into the feature selection process, thus reducing information loss.

Given a dataset of N features denoted as $\mathcal{X} = \{\mathbf{f}_1, \ldots, \mathbf{f}_i, \ldots, \mathbf{f}_N\} \in \mathbb{R}^{M \times N}$, $\mathbf{f}_i$ represents the i-th vectorial feature and has M samples as $\mathbf{f}_i = \{f_{i1}, \ldots, f_{ia}, \ldots, f_{ib}, \ldots, f_{iM}\}^T$. We transform each feature $\mathbf{f}_i$ into a graph-based feature $\mathbf{G}_i(V_i, E_i)$, where each vertex $v_{ia} \in V_i$ indicates the a-th sample f_{ia} of $\mathbf{f}_i$, each pair of vertices v_{ia} and v_{iv} is connected by a weighted edge $(v_{ia}, v_{ib}) \in E_i$, and the weight $\omega(v_{ia}, v_{ib})$ of (v_{ia}, v_{ib}) is the Euclidean distance between f_{ia} and f_{ib}, i.e.,

$$\omega(v_{ia}, v_{ib}) = \| f_{ia} - f_{ib} \|_2 . \tag{4}$$

Similarly, if the sample values of the target feature $\mathbf{Y} = \{y_1, \ldots, y_a, \ldots, y_b, \ldots, y_M\}^T$ are continuous, its graph-based feature $\hat{\mathbf{G}}(\hat{V}, \hat{E})$ can be computed using Eq. (4) and each vertex $\hat{v}_a$ represents the a-th sample y_a. However, for classification problems, the sample of the target feature Y is the class label c and thus takes the discrete value $c = 1, 2, \ldots, C$, i.e., the samples of each feature $\mathbf{f}_i$ belong to the C different classes. In this case, we propose to compute the graph-based target feature $\hat{\mathbf{G}}_i(\hat{V}_i, \hat{E}_i)$ for each feature $\mathbf{f}_i$, where the weight $\omega(\hat{v}_{ia}, \hat{v}_{ib})$ of each edge $(\hat{v}_{ia}, \hat{v}_{ib}) \in \hat{E}_i$ is

$$\omega(\hat{v}_{ia}, \hat{v}_{ib}) = \| \mu_{ia} - \mu_{ib} \|_2, \tag{5}$$

where μ_{ia} is the mean value of all samples in $\mathbf{f}_i$ from the same class c.

Note that, constructing the graph-based feature from the original vectorial feature is an open problem. In fact, in addition to the distance measure employed in this paper, one could employ a number of alternative measures, e.g., covariance, cosine similarity, etc. Moreover, instead of a complete graph, one may want to define a sparser graph.

3.2   The Information Theoretic Criterion for Feature Selection

We propose to use the following information theoretic criterion to measure the joint relevance of different pairwise feature combinations with respect to either the continuous or discrete target feature. For a set of N features $\mathbf{f}_1, \ldots, \mathbf{f}_i, \ldots, \mathbf{f}_j, \ldots, \mathbf{f}_N$ and the associated continuous target feature $\mathbf{Y}$, the relevance degree of a feature pair $\{\mathbf{f}_i, \mathbf{f}_j\}$ is

$$W_{i,j} = I_S(\mathbf{G}_i, \hat{\mathbf{G}}) \times I_S(\mathbf{G}_j, \hat{\mathbf{G}}) \times I_D(\mathbf{G}_i, \mathbf{G}_j), \tag{6}$$

where $\mathbf{G}_i$ and $\hat{\mathbf{G}}$ are the graph-based features of $\mathbf{f}_i$ and $\mathbf{Y}$, I_S is the JSD based information theoretic similarity measure defined in Eq. (3), and I_D is the JSD based information theoretic dissimilarity measure defined in Eq. (2). The above relevance measure consists of three terms. The first and second terms $I_S(\mathbf{G}_i, \hat{\mathbf{G}})$ and $I_S(\mathbf{G}_j, \hat{\mathbf{G}})$ are the relevance degrees of individual features $\mathbf{f}_i$ and $\mathbf{f}_j$ with respect to the target feature $\mathbf{Y}$, respectively. The third term $I_S(\mathbf{G}_i, \mathbf{G}_j)$ measures the non-redundancy between the feature pair $\{\mathbf{f}_i, \mathbf{f}_i\}$. Therefore, $W_{\mathbf{f}_i, \mathbf{f}_j}$ is large if and only if both $I_S(\mathbf{G}_i, \hat{\mathbf{G}})$ and $I_S(\mathbf{G}_j, \hat{\mathbf{G}})$ are large (i.e., both $\mathbf{f}_i$ and $\mathbf{f}_j$ are informative themselves with respect to the target feature $\mathbf{Y}$) and $I_D(\mathbf{G}_i, \mathbf{G}_j)$ is also large (i.e., $\mathbf{f}_i$ and $\mathbf{f}_j$ are not redundant).

For classification problems, the samples of the target feature $\mathbf{Y}$ take the discrete value c and $c = 1, 2, \ldots, C$. In this case, we compute the individual graph-based target feature $\hat{\mathbf{G}}_\mathbf{i}$ for each feature $\mathbf{f}_i$, and the relevance measure defined in Eq. (6) can re-written as

$$W_{i,j} = \{S(\mathbf{f}_i)I_S(\mathbf{G}_i, \hat{\mathbf{G}}_i)\} \times \{S(\mathbf{f}_j)I_S(\mathbf{G}_j, \hat{\mathbf{G}}_j)\} \times \{I_D(\mathbf{G}_i, \mathbf{G}_j)\}, \tag{7}$$

where $S(\mathbf{f}_i)$ is the Fisher score of feature $\mathbf{f}_i$ [13] and is defined as

$$S(\mathbf{f}_i) = \sum_{c=1}^{L} n_l(\mu_l - \mu)^2 / \sum_{c=1}^{C} n_c \sigma_c^2, \tag{8}$$

where μ_c and σ_c^2 are the mean and variance of the samples belonging to the c-th class in feature $\mathbf{f}_i$, μ is the mean of feature $\mathbf{f}_i$, and n_c is the sample number of the c-th class in feature $\mathbf{f}_i$. For Eq. (8), the Fisher score $S(\mathbf{f}_i)$ indicates the quality of the graph-based target feature $\hat{\mathbf{G}}_i$ for $\mathbf{f}_i$, i.e., a higher Fisher score $S(\mathbf{f}_i)$ means a better target feature graph $\hat{\mathbf{G}}_i$. This follows the definition of Eq. (5). More specifically, the graph-based target feature $\hat{\mathbf{G}}_i$ of original vectorial feature $\mathbf{f}_i$ is preferred, if the distances between samples in different classes are as large as possible and the distances between data points in the same class are as small as possible. Similar to Eq. (6), the three terms of Eq. (7) have the same corresponding theoretical significance.

3.3   Determination of the Most Informative Feature Subset

We adaptively determine the most informative subset of features by solving a quadratic program problem [17]. More specifically, for a set of N features

$\mathbf{f}_1, \ldots, \mathbf{f}_i, \ldots, \mathbf{f}_j, \ldots, \mathbf{f}_N$ and the target feature $\mathbf{Y}$, we commence by transforming each feature into a graph-based feature. Moreover, based on the graph-based features, we construct a feature informativeness matrix $\mathbf{W}$, where each element $W_{i,j} \in \mathbf{W}$ represents the information theoretic measure between a feature pair $\{\mathbf{f}_i, \mathbf{f}_j\}$ based on Eq. (6) (for $\mathbf{Y}$ is continuous) or Eq. (7) (for $\mathbf{Y}$ is discrete). As we have stated in Sect. 3.2, $W_{\mathbf{f}_i, \mathbf{f}_j}$ is large if and only if both $\mathbf{f}_i$ and $\mathbf{f}_j$ are informative themselves with respect to the target feature $\mathbf{Y}$, and $\mathbf{f}_i$ and $\mathbf{f}_j$ are not redundant. Therefore, we locate the most informative feature subset by finding the solution of the following quadratic program problem [17]

$$\max \; f(\mathbf{a}) = \frac{1}{2}\mathbf{a}^T\mathbf{W}\mathbf{a} \tag{9}$$

subject to $\mathbf{a} \in \mathbb{R}^N$, $\mathbf{a} \geq 0$ and $\sum_{i=1}^{N} a_i = 1$. The solution vector $\mathbf{a}$ to the above quadratic program is an N-dimensional vector. When $a_i > 0$, the i-th feature $\mathbf{f}_i$ belongs to the most informative feature subset. Therefore, the number of the selected features n can be determined by counting the positive components of vector $\mathbf{a}$. Pavan and Pelillo [17] have shown that the local maximum of $f(\mathbf{a})$ can be solved using the following equation

$$a_i(t+1) = a_i(t)\frac{(\mathbf{W}\mathbf{a}(t))_i}{\mathbf{a}(t)^T\mathbf{W}\mathbf{a}(t)} . \tag{10}$$

where $a_i(t)$ corresponds to the i-th feature $\mathbf{f}_i$ at iteration t of the update process. According to the value of the element in $\mathbf{a}$, all features $\mathbf{f}_1, \ldots, \mathbf{f}_N$ fall into two disjoint subsets, i.e., $\mathbf{S}_1(a) = \{\mathbf{f}_i \mid a_i > 0\}$ and $\mathbf{S}_2(a) = \{\mathbf{f}_j \mid a_j = 0\}$. Clearly, the set $\mathbf{S}_1$ that has nonzero variables is the selection of the most informative feature subset. The features in $\mathbf{S}_1$ have both low redundancy and strong discriminative power.

3.4   Complete Feature Ranking

The proposed feature selection method aims to adaptively select a compact most informative feature subset that falls into the subset $\mathbf{S}_1(a) = \{\mathbf{f}_i \mid a_i > 0\}$. We can rank the feature $\mathbf{f}_i \in \mathbf{S}_1$ by evaluating the values of their indicators a_i. A higher indicator a_i means a more informative feature. Moreover, we can also rank the features contained in the unselected feature subset $\mathbf{S}_2(a) = \{\mathbf{f}_j \mid a_j = 0\}$ based on the selection method in [15]. More specifically, we compute the reward of each feature $\mathbf{f}_j \in \mathbf{S}_2$ as

$$r_j = \sum_{\mathbf{f}_i \in \mathbf{S}_1, a_i > 0} W_{i,j}a_i, \tag{11}$$

which summarizes the pairwise informativeness between the feature $\mathbf{f}_j \in S_2$ and each feature $\mathbf{f}_i \in \mathbf{S}_1$. A higher r_j means a more informative feature in $\mathbf{S}_2$, thus providing a measure to rank the features in $\mathbf{S}_2$. Based on the feature ranking of $\mathbf{S}_1$ and $\mathbf{S}_2$, we can obtain a Complete Feature Ranking List L, from 1 to a user-specified number.

4   Experimental Evaluations

To validate the effectiveness of the proposed feature selection approach, we perform the following experimental evaluation on a P2P dataset collected from the Peer-to-Peer (P2P) lending sector in China. The reasons for using this dataset are twofold. **First**, P2P lending data are usually high-dimensional, highly correlated, and unstable, thus representing a challenge for traditional statistical and machine learning techniques. To better analyze the P2P data, the sample relationship of the P2P data encapsulating significant information should be incorporated, when designing feature selection methods. Unfortunately, most existing feature selection methods ignore the sample relationships and may cause significant information loss. By contrast, our proposed adaptive feature selection method is able to encapsulate the sample relationship of P2P data and overcome these shortcomings. **Second**, the P2P lending industry in China has developed rapidly since 2007, with more than 3,000 P2P lending platforms and an accumulative loan amount of 12 trillion by 2015. It is of great significance to develop an effective decision aid for the credit risk analysis of the P2P platforms.

The **P2P dataset** is collected from a reputable P2P lending portal in China[1], which consists of the most popular 200 platforms (i.e., 200 samples) until Aug 2014. For each platform, we collect 19 features including (1) transaction volume, (2) total turnover, (3) average annualized interest rate, (4) total number of borrowers, (5) total number of investors, (6) online time, which refers to the foundation year of the platform, (7) operation time, i.e., number of months since the foundation of the platform, (8) registered capital, (9) weighted turnover, (10) average term of loan, (11) average full mark time, i.e., tender period of a loan raised to the required full capital, (12) average amount borrowed, i.e., average loan amount of each successful borrower, (13) average amount invested, which is the average investment amount of each successful investor, (14) loan dispersion, i.e., the ratio of the repayment amount to the total capital, (15) investment dispersion, the ratio of the invested amount to the total capital, (16) average times of borrowing, (17) average times of investment, (18) loan balance, and (19) popularity.

4.1   Identification of the Most Influential Factors for Credit Risk

We evaluate the performance of the proposed feature selection approach with respect to **continuous target features**. Specifically, we use the proposed method to perform credit risk evaluation of the P2P lending platforms. As it is difficult to obtain sufficient data of the platforms which encountered problem, we use the annualized average interest rate as an indicator of the credit risk of the P2P lending platforms. In finance, interest rate is the amount charged, expressed as a percentage of principal, by a lender to a borrower for the use of assets. When the borrower is a low-risk party, they will usually be charged a low interest rate. On the other hand, if the borrower is considered high risk, the interest rate

---

[1] See the website http://www.wdzj.com/ for more details.

charged will be higher. Likewise, a higher annualized average interest rate of the P2P lending platforms often indicates greater likelihood of default, i.e., higher credit risk of the platforms. Identifying the most relevant features to the interest rate can help investors effectively manage the credit risks involved in P2P lending. Therefore, in our experiment, we set the average annualized interest rate as the target feature which takes continuous values. Our purpose is to identify the most informative subset of features for the credit risk of the P2P platforms by using the proposed feature selection method. To further strengthen our findings, we also compare the proposed adaptive feature selection method associated with the SSRW (AFS-RW) with three alternative methods. These include correlation analysis (CA) and multiple linear regression (MLR), which are simple but widely applied. Furthermore, we also compare the proposed method to the most relevant graph-based feature selection method associated with the SSRW (FS-RW) [11], since it can also accommodate the continuous target feature.

Table 1 presents a comparison of the results obtained using these methods. For each method, we display the top 10 features in terms of correlation to the average annualized interest rate. Because the number of the most informative features adaptively located by AFS-RW is six, we display these results in bold. It is worth noting that the most influential factors located by the proposed AFS-RW method is in general different from the remaining three methods used for comparison. This is due to the unique characteristics of the proposed feature selection method which encourages the most informative and least redundant features to be selected. For instance, AFS-RW identifies average full mark time, transaction volume, and average amount borrowed as the most informative features. This is reasonable because a longer full mark time of the loan often reflects a higher level of credit risk and a higher amount of total transaction volume and

Table 1. Comparison of four methods

Ranking	AFS-RW	FS-RW	Correlation analysis	Multiple linear regression
1#	**Average full marktime**	Registered capital	Popularity	Loan dispersion
2#	**Transaction volume**	Operation time	Loan balance	Investment dispersion
3#	**Average amount borrowed**	Average amount invested	Average times of investment	Online tim
4#	**Loan balance**	Loan dispersion	Average times of borrowing	Popularity
5#	**Investment dispersion**	Average times of investment	Investment dispersion	Operation time
6#	**Total number of borrowers**	Online time	Loan dispersion	Average times of borrowing
7#	Average times of borrowing	Average term of loan	Average amount invested	Total number of borrowers
8#	Total turnover	Total number of investors	Average amount borrowed	Loan balance
9#	Average amount invested	Investment dispersion	Average full mark time	Transaction volume
10#	Weighted turnover	Popularity	Average term of loan	Weighted turnover

a higher level of the average amount borrowed indicate a higher preference of both the borrowers and investors for the P2P lending platform due to a higher degree of security. Also, AFS-RW and CA consider loan balance as a relevant feature. This is also reasonable because a higher amount of loan balance often indicates a higher level of credit risk and can result in a higher interest rate. In addition, the total number of borrowers reflects the borrowers preference for the P2P lending platforms and is a significant influential factor. A platform with a relatively low average annualized interest rate is often more attractive to the borrowers because this indicates both a lower transaction cost and a lower credit risk of the platform. However, only the proposed AFS-RW method is able to select this factor, whereas the remaining three methods rank this factor much lower. These results demonstrate the effectiveness of the proposed method for identifying the most influential factors for credit risk of P2P lending platforms.

4.2 Classification for the Credit Rating of the P2P Lending Platforms

We evaluate the performance of the proposed feature selection approach with respect to **discrete target features**. Specifically, we aim to locate the most informative subset of features for the credit rating of the P2P platforms in China, which takes discrete values and is collected from the Report on the Development of the P2P lending industry in China, 2014–2015, issued by the *Financial Research Institute of the Chinese Academy of Social Sciences*. In this Report, only 104 platforms are included due to the strict evaluation criteria involved, among which only 42 platforms belong to the 200 platforms used in the above P2P dataset. Thus, we use the 42 platforms (i.e., samples) for the evaluation. We set the credit ranking for these platforms as the discrete target feature, and aim to locate the most informative feature subset using the proposed approach.

To evaluate the effectiveness of the features selected by the proposed approach, we set the discrete credit ranking targets as classification labels. Since there are only 42 samples and these need to be classified into four classes, it is a very challenging classification problem. In the experiment, we randomly select 50% samples as training data and the remainder as testing data. By repeating this selection process 10 times, we obtain 10 random partitions of the original data. For each partition, we identify the most relevant features via the proposed method based on the train data, and perform a 10-fold cross-validation using a C-Support Vector Machine (C-SVM) to evaluate the classification accuracy associated with the selected features based on the testing data, i.e., we use 9 folds for training and 1 fold for testing. For the C-SVM on each partition, we repeat the process 10 times and compute the average classification accuracy. Finally, we compute the average classification accuracy over the 10 partitions. To further evaluate our study, we compare the proposed method (AFS-QW) with several alternative feature selection methods. These alternative methods include: (1) the Fisher Score method (FS) [13], (2) the Mutual Information based method (MI) [19], and (3) most relevant graph-based feature selection method

(FS-RW) [11]. The classification accuracy of each method is shown in Fig. 1 as a function of the number of features selected.

Figure 1 indicates that the proposed method AFS-RW achieves the best classification accuracy (34.50%) while requiring the lowest number of features, i.e., 3 adaptively selected features. In contrast, FS and MI both require 4 features to generate best classification accuracies. Like the proposed method, the FS-RW also achieves best accuracy with 3 features. However, only the proposed method can adaptively determine the most informative feature subset. Finally, recall that there are only 42 samples divided into 4 class for the evaluation, making this classification task very challenging. Thus, these results demonstrates the effectiveness of the proposed method.

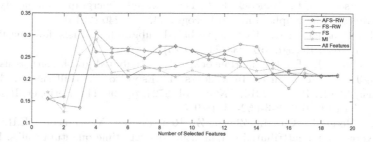

Fig. 1. Accuracy vs. number of selected features for different feature selection methods.

5   Conclusion

In this paper, we have proposed an adaptive feature selection method, based on a new information theoretic criterion between graph-based features. Unlike most existing information theoretic feature selection methods, our approach has two advantages. First, it is based on graph-based features and thus incorporates the relationships between feature samples into the feature selection process. Second, it can accommodate both continuous and discrete target features. Experiments on the analysis of P2P lending platforms in China demonstrate the effectiveness of the proposed feature selection method.

We will extend our method in a number of ways. First, in our previous works [4,5], we have developed a number of quantum Jensen-Shannon kernels using both the continuous-time and discrete-time quantum walks. It is interesting to extend the proposed feature selection method using the classical Jensen-Shannon divergence to that using its quantum counterpart. Second, we will also use our previous graph kernel measures as the graph similarity measures for our feature selection frameworks [2,6,7]. We will explore the performance of our feature selection method associated with different graph kernels. Third, the proposed feature selection method only considers the relationship between pairwise features, i.e., it only evaluates the two-order relationship between features. Our future work will extend the proposed method into a high-order feature selection method by establishing higher order relationship between features. Finally, it is interesting to establish hypergraph-based features [1] and thus develop a new hypergraph-based feature selection method.

Acknowledgments. This work is supported by the National Natural Science Foundation of China (Grant no. 61602535 and 61503422), the Open Projects Program of National Laboratory of Pattern Recognition, the Young Scholar Development Fund of Central University of Finance and Economics (No. QJJ1540), and the program for innovation research in Central University of Finance and Economics.

References

1. Bai, L., Escolano, F., Hancock, E.R.: Depth-based hypergraph complexity traces from directed line graphs. Pattern Recogn. **54**, 229–240 (2016)
2. Bai, L., Hancock, E.R.: Fast depth-based subgraph Kernels for unattributed graphs. Pattern Recogn. **50**, 233–245 (2016)
3. Bai, L., Rossi, L., Bunke, H., Hancock, E.R.: Attributed graph Kernels using the Jensen-Tsallis q-differences. In: Calders, T., Esposito, F., Hüllermeier, E., Meo, R. (eds.) ECML PKDD 2014. LNCS, vol. 8724, pp. 99–114. Springer, Heidelberg (2014). doi:10.1007/978-3-662-44848-9_7
4. Bai, L., Rossi, L., Cui, L., Zhang, Z., Ren, P., Bai, X., Hancock, E.R.: Quantum Kernels for unattributed graphs using discrete-time quantum walks. Pattern Recogn. Lett. **87**, 96–103 (2017)
5. Bai, L., Rossi, L., Torsello, A., Hancock, E.R.: A quantum Jensen-Shannon graph Kernel for unattributed graphs. Pattern Recogn. **48**(2), 344–355 (2015)
6. Bai, L., Rossi, L., Zhang, Z., Hancock, E.R.: An aligned subtree Kernel for weighted graphs. In: Proceedings of ICML, pp. 30–39 (2015)
7. Bai, L., Zhang, Z., Wang, C., Bai, X., Hancock, E.R.: A graph Kernel based on the Jensen-Shannon representation alignment. In: Proceedings of IJCAI, pp. 3322–3328 (2015)
8. Battiti, R.: Using mutual information for selecting features in supervised neural net learning. IEEE Trans. Neural Netw. **5**(4), 537–550 (1994)
9. Bonev, B., Escolano, F., Cazorla, M.: Feature selection, mutual information, and the classification of high-dimensional patterns. Pattern Anal. Appl. **11**(3–4), 309–319 (2008)
10. Brown, G.: A new perspective for information theoretic feature selection. In: Proceedings of AISTATS, pp. 49–56 (2009)
11. Cui, L., Bai, L., Wang, Y., Bai, X., Zhang, Z., Hancock, E.R.: P2P lending analysis using the most relevant graph-based features. In: Robles-Kelly, A., Loog, M., Biggio, B., Escolano, F., Wilson, R. (eds.) S+SSPR 2016. LNCS, vol. 10029, pp. 3–14. Springer, Cham (2016). doi:10.1007/978-3-319-49055-7_1
12. Han, J., Sun, Z., Hao, H.: Selecting feature subset with sparsity and low redundancy for unsupervised learning. Knowl.-Based Syst. **86**, 210–223 (2015)
13. He, X., Cai, D., Niyogi, P., Laplacian score for feature selection. In: Proceedings of NIPS, pp. 507–514 (2005)
14. Kwak, N., Choi, C.-H.: Input feature selection by mutual information based on Parzen window. IEEE Trans. Pattern Anal. Mach. Intell. **24**(12), 1667–1671 (2002)
15. Liu, S., Liu, H., Latecki, L.J., Yan, S., Xu, C., Lu, H.: Size adaptive selection of most informative features. In: Proceedings of AAAI (2011)

16. Malekipirbazari, M., Aksakalli, V.: Risk assessment in social lending via random forests. Expert Syst. Appl. **42**(10), 4621–4631 (2015)
17. Pavan, M., Pelillo, M.: Dominant sets and pairwise clustering. IEEE Trans. Pattern Anal. Mach. Intell. **29**(1), 167–172 (2007)
18. Peng, H., Long, F., Ding, C.H.Q.: Feature selection based on mutual information: criteria of max-dependency, max-relevance, and min-redundancy. IEEE Trans. Pattern Anal. Mach. Intell. **27**(8), 1226–1238 (2005)
19. Pohjalainen, J., Räsänen, O., Kadioglu, S.: Feature selection methods and their combinations in high-dimensional classification of speaker likability, intelligibility and personality traits. Comput. Speech Lang. **29**(1), 145–171 (2015)
20. Yang, H., Moody, J.: Feature selection based on joint mutual information. In: Proceedings of AIDA, pp. 22–25 (1999)
21. Zhang, Z., Hancock, E.R.: Hypergraph based information-theoretic feature selection. Pattern Recogn. Lett. **33**(15), 1991–1999 (2012)

Author Index

Abu-Aisheh, Zeina   197
Algabli, Shaima   143

Bai, Lu   59, 276
Berton, Gottfried   39
Biasotti, Silvia   13
Blumenthal, David B.   211
Bougleux, Sébastien   118
Brun, Luc   118
Bunke, Horst   222

Carletti, Vincenzo   128
Conte, Donatello   49
Cui, Hongxia   187
Cui, Lixin   59, 276
Curado, Manuel   13, 165

Deville, Romain   177
Di Fabio, Barbara   23
Ding, Chris   3
Drees, Dominik   73
Dupé, François-Xavier   39

Escolano, Francisco   13, 165

Fiorucci, Marco   165
Fischer, Andreas   83, 222, 242
Foggia, Pasquale   128
Fornés, Alicia   107
Fromont, Elisa   177

Gamper, Johann   211
Gaüzère, Benoit   118

Hancock, Edwin R.   13, 59, 94, 255, 276
He, Zhouqin   3
Hou, Jian   187

Jeudy, Baptiste   177
Jiang, Bo   3

Jiang, Xiaoyi   73, 232
Jiao, Yuhang   276

Landi, Claudia   23
Lladós, Josep   107
Luo, Bin   3

Malfait, Nicole   39
Martineau, Chloé   49
Moreno-García, Carlos Francisco   232

Pelillo, Marcello   165

Ramel, Jean-Yves   197
Raveaux, Romain   49, 197
Riba, Pau   107
Riesen, Kaspar   83, 222, 242
Rossi, Luca   59, 154, 276

Saggese, Alessia   128
Santacruz, Pep   143
Scherzinger, Aaron   73
Schiavinato, Michele   266
Serratosa, Francesc   143, 232
Solnon, Christine   177
Stauffer, Michael   83, 242

Takerkart, Sylvain   39
Torcinovich, Alessandro   165
Torsello, Andrea   154, 266
Tschachtli, Thomas   242

Vento, Mario   128
Venturini, Gilles   49

Wang, Jianjia   94, 255
Wilson, Richard C.   94, 255

Xiao, Yun   3

Printed in the United States
by Baker & Taylor Publisher Services